To my parents
June, 1989

ACTIVATION AND FUNCTIONALIZATION OF ALKANES

ACTIVATION AND FUNCTIONALIZATION OF ALKANES

Edited by

Craig L. Hill

Emory University, Atlanta, Georgia

WILEY

A Wiley-Interscience Publication

JOHN WILEY & SONS

New York Chichester Brisbane Toronto Singapore

Library of Congress Cataloging in Publication Data

Activation and functionalization of alkanes / Craig L. Hill,
editor.
     p.   cm.
  "A Wiley-Interscience publication."
  Includes bibliographies.
  ISBN 0-471-60016-4
  1. Paraffins--Reactivity. 2. Catalysis. I. Hill, Craig L.
QA305.H8A20 1900
547' .4110459--dc19   88-31320
                                   CIP

Printed in the United States of America

10 9 8 7 6 5 4 3 2 1

Contributors

Derek H. R. Barton • Department of Chemistry, Texas A&M University, College Station, Texas

Pierrette Battioni • Laboratoire de Chimie et Biochimie Pharmacologiques et Toxicologiques, Université René Descartes, Paris, France

Robert H. Crabtree • Department of Chemistry, Yale University, New Haven, Connecticut

Joseph D. Druliner • Central Research and Development Department, E. I. du Pont de Nemours & Company, Wilmington, Delaware

Omar Farooq • Department of Chemistry, University of Southern California, Los Angeles, California

Norman Herron • Central Research and Development Department, E. I. du Pont de Nemours & Company, Wilmington, Delaware

Craig L. Hill • Department of Chemistry, Emory University, Atlanta, Georgia

William D. Jones • Department of Chemistry, University of Rochester, Rochester, New York

Daniel Mansuy • Laboratoire de Chimie et Biochimie Pharmacologiques et Toxicologiques, Université René Descartes, Paris, France

Mario J. Nappa • Chemicals and Pigments Department, E. I. du Pont de Nemours & Company, Chambers Works, Deepwater, New Jersey

George A. Olah • Department of Chemistry, University of Southern California, Los Angeles, California

Nubar Ozbalik • Department of Chemistry, Texas A&M University, College Station, Texas

Ian P. Rothwell • Department of Chemistry, Purdue University, West Lafayette, Indiana

Alexander E. Shilov • Institute of Chemical Physics, U.S.S.R. Academy of Sciences, Chernogolovka, U.S.S.R.

G. K. Surya Prakash • Department of Chemistry, University of Southern California, Los Angeles, California

Kenneth S. Suslick • School of Chemical Sciences, University of Illinois at Urbana-Champaign, Urbana, Illinois

Chadwick A. Tolman • Central Research and Development Department, E. I. du Pont de Nemours & Company, Wilmington, Delaware

Preface

Methods of activating or functionalizing alkanes have been numerous and well publicized in the past few years; as the most abundant class of organic compounds, alkanes constitute one of the greatest potential resources for chemical energy and for precursor compounds in organic synthesis. Both the abundance of alkanes and their extremely low reactivity have greatly interested a wide spectrum of chemical scientists. In the past five years investigators from several distinct subdisciplines of chemistry have made substantive contributions to the study of alkane activation or functionalization. The investigators who have concentrated on the development of liquid phase systems include a number of prominent organometallic chemists, bioorganic-bioinorganic chemists, and others. Paralleling this development in liquid phase methods have been equally strong experimental efforts by physical chemists and chemical physicists, the latter investigators examining both gas phase and solid state processes. In the early 1980s, when most of these programs were first developing, there appeared to be little communication among investigators in the different subdisciplines. For example, the chemical physicists had little awareness of the results and facts generated from studies directed toward liquid phase activation of alkanes by coordinatively unsaturated organometallic species. Similarly, the organometallic chemists had little awareness of the research on alkane activation in the gas phase, in spite of the fact that such awareness would have had widespread benefits for the conceptualization and understanding of the organometallic systems. Although cross-disciplinary communication and fertilization have improved substantially in the past five years, much more is possible and desirable. A factor present in 1989 that exacerbates the interdisciplinary communication problem is common to areas of research that spawn fundamentally new ideas, concepts, or methods: proliferation of the pertinent literature. In 1980 there were only a few papers that addressed new systems or investigations of alkane carbon-hydrogen bond cleavage of any kind. There are now on the order of 10^3 such publications.

This volume does not attempt to cover all areas of chemical research that address the activation or cleavage of alkane carbon-hydrogen bonds. To do this in a sensible and pragmatic manner within a single volume at this stage would be difficult, if not impossible. This book concentrates on the systems that are conceptually and practically closest to the first highly selective catalytic high-conversion alkane functionalization processes to be developed in the future -- homogeneous processes, and largely those in the liquid phase. The goals of this book are twofold:

1. To provide in one volume discussions of most of the principal approaches to transforming alkanes so that the energetic and mechanistic features as well as the preparative potential of these processes can be examined in relation to one another.

2. To bring together in one volume most of the principal methods for activating or functionalizing alkanes that are currently of interest to preparative chemists or should ultimately lead to processes that are of interest to these investigators.

The book begins with a chapter by A. E. Shilov of the U.S.S.R. Academy of Sciences, one of the early pioneers of alkane activation in the liquid phase. His chapter addresses both the early work and more recent biomimetic approaches from his laboratory and those of his colleagues. The second chapter by G. A. Olah et al. outlines the activation of alkanes by superacid and other electrophilic systems, work that predates most of the alkane chemistry effected by organometallic, metalloporphyrin, and other systems, but which remains unique and yet in many ways complementary to all other homogeneous alkane activation or functionalization processes in the literature. In Chapters III, IV, and V, some of the most prominent investigators in their respective areas describe three increasingly distinct types of organometallic systems that activate alkanes. Nearly all the work in these chapters involves research done since the first reports of nonradical C-H bond activation facilitated by transition metal complexes reported by R. H. Crabtree and R. G. Bergman in early 1982. R. H. Crabtree in Chapter III discusses alkane activation by the polyhydride complexes, including those few organometallic compounds that catalytically transform alkanes, as well as a new highly catalytic and selective approach involving photosensitized Hg vapor. In Chapter IV, W. D. Jones discusses the alkane chemistry seen in the cyclopentadienyl rhodium, iridium, rhenium, and related systems. The stoichiometric alkane activation reactions effected by these complexes have produced a wealth of important energetic and mechanistic information. In Chapter V, I. P. Rothwell reviews C-H bond activation in the d^0 transition metal systems. Some of these systems investigated primarily by the groups of I. P. Rothwell, P. L. Watson, and T. J. Marks have generated substantial new chemistry, including the first reported organotransition metal species capable of activating methane. Chapters VI and VII and a portion of Chapter X address the homogeneous catalytic functionalization of alkanes by metalloporphyrins. Most of the molecular features of hydrocarbon transformation by these compounds are distinct from those seen in the organometallic systems. In Chapter VI, D. Mansuy and P. Battioni describe some of the alkane functionalization experiments involving the enzyme cytochrome P-450, the principal catalyst for hydrocarbon oxidation in the biosphere, as well as cytochrome P-450 metalloporphyrin model compounds, using two of the most biologically and industrially available oxidants, hydrogen peroxide and dioxygen. In Chapter VII, K. S. Suslick focuses on the modification of metalloporphyrin catalysts to facilitate the shape-selective hydroxylation of alkanes. Chapters VIII and IX describe hydrocarbon oxidation systems unlike any others in the literature. In Chapter VIII, C. L. Hill reviews the literature on oxo transfer to alkanes and describes alkane oxygenation by transition metal-substituted polyoxometalates, complexes that effectively represent homogeneous inorganic metalloporphyrin analogs that are thermodynamically, not just kinetically, resistant to oxidation. The new catalytic selective photochemical methods for the functionalization of alkanes based on polyoxometalates developed by the Hill group are not reviewed here but will be reviewed elsewhere in the near future. In Chapter IX, D. H. R. Barton and N. Ozbalik address several aspects of the "Gif" system for C-H functionalization. Some of the mechanistic attributes of this unique chemistry as well as its application to synthetic problems are presented. In the final chapter, C. A. Tolman, J. D. Druliner, M. J. Nappa, and N. Herron, all contributors to problems associated with alkane oxidation or functionalization themselves, collectively address the substantial alkane oxidation research, practical and exploratory alike, that has garnered serious attention for more than two decades in Du Pont's Central Research & Development Department.

Although the methodologies and the basic scientific information pertaining to the activation and functionalization of alkanes as well as unactivated C-H bonds in more complex molecules is still advancing rapidly and minor aspects of this book will be dated by publication time, the in-depth discussions of mechanism, structure-reactivity information, and other aspects of this chemistry should ensure the lasting value of this monograph. Given the continuing advances in the subject areas of this book and the consequent limited shelf life of a portion of the research, we chose to produce the entire book in a camera-ready but print quality format using a laserwriter printer system in my office rather than using the normal typesetting procedure. The former procedure although faster, in principle, than the latter, proved to be a very substantial amount of work -- translation: a full time job for a highly competent secretary for one year. It is with this fact articulated that I give my tremendous gratitude to my two secretaries that worked on this project, Deborah Finn and Nithya Raghunathan. Without them the entire project would have been impossible. Deborah Finn started this opus and contributed a great deal to the editing, organizing, and other tasks in the early stages. Nithya Raghunathan took over most of the aspects of production for a period of 7 months making mistakes so rarely that it became a point of some amusement when she made one. It is unusual to have one secretary as professional, dedicated, and competent as either Deborah or Nithya; it is rare indeed to have been

blessed with two of them. Again, they have my admiration and deep
felt thanks.

Atlanta, GA 1989   CRAIG L. HILL

Contents

Chapter I

HISTORICAL EVOLUTION OF HOMOGENEOUS ALKANE ACTIVATION SYSTEMS

A. E. Shilov

Institute of Chemical Physics
U.S.S.R. Academy of Sciences
Chernogolovka 142 432, U.S.S.R.

I. Introduction

Since the end of the 1960s a new problem has arisen in the field of homogeneous catalysis, that of homogeneous activation of saturated hydrocarbons by metal complexes. The development of coordination chemistry and catalysis had lead to increased success in the activation of various molecules by metal complexes, such as hydrogen, olefins, aromatics, carbon monoxide, and molecular nitrogen. "Activation" of a certain molecule by a metal complex, although a rather vague concept, usually means that the molecule or its part becomes a ligand in the coordination sphere of the complex and then undergoes a subsequent chemical transformation.

For some time saturated hydrocarbons remained outside of this process, which involved all the other classes of organic and inorganic substances. Meanwhile, an increasingly urgent problem involved the great resources of natural gas that necessitated the creation of new selective processes involving methane, which constitutes its primary part.   At the same time, the activation of methane and its analogs is an interesting theoretical problem, since the absence of double or triple bonds, lone electron pairs, and the strength of covalent C-H and C-C σ-bonds in alkanes constitute natural difficulties in the search for desirable systems.

In 1968 J. Halpern[1], one of the leading specialists in homogeneous catalysis, described the task as "to develop a successful approach for activation of C-H bonds, particularly saturated hydrocarbons, this problem being at present one of the most important and challenging in the entire field of homogeneous catalysis."

The chemical inertness of alkanes is well known and is reflected in one of their old names, "paraffins" (from the Latin *parum affinis* - without affinity). The alkane reactions known in chemistry usually require particularly active particles, e.g., strong oxidants, superacids, free atoms, radicals, and carbenes, or proceed at high temperatures or require other sources of energy (as in radiation chemistry.)

In the 1960s it was possible to suggest the existence of other more selective reactions of alkanes under comparatively mild conditions. The bases for such suggestions are summarized as follows:

1. The σ-bond in molecular hydrogen is not weaker than the σ-C-H bond in alkanes. However, multiple examples of homogeneous H_2 activation are known, for example in H-D exchange or hydrogenation. Therefore an analogy with H_2 activation could be used in the search for alkane catalytic reactions in solution.*

2. A number of metal complexes (ML_n or M) are capable of reacting with substances containing "activated" C-H bonds, e.g., C-H bonds in aromatic hydrocarbons or in the α-position to double bonds. The C-H bond is being cleaved in this reaction and an M-C bond is formed. Therefore the C-H bond energy, which in aromatic compounds is even higher than in alkanes, does not prevent the reaction.

3. A "nonactivated" aliphatic C-H bond may be involved in the reaction when it is present in a suitable position in a ligand of the complex coordination sphere. The result of the reaction is often cyclometallation. This reaction indicates the possibility of a similar reaction (perhaps at high temperatures) of alkanes.

4. Since the 1930s facile alkane reactions at the surface of metals and oxides have been known, in which the C-H bond is definitely cleaved, e.g., in H-D exchange with D_2. Similar reaction may be visualized in homogeneous solutions, at least for polynuclear complexes.

5. Enzymatic oxidation is known including hydroxylation of C-H bonds in saturated hydrocarbons catalyzed by metal enzymes. For example, the methane monooxygenase catalyzes methane oxidation primarily to methanol. Mechanistic investigation of these processes may help to use a biomimetic approach to create similar purely chemical catalytic systems.

Beginning in the 1960s, investigations directed to the activation of alkanes and to the discovery of new catalytic reactions involving the

---

*The analogy of alkanes with other compounds from the viewpoint of the search for their activation is considered more thoroughly by Shilov.[2] More references on different sources can also be found there.

participation of metal complexes were conducted along several main routes.

II. Homogeneous Activation of Alkanes by Platinum(II) Complexes

In the search for a successful approach to C-H bond activation in alkanes (RH) and using the analogy with H_2 activation we could suggest an oxidative addition of an alkane to the coordinatively unsaturated complex of a transition metal:

$$L_nM \ + \ RH \ \rightleftharpoons \ L_nM \Big\langle {{}^R \atop {}_H}$$

The reaction may be reversible and shifted to the left, taking into account the strong C-H bond and comparatively weak M-R and M-H bonds.

Therefore there must be some indication that the oxidative addition does take place, even though the equilibrium is shifted to the left. The simplest consequence of equilibrium 1 may be H-D exchange with D_2, solvent, or other substances present in the solution. In particular H-D exchange may be expected with the solvent protons if the alkyl hydride complex formed primarily has sufficiently strong acid properties.

$$L_nM \Big\langle {{}^R \atop {}_H} \ \rightleftharpoons \ L_nM\text{-}R^- + H^+ \tag{1}$$

At the end of the 1960s no such examples were known. It was natural to start with the simplest systems. Therefore, our attention was attracted to the works by Garnett and Hodges, who had shown that ordinary platinum(II) complexes, $PtCl_4^{2-}$, catalyze the H-D exchange of aromatic hydrocarbons with the solvent in a water-acetic acid mixture.[3] Hydrogen atoms involved in the exchange with the protic solvent had been shown to include not only those of the aromatic ring but also of its side chain, including ß-H atoms, e.g., H atoms of the methyl group in ethylbenzene. Garnett and Hodges[3] postulated the formation of π-complexes of Pt(III) with the aromatic ring as a necessary precondition of C-H bond activation. However, another alternative should be taken into account, i.e., direct attack of the Pt(II) complex at the nonactivated C-H bond. To our surprise this first and simplest system did indeed work: methane and ethane revealed their ability to exchange catalytically their H atoms for deuterium of the solvent at 90-120°C in the presence of a chloride complex of bivalent platinum.[4] Moreover, it became clear afterward that the species present in the water-acetic acid solution of the well known Pt(II) complex (K_2PtCl_4) contains the most active catalysts of the H-D exchange. The use of other metals with Cl⁻ or other ligands with Pt(II) leads to either a decrease in the reaction rate or to a complete stoppage of the reaction.

Catalytic H-D exchange of alkanes with water in the presence of Pt(II) complexes could have been discovered much earlier, e.g., in 1930s, when H-D exchange of alkanes with D_2 was found at metal surfaces. Apparently the chemical inertness of alkanes inhibited the search for reactions catalyzed by simple Pt(II) complexes.

According to Reaction 1 the mechanism of H-D exchange has to involve alkyl platinum derivatives. In view of the reactivity of alkyl groups bound to metal, other reactions could be expected under the same conditions as those for H-D exchange. Such reactions soon were found.[5] Platinum(IV) complexes turned out to be capable of oxidizing alkanes in the presence of Pt(II) to chlorides and (as found later) alcohols. For example, methane reacts according to the equations:

$$H_2PtCl_6 + CH_4 \xrightarrow[\text{Pt(II)}]{} H_2PtCl_4 + CH_3Cl + HCl$$

$$H_2PtCl_6 + CH_4 + H_2O \xrightarrow[\text{Pt(II)}]{} H_2PtCl_4 + CH_3OH + 2HCl$$

Dehydrogenation proceeds in the case of cyclohexane:

$$3H_2PtCl_6 + C_6H_{12} \xrightarrow[\text{Pt(II)}]{} C_6H_6 + 6HCl + 3H_2PtCl_4$$

These were the first examples of alkane homogeneous functionalization catalyzed by electron-rich complexes (see below).

In the presence of the external oxidant (dioxygen) and electron transfer agent ($CuCl_2$, quinones, heteropolyacids) the system H_2PtCl_6-H_2PtCl_4 can become catalytic with respect to both platinum compounds. However, the turnover number is small, primarily because of secondary reactions, e.g., oxidation of methanol in the case of methane.

Further investigations of alkane H-D exchange and oxidation in the presence of platinum complexes were conducted in several research groups by Garnett et al. in Australia, Webster and Hodges in Great Britain[6] and Rudakov[7] and Shilov[2] the U.S.S.R. In all cases the results obtained were similar and mutually complementary. They can be summarized as follows:

1. H-D exchange and oxidation by H_2PtCl_6 involve the same intermediate, which is presumably the alkylated complex of Pt(II): $RPtClL_2$ (L is the H_2O or HOAc). The result is that when the formation of this intermediate is a rate-determining step [temperature below 100°C, high Pt(IV) concentrations] the rate of H-D exchange in the absence of Pt(IV) coincides with that of oxidation under the action of H_2PtCl_6.

2. Among the Pt(II) complexes both positive, $PtClL_3^+$, and negative, $PtCl_3L^-$, ions are active toward alkanes as well as uncharged $PtCl_2L_2$ (Table I). Symmetrical PtL_4^{2+} and $PtCl_4^{2-}$ are inactive.

Table I. Rate Constants (mol^{-1} liter sec^{-1}) of Reactions of PtL_4^{2+} (k_0), $PtClL_3^+$ (k_1), $PtCl_2L_2$ (k_2), $PtCl_3L^-$ (k_3), and $PtCl_4^{2-}$ (k_4)

Hydrocarbon	k_0	k_1	k_2	k_3	k_4
Isobutane	0	1.9±0.1	3.8±1.1	3.7±0.1	0
Cyclohexane	0	5.3±0.2	9.5±2.4	12±0.3	0
Benzene	9.1±1.4	66±31	12±4	18±2	0

Source: From Rudakov et al.[7b]

Introduction of "softer" ligands in the platinum coordination sphere as phosphines or dimethyl sulfoxide (DMSO) slows down the reaction. The order of reactivity in reaction of alkanes is inverse to the trans-effect observed for substitution reactions in Pt(II) complexes.

3. With respect to the series of alkanes, platinum(II) complexes behave as moderate electrophiles (the rate increases in the order $CH_4 < C_2H_6 < C_3H_8$). Steric factors play an important role. Isolated methyl groups are the most active. Methyl and methylene groups situated next to bulky *tert*-butyl group display very little activity. Tertiary C-H bond is not active at all.

Kinetic investigation leads to the following reactions:

$$PtCl_4^{2-} \; \underset{}{\overset{L}{\rightleftharpoons}} \; PtCl_3L^- + Cl^- \qquad (2)$$

$$PtCl_3L^- \; \underset{}{\overset{K}{\rightleftharpoons}} \; PtCl_2L_2 + Cl^- \qquad (3)$$

$$PtCl_2L_2 + RH \; \underset{k_{-1}}{\overset{k_1}{\rightleftharpoons}} \; RPtClL_2 + H^+ + Cl^- \qquad (4)$$

$$RPtClL_2 + PtCl_6^{2-} \; \overset{k_2}{\longrightarrow} \; RPtCl_5^{2-} + PtCl_2L_2 \qquad (5)$$

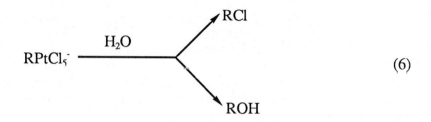

$$\text{RPtCl}_5^- \xrightarrow{\text{H}_2\text{O}} \begin{array}{c} \text{RCl} \\ \\ \text{ROH} \end{array} \tag{6}$$

The following kinetic equation corresponds to this scheme, provided the stationary state is reached:

$$(k_1[\text{PtCl}_2\text{L}_2][\text{RH}] = k_{-1}[\text{RPtClL}_2][\text{H}^+][\text{Cl}^-] + k_2[\text{RPtCl}_2\text{L}_2][\text{PtCl}_6^{2-}])$$

$$-\frac{d[\text{Pt(IV)}]}{dt} = \frac{k_1 k_2 \text{K}[\text{RH}][\text{Pt(II)}][\text{Pt(IV)}]}{([\text{Cl}^-] + \text{K})(k_{-1}[\text{H}^+][\text{Cl}^-] + k_2[\text{Pt(IV)}])}$$

In the absence of Pt(IV) Reaction 4 leads to the H-D exchange with the solvent. At the addition of Pt(IV) the H-D exchange rate decreases, the sum of the rates [exchange and oxidation by Pt(IV)] remaining constant.[7a] This is the evidence for the same intermediate (RPtClL$_2$) formed with the constant rate $k_1[\text{PtCl}_2\text{L}_2]$ [RH].

Kinetic curves for methane oxidation are shown in Figure 1.[8] The product is the complex of Pt(II), which is the catalyst of the process, and the reaction rate initially increases with time. Methyl chloride and methanol are seen to form in parallel reactions, approaching quantitative yield per oxidant.

To prove Reactions 2 - 6, it was essential to provide evidence for the formation of the alkyl platinum intermediate. Some indirect data supported this view. For aromatic compounds arylplatinum(IV) complexes are sufficiently stable to be isolated and in some cases structural analysis has been made. Platination of aromatic compounds by Pt(IV) turned out to be a very effective way to synthesize these complexes (see ref. 9 for a review).

For alkanes, e.g., methane, the reaction temperature is higher and expectations of observing the intermediate in water as a solvent were not particularly high. However, such an intermediate was detected, first by chemical analysis (Fig. 1) and then directly by NMR spectroscopy (Fig. 2). Moreover, kinetic analysis has shown that the entire process proceeds via the alkylated Pt(IV) complex CH$_3$PtCl$_5^{2-}$.

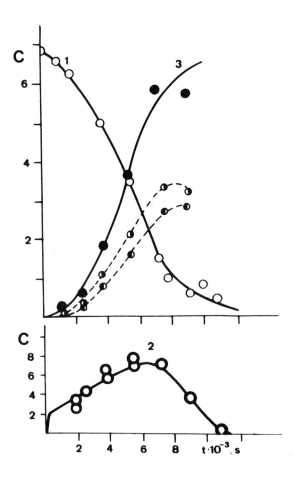

FIGURE 1. The kinetic data on Pt(IV) consumption (curve 1, O, [Pt(IV)] x 10^2, M); methylplatinum(IV) complex behavior (curve 2, $\mathbf{O}$, [Pt(IV)-CH$_3$] x 10^5, M), the total of products accumulation (curve 3, $\bullet$, [CH$_3$OH] + [CH$_3$Cl]) x 10^2, M), and the accumulation of methanol (⦶, [CH$_3$OH] x 10^2, M) and methyl chloride (⦶, [CH$_3$Cl] x 10^2, M) in the reaction of methane with an aqueous solution of platinum chloride complex; 120°C, [CH$_4$] = 0.1 M, [K$_2$PtCl$_4$]$_0$ = 1.5 x 10^{-2} M.

Its decomposition rate was measured separately, the products being the same as for the CH$_4$ oxidation, i.e., Pt(II) and CH$_3$Cl together with CH$_3$OH. Extrapolating the temperature dependence of the rate constant to the temperature of the methane reaction with H$_2$PtCl$_6$ in the presence of Pt(II) and knowing the experimental value of CH$_3$PtCl$_5{}^{2-}$ concentration in the reaction mixture, it was possible to calculate the reaction rate for decomposition of this complex under the reaction conditions. The calculated reaction rate was found to coincide quantitatively with the observed overall reaction rate of Pt(IV) consumption.

This kinetic investigation confirming the reaction mechanism was also very convincing evidence for the homogeneity of the reaction. Some doubts were expressed as regards this subject in the literature (for example, see ref. 10). Apparently the simplicity of the system aroused some distrust as to the correctness of the proposed interpretation. Moreover, under the conditions of H-D exchange some precipitation of platinum black is indeed often observed, particularly in a water solution (the system is more stable in a water-acetic acid mixture) and in the presence of some impurity. However, the formation of free platinum was shown to suppress the reaction rather than increase its rate. The results were reproducible in a homogeneous solution.

In the presence of Pt(IV) the reaction solution remains homogeneous for a long time, spectral investigation showing only the reacting Pt(IV) complex and Pt(II) complex present in solution as they are being formed. The kinetic results show that the reaction rate is determined by the concentrations of mononuclear Pt(II) and Pt(IV) complexes and, as previously mentioned, *the reaction rate is equal to the rate of transformation of a perfectly soluble alkyl Pt(IV) complex.* This seems to be sufficient evidence for the homogeneity of the reaction and, thus, for the homogeneity of the H-D exchange.

The mechanism of the initial elementary interaction of alkane molecules with Pt(II) complexes deserves special attention. The absence of evidence for alkyl hydride complex formation in Reaction 4 has made some authors (e.g., Crabtree[11]) consider the reaction as an electrophilic substitution of proton in the alkane molecule by the Pt(II) complex. Though definitions are always rather relative and subjective, and Reaction 4 may be indeed formally called electrophilic substitution by outward resemblance, detailed consideration shows that the term "electrophilic" is very misleading and inconvenient with respect to Pt(II) complex reactions with alkanes. In the first place, for an electrophilic reaction, the reaction of Pt(II) complexes with weak nucleophiles such as alkanes is strikingly insensitive toward electrophilic properties of the reaction species. As mentioned, the reaction rate is not very much different for ions of opposite charge ($PtClL_3^+$ and $PtCl_3L^-$), whereas the noncharged complex $PtCl_2L_2$ is often the most active. Second, if alkanes including methane are considered as nucleophiles replacing Cl^- in Pt(II) complexes, then in spite of some misgivings about calling alkanes nucleophiles in water solution, bear in mind that the influence of the ligand nature on [Pt(II)] complexes is opposite to that observed for nucleophilic substitution.*

The real mechanism of alkane reactions with Pt(II) complexes is

---

* For Pt(II) complexes even for substitution by typical nucleophiles the term "nucleophilic" is not suitable for a description of the mechanism. Other terms, i.e., associative, dissociative, and interchange are proposed (see ref. 12). It is even more inconvenient with respect to methane as a nucleophile, in particular when the electrophile reacting with CH_4 is $PtCl_3L^-$.

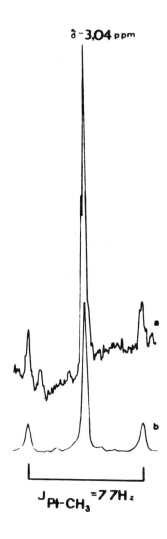

FIGURE 2. ^{1}H-NMR spectra of the Pt(IV)-CH$_3$ complex obtained in (*a*) the PtCl$_6^{2-}$ (0.32 *M*) + PtCl$_4^{2-}$ (0.24 *M*) + CH$_4$ (0.1 *M*) system, reaction time 30 min, 120°C, number of scans 4 x 10^3; (*b*) the reaction of CH$_3$I and PtCl$_4^{2-}$; [Pt(IV)-CH$_3$] = 2 x 10^{-2} *M*, number of scans 1 x 10^2, solvent D$_2$O.

apparently close to the oxidative addition of C-H bonds[1] to electron-rich species. The driving force of the reaction is the formation of a strong Pt-C bond and also initially a Pt-H bond. Nucleophilic or electrophilic properties of reacting particles contribute only in a minor way to the interaction determining the rate of the process.

According to the theoretical results obtained by Saillard and Hoffmann,[13] a linear approach of the methane C-H bond to the metal atom of the complex (A) is a favorable rather than angular one (B) for oxidative addition. This approach was first suggested[14] for alkane

$$\underset{(A)}{\overset{\diagdown}{\diagup}C\text{-}H\cdots M} \qquad\qquad \underset{(B)}{\overset{\overset{\diagdown \,|\,\diagup}{C}}{\underset{H}{|}}\overset{\diagup}{\diagdown}M}$$

(A)   (B)

reaction with Pt(II) complexes by analogy with CH_2 insertion into the methane C-H bond.

Accordingly, at the approach of the methane molecule to the platinum complex, it is the orbital corresponding to the Pt-H bond that starts to be occupied first, platinum and carbon atoms repulsing each other.[15] The development of the process, which at first follows the oxidative addition path, leads to an increase in the positive charge on the H atom. In the presence of water in the coordination sphere of platinum or in solution the reaction may change its path, not reaching a stable alkyl hydride complex but transferring a proton to the water molecule. It should be noted that the reverse process of Pt-C bond hydrolysis depending on the ligands may proceed both via oxidative addition of proton to form alkyl hydride complex and via a three-center transition state with synchronous C-H bond formation and Pt-C bond cleavage.[16]

In reactions discovered in 1979 by Crabtree et al.[17] and also by Baudry et al.[18] in 1980 of alkanes with cationic iridium(I) complexes and hydrido-rhenium complexes, respectively, with participation of 3,3-dimethylbutene-1, the oxidative addition of alkane might also proceed without alkyl hydride complex formation. The hydrogen atom of the C-H bond may be transferred instead of the olefin molecule (see Chapter III).

$$L_n Ir + RH + \quad \diagup\!\!\!\!\diagup\!\!\bigvee \quad \longrightarrow \quad L_n Ir \overset{\diagup R}{\underset{CH_2CH\text{-}C(CH_3)_3}{\diagdown}}$$

$$\longrightarrow \quad L_n Ir(R(\text{-}H)) \;+\; C_2H_5C(CH_3)_3$$

Coordinatively unsaturated Ir(I) and Rh(I) complexes were found in 1982 to oxidatively add alkanes to form alkyl hydrides in hydrocarbon solutions.[19,20] In these cases, photochemical elimination of H_2 induces formation of sufficiently active particles to react with C-H at low temperatures. (See Chapter IV by W. D. Jones in this volume.) These definitive oxidative addition reactions are in many respects similar to alkane reactions with platinum complexes from the viewpoint of the weak dependence of polar factors and sensitivity of steric hindrances and M-C bond energy. Presumably all these reactions must be considered in the ever expanding area of electron-rich species interactions with σ-bonds, C-H bond in alkanes being included with increasing number of examples.

Square planar Pt(II) complexes, however, remain unique, since as a result of particularly high Pt-C bond energy they are apparently more active toward alkanes in water solution than complexes of any other metal.

III. Alkane Reactions with Hydrides and Organometallics

Besides heterolytic exchange with protic solvent according to Reaction 1, H-D exchange between alkanes and D_2 might be suggested via polyhydrido complexes with coordinatively unsaturated complexes as intermediates, e.g.:

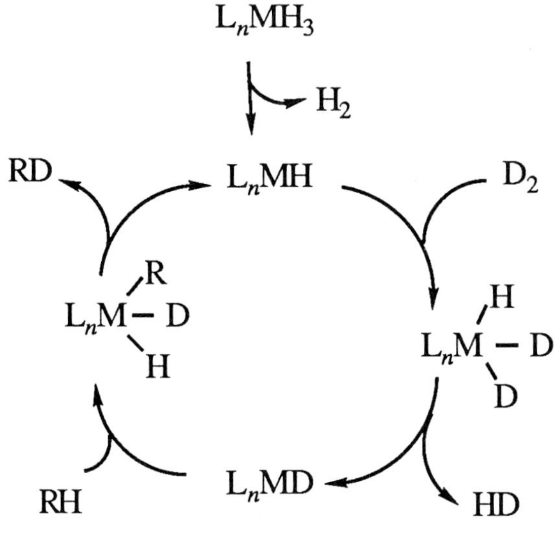

Accordingly H-D exchange of methane with D_2 was observed in 1969 in our work on benzene solution of $(Ph_3P)_3CoH_3$ at room temperature[4] at the same time the first reactions of alkanes with Pt(II) solutions were found.

The reaction is slow and because Pt(II) reactions looked more promising to us at the time, we decided to concentrate our efforts on them. It later turned out that the chemistry of alkanes reacting with coordinatively unsaturated metal complexes formed in photochemical or thermal decomposition of di- and polyhydrides was very rich.

In the works of M. L. H. Green, R. Crabtree, H. Felkin, W. D. Jones, R. Bergman, W. A. G. Graham, and others this approach was successfully used for to discover and study new and exceptionally interesting transformations of alkanes (see Chapters III and IV in this volume).

The natural obstacle in the search for agents to activate methane and its homologs with the participation of hydride and alkyl derivatives of transition metals is the choice of a suitable solvent (except for hydrocarbons there are few solvents that are chemically inert to such complexes), but if a hydrocarbon is used as a solvent, it usually is more reactive towards C-H activating particles than methane.

Therefore the results that Grigoryan et al.[21] began to obtain in 1975 were somewhat unexpected. It was shown that methane can be activated in Ziegler-Natta catalytic systems in hydrocarbon media. For example, CD_4 was shown to be capable of exchanging deuterium atoms with the hydrogen of methyl groups in such systems as $TiCl_4$ + $AlMe_2Cl$; $(\eta^5\text{-}C_5H_5)_2TiCl_2$ + $AlMe_2Cl$; VCl_3 + $AlMe_2Cl$. The reaction was carried out in heptane solution at 20-50°C and methane pressure of 0.3-4 atm. The products were found to be CHD_3, CH_2D_2, and CH_4. The authors suggested carbene complex as an intermediate:

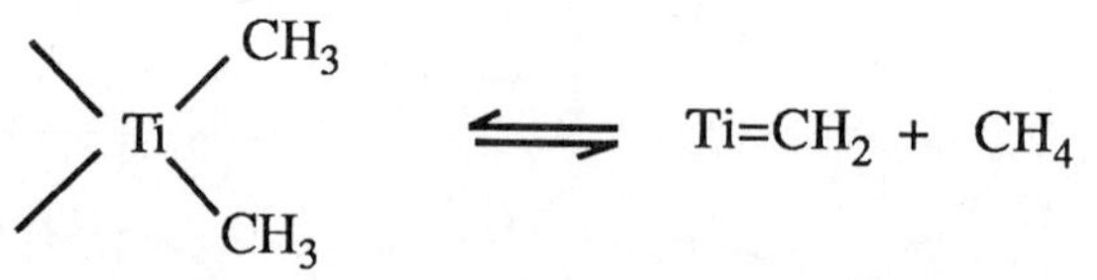

$$\text{Ti}(CH_3)_2 \;\rightleftharpoons\; Ti{=}CH_2 \;+\; CH_4$$

or in a binuclear complex

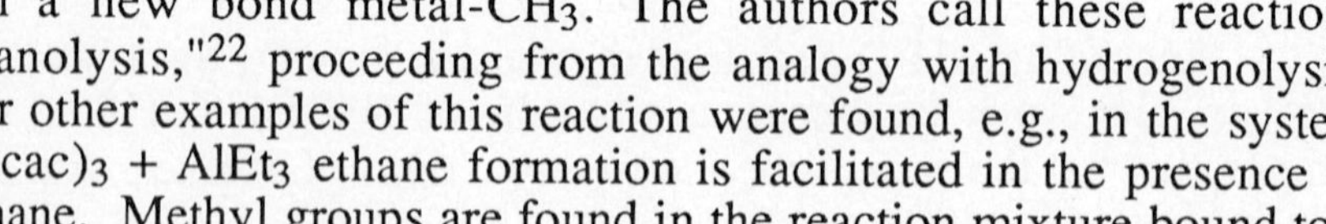

$$\text{Ti}(CH_2)Ti \;+\; 2\,CD_4 \;\longrightarrow\; CH_2D_2 \;+\; 2\,Ti(CD_3)_2$$

Thus, methane is capable of cleaving to a metal-carbon bond to form a new bond metal-CH_3. The authors call these reactions "alkanolysis,"[22] proceeding from the analogy with hydrogenolysis. Later other examples of this reaction were found, e.g., in the system $Fe(acac)_3$ + $AlEt_3$ ethane formation is facilitated in the presence of methane. Methyl groups are found in the reaction mixture bound to a metal releasing methane at hydrolysis (CD_3H in the case of CD_4). Therefore the following reaction may be suggested:

$$L_nFeC_2H_5 \;+\; CH_4 \;\longrightarrow\; L_nFeCH_3 \;+\; C_2H_6 \qquad (7)$$

H-D exchange between CH_4 and C_2D_4 was detected in the presence of $(\eta^5\text{-}C_5H_5)_2V$ in benzene solution.[23] According to the authors, the reaction mechanism involves oxidative addition with olefin participation:

$$\overset{\diagdown}{\underset{\diagup}{V}} \longleftarrow \overset{CH_2}{\underset{CH_2}{\parallel}} + CH_4 \rightleftharpoons \overset{\diagdown}{\underset{\diagup}{V}} \overset{CH_3}{\underset{C_2H_5}{\diagup}}$$

Thus, the reaction suggested involves the reverse process of alkyl disproportionation in the metal coordination sphere that was reported earlier.[24]

The reaction mechanisms could cast some doubts. Unexpectedly methane is capable to enter various reactions with different metal complexes in different oxidation states.

However, all these mechanisms were subsequently confirmed in quite definite systems: P. Watson, T. J. Marks, J. Bercaw, and I. P. Rothwell found the reaction type (7) with the d^0 complexes Lu, Y, Th, and Sc (see Chapter V). Coreaction of olefin in oxidative addition of alkane already mentioned above was found by Crabtree and Felkin and their co-workers (see Chapter III).

Thus, methane chemistry includes many new reactions in solution of organometallic compounds. In these reactions methane is often (particularly with 3d elements) much more reactive than other hydrocarbons, which can be used as solvents (toluene, heptane, etc.). Apparently, steric factors play a particularly important role in these processes.

Another interesting methane reaction also observed for the first time by Enikolopyan and Grigoryan[21] is the addition of methane to a multiple bond, which can be called hydromethylation.

Hydromethylation of ethylene was first reported for the system $Ti(OBu)_4 + AlEt_3$ in benzene,[25] the product being propane, though with rather small yields. Subsequently hydromethylation of acetylene was observed, producing propylene for the complexes of iron,[26] nickel,[27] and platinum.[28]

It may be suggested that the reaction mechanism involves recombination of alkyl groups in the coordination sphere of the metal, e.g., for acetylene

$$M \longleftarrow \overset{CH}{\underset{CH}{\parallel}} + CH_4 \longrightarrow M\overset{CH_3}{\underset{CH=CH_2}{\diagup}} \longrightarrow M + C_3H_6$$

Nevertheless, the detailed mechanism of hydromethylation remains unclear; perhaps in the system reported so far the reaction is not homogeneous. These problems still remain to be solved.

Hydromethylation of carbon monoxide (methane carbonylation) was shown recently[29] to proceed with the formation of acetaldehyde in the system Cp_2TiCl_2 - $AlMe_2Cl$ in benzene or toluene at 20-50°C.

The thermodynamics of the reaction

$$CH_4 + CO \longrightarrow CH_3COH$$

is not favorable ($\Delta G°_{298} = +13.2$ kcal/mol) and the process was conducted at high pressure (1500 - 3000 atm). In the presence of ethylene the formation of methyl ethyl ketone was observed apparently according to the scheme:

$$CH_4 + CO + C_2H_4 \longrightarrow CH_3COC_2H_5$$

It should be mentioned that with longer chain hydrocarbons insertion of CO into the C-H bond becomes more favorable. Recently catalytic formation of aldehyde was reported for carbonylation of cyclohexane[30] and pentane[31] in the presence of $RhCl(CO)(PR_3)_2$ (P_{CO} = 1 atm, UV irradiation), the reaction being regioselective in the latter case.

Thus, a number of new reactions of methane and other alkanes have been discovered in solutions of organometallic and hydrido complexes of metals. The steric hindrances produce very pronounced effects, and as a consequence of their minimum significance for methane, hydrocarbon solutions can be used for methane activation. High Lewis acidity facilitates methane reactions with $d°$ high-valent metal complexes by replacement of alkyl groups on the metal with CH_3. Oxidative addition in the case of low-valent complexes is facilitated as a result of coupled reactions of the type of hydride or H atom addition to an unsaturated molecule present in the coordination sphere of the metal.

IV.   Models of Biological Alkane Oxidation: Activation of Dioxygen

As previously mentioned, the construction of chemical models of biological alkane oxidation is another approach in the search for new ways of alkanes activation and functionalization. Nature is known to overcome successfully the difficulties in activation of chemically inert hydrocarbon molecules, molecular oxygen being an oxidant.

Monooxygenases are enzymes catalyzing oxidation in coupled processes, which involve C-H bond hydroxylation (or epoxidation of olefins) and simultaneous oxidation of NADH or NADPH according to the equation:

$$RH + O_2 + 2e + 2H^+ \longrightarrow ROH + H_2O$$

The cytochrome *P*-450-dependent monooxygenases have become particularly well known since 1958.[32] They contain an iron porphyrin complex in the enzyme active center. Cytochrome *P*-450 is the base for various monoxygenases functioning in different living organisms from bacteria to mammals. There are also bacterial ω-hydroxylases that catalyze selective hydroxylation of alkane methyl groups. Finally, methane monooxygenase hydroxylates methane and its analogs, methane being the most active among the other alkanes. Iron is again present in the reaction center but in the two latter cases it resides in nonheme surrounding.[33]   Ketoglutorate-dependent diooxygenase should also be mentioned; it hydroxylates the C-H bond with ketoglutorate participation according to the following equation.[34]

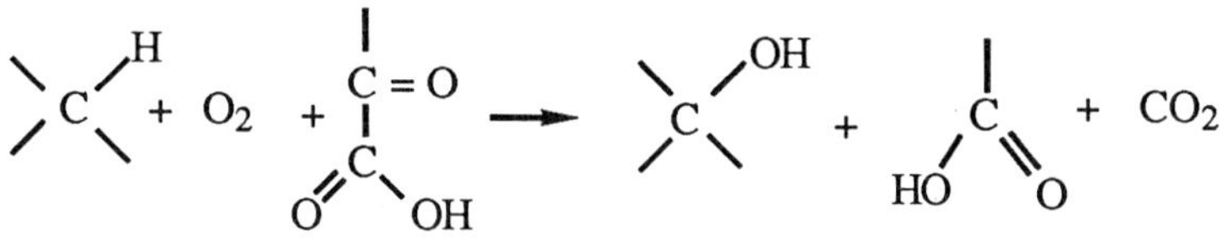

The enzyme active center again contains iron in this case. Experiments with labeled dioxygen $^{18}O_2$ have proved that one oxygen atom of the O_2 molecule is inserted into the product and the other into the succinate molecule formed from α-ketoglutarate. Participation of iron in all cases mentioned suggests a mechanism similar to that studied in detail in the case of cytochrome *P*-450.[35]   The structure of the enzyme was recently determined by an X-ray crystallography.[36] (See Chapter VI for oxygenation by cytochrome *P*-450.)
The mechanism of cytochrome *P*-450 action may be presented by the following scheme (P = porphyrin).

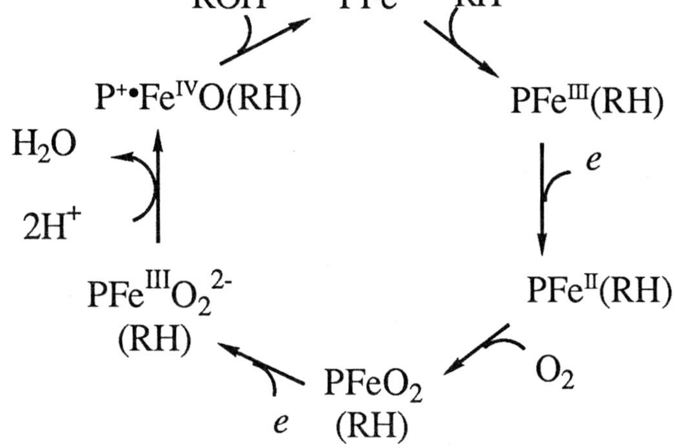

The active species hydroxylating the C-H bond is an $Fe^{IV}O$ complex of a porphyrin cation-radical $P^{+\bullet}$[37] This conclusion is supported by the fact that the active particle can be produced at cytochrome *P*-450 by the action of single oxygen atom donors such as hydroperoxides, peroxy acids, iodosylbenzene, or amine *N*-oxides[38] (see Chapters VI-VIII). The coupling with the electron donor (e.g., NADH) for O_2 utilization is necessary to increase the energy for four-electron oxidation by O_2 to form high potential species:

$$PFe^{III} + O_2 + 2e + 2H^+ \longrightarrow P^{+\bullet}Fe^{IV}O + H_2O$$

The analysis of two-electron redox potential E_o of $Fe^{V}O$ formed in the reaction with O_2 (without participation of a porphyrin ring) shows that E_o may be for example twice as high as $E_o = +1.229$ V of the four-electron redox process with O_2

$$O_2 + 4e + 4H^+ \rightleftharpoons 2H_2O$$

(see pp. 130-131 in ref. 2).

Donation of electron density from the porphyrin ring evidently decreases the potential of the oxidant but facilitates its formation and makes it less dangerous for the surrounding. Thus, the energy of the donor oxidation is used to create a strong oxidant capable of hydroxylating a strong C-H bond. C-H activation in the sense of its involvement in the coordination sphere of the complex apparently does not take place in this case. It is only dioxygen that is chemically activated. The alkane molecule enters the hydrophobic cavity next to the reaction center and is constrained in the neighborhood of the active oxygen species.

Analogs of the oxidized iron(IV) complex have been known in chemistry for many years. They are high-valent metal compounds, e.g., chromic acid, chromyl chloride, and paramanganate, which are traditional oxidants for organic substances including alkanes. Their oxidizing properties are usually considerably strengthened in concentrated acids such as sulfuric acid.[39]

Such oxidants attack the C-H bond primarily by abstracting an H atom and forming a free radical. The latter may interact with the MOH group without leaving the solvent cage to form an alcohol. The result is an insertion of an O atom (oxene) in the C-H bond, and the process may really develop without the formation of kinetically independent free radicals.

The oxenoid mechanism (insertion of oxene) was first proposed by Hamilton.[40] However, even at present this problem is not completely solved and there is some evidence for the intermediate radical in the enzymatic process.[41]

For ketoglutarate-dependent dioxygenase the following mechanism may be suggested[*]:

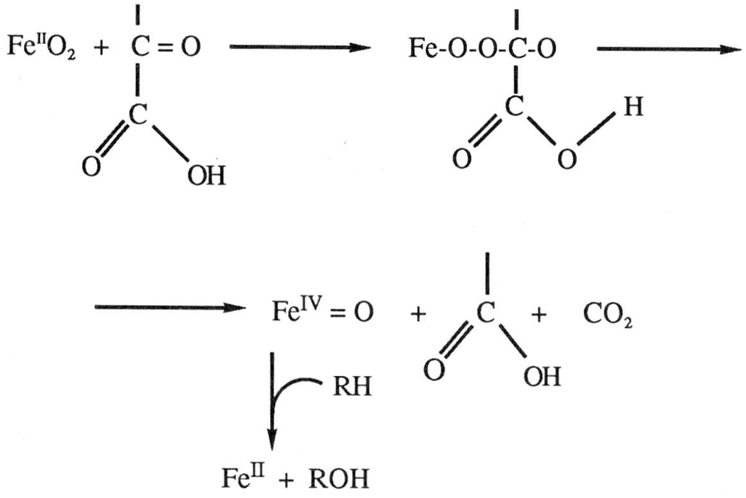

The formation of a stable CO_2 molecule in the concerted process involving a homolytic O-O bond cleavage in the transition state must help to produce high-valent iron(IV) species.

The mechanism of methane monooxygenase is also becoming increasingly clear as the result of intense studies. From the data received by Woodland et al.[42] it follows that the active center is in this case similar to that of a biological O_2 carrier, i.e., hemerythrin, which contains two iron atoms. The interaction of O_2 with hemerythrin according to recent data[43] may be presented as follows:

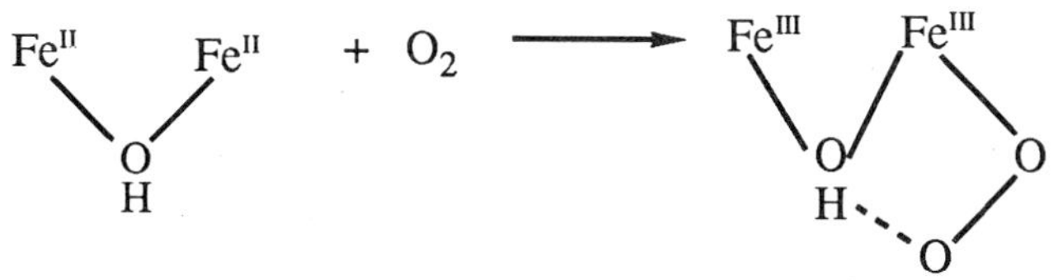

This structure applied to methane monooxygenase helps to suggest the following mechanism for C-H hydroxylation of methane:

---

[*] See p. 102 in ref. 2.

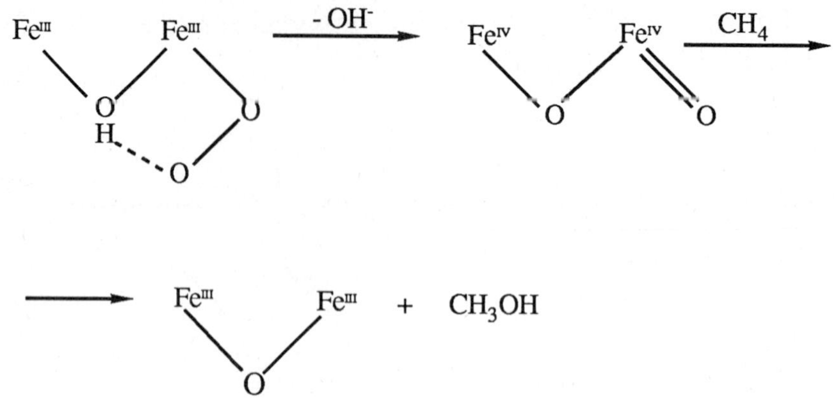

$$Fe^{III} \quad Fe^{III} \xrightarrow{\;-OH^-\;} Fe^{IV} \quad Fe^{IV} \xrightarrow{\;CH_4\;}$$

(with bridging O, H····O structure on the left; bridging O and =O on the right)

$$\longrightarrow Fe^{III} \quad Fe^{III} \; + \; CH_3OH$$

(with bridging O)

Though future investigations may correct these hypothetical mechanisms, at present the active intermediate attacking the C-H bond may in different oxygenases be represented in a noncontradictory way as containing the same $Fe^{IV}O$ fragment but bound to different neighboring groups. In the latter case, the presence of one more Fe^{IV} next to the active center naturally makes it a particularly active particle because of the electrophilic coaction of Fe^{IV} and because of the high redox potential of the transition $Fe^{IV}Fe^{IV} = O + 2e^- + 2H^+ \rightleftharpoons 2Fe^{III} + H_2O$ as compared with $Fe^{IV} = O + 2e^- + 2H^+ \rightleftharpoons Fe^{II} + H_2O$. Methane may be sufficiently reactive toward this particle, however, other more nucleophilic hydrocarbons must be even more reactive. We have seen that steric hindrance can be used to make methane the most active alkane.

The question about methane activation by methane monooxygenase still remains to be resolved, however. It may be that the hydrophobic cavity formed by protein surroundings is stabilized at the introduction of methane and more with this alkane than with other hydrocarbons (some kind of clathrate). This may be reached if the size of the cavity is more suitable for methane than for molecules of other alkanes.

The attempts to mimic cytochrome *P*-450 and other oxygenases started long ago. At the oxidation of simple metal salts such as Fe^{2+}, Cu^+, and Sn^{2+} by dioxygen in the presence of alkanes, coupled oxidation of the latter is observed to produce mainly alcohols and ketones, sometimes even with quite noticeable yields. However, the mechanism of these reactions usually involves free radicals (see ref. 2):

$$M + O_2 \xrightarrow{H^+} M^+ + HO_2\bullet$$

$$M + HO_2\bullet \xrightarrow{H^+} M^+ + H_2O_2$$

$$H_2O_2 + M \longrightarrow M^+ + OH^- + OH\bullet$$

Hydroxyl radicals react with the alkane

$$\bullet OH + RH \longrightarrow H_2O + \bullet R$$

Alkyl radicals react with the O_2 to form finally alcohol:

$$R\bullet + O_2 \longrightarrow RO_2\bullet$$

$$RO_2\bullet + M \xrightarrow{H^+} RO_2H + M^+$$

$$RO_2H + M \longrightarrow RO\bullet + OH^- + M^+$$

$$RO\bullet + RH \longrightarrow ROH + R\bullet \quad \text{etc.}$$

Naturally, the selectivity (e.g., the ratio prim:sec:tert) corresponds to that of hydroxyl attack on C-H bonds and there is no retention of configuration (e.g., in the case of *cis*-dimethylcyclohexane), if the tertiary C-H bond is hydroxylated.

An important step toward the creation of a hydroxylating complex similar to the active center of cytochrome *P*-450 was made in 1979 by Groves et al.,[44] who used iodosobenzene as an oxygen atom donor for iron porphyrins, which, accepting the oxygen atom, are transformed to an active form capable of epoxidizing olefins and hydroxylating alkanes. Iodosobenzene was also used for the Mn^{III} porphyrin complex to create an active species containing manganese.[45]

Full retention of the configuration was observed during the epoxidation of *cis*- and *trans*-stilbene by iodosobenzene with dimethylferroporphyrin as a catalyst demonstrating stereoselective oxene transfer similar to the enzymatic epoxidation. C-H bonds in alkanes are hydroxylated to form alcohols. At the same time the absence of stereo-selectivity at the epoxidation by chloroaqua(tetraphenylporphyrinato) manganese(III) with iodosobenzene shows that free radicals are formed in this case.[46]

The first observation of a high-valent iron porphyrin complex that is apparently a close model of the intermediate in enzymatic cytochrome *P*-450-catalyzed hydroxylation was made by Groves et al.[47] The complex was obtained by the oxidation of chloro-5,10,15,20-tetramesitylporphyrinate iron(III) by *m*-chloroperbenzoic acid in the mixture of methylene chloride with methanol. It was characterized as a complex of $Fe^{IV}O$ with a porphyrin π-cation radical (Fig. 3).

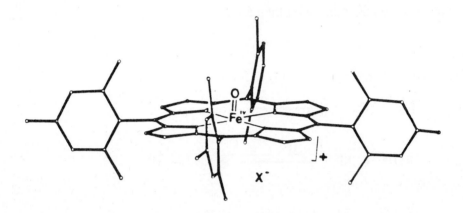

FIGURE 3.   Active center in iron porphyrin hydroxylating systems.[47]

It was demonstrated in a number of papers that C-H bond hydroxylation in a system containing a porphyrin complex with an oxygen donor (e.g., iodosobenzene) proceeds with unusual selectivity and sensitivity to steric hindrances created by both the C-H bond surrounding in an alkane and different groups attached to the porphyrin ring[48-51] (see Chapter VII). By increasing the electrophilic properties of the porphyrin complex by the introduction of nitro groups it is possible to hydroxylate the most inert hydrocarbons including methane[52] (Table II). However, the reactivity of the latter toward the porphyrin complex is still low.

To put into action a mechanism similar to that of the enzymatic process with molecular oxygen, it is first necessary to prevent the formation of $HO_2(O_2^-)$ and then OH radicals in the reaction of the O_2 complex with iron porphyrinate and also bimolecular reactions of the complexes with each other. In 1975 Collman et al.[53] synthesized the now well-known picket-fence complex of iron porphyrin with O_2. Bulky groups on the porphyrin ring stabilize the dioxygen complex preventing its reactions with other particles.

Table II. Oxidation of Alkanes in a Catalytic system in 1 ml Benzene[a]

Alkane	P (atm)	Products	x $10^5/M$	Relative reactivity per one C-H bond
Methane	80	Methanol	1 ± 0.5	0.08
Ethane	20	Ethanol	20 ± 5	(1)
Propane	5	1-Propanol	16 ± 5	0.8
		2-Propanol	100 ± 10	15
Hexane		1-Hexanol	15 ± 5	0.8
		2-Hexanol	180 ± 10	14
		3-Hexanol	70 ± 10	5
Cyclohexane		Cyclohexanol	1200 ± 100	30

[a] [TNPP FeCl] = $1mM$, [alkane] = $2\ M$, [PhIO] = 68 mM, 20°C, 3 hr.

The complex itself and other similar O_2 complexes do not hydroxylate C-H bonds. The problem that arises is how to transform O_2 complexes into the active species, i.e., to perform the reaction

$$PFe^{II}O_2 \xrightarrow{\ e,\ 2H^+\ } P^{+\bullet}Fe^{IV}O + H_2O$$

The cysteine-S^- group plays the role of the fifth axial ligand in the porphyrin complex in cytochrome P-450. Presumably S^- decreases O-O bond order and faciliates further O^{2-} abstraction to form water.

The analogs of the porphyrin complex with -S^- in the axial position were synthesized and then the complex $RS^-\cdots PFe^{II}\cdots O_2$ was synthesized, which is the analog of the precursor of the active species $P^{+\bullet}Fe^{IV}O$.[54]

However, the further conversion of the complex into an active one is not easy to perform in nonbiological conditions. The reaction with protons apparently leads to free radicals.

There is an indication that the oxygen atom (perhaps O^{2-}) initially inserts itself into the organic molecule of an enzyme effector. At the introduction of dihydrolipoic acid as such an effector into the functioning enzyme one of the dioxygen $^{18}O_2$ atoms turned out initially in the carboxyl group of the dihydrolipoic acid.[55]

Proceeding from this, Khenkin and Shteinman[56] used the addition of a strong acylating agent, acetic anhydride, in aprotic media (acetonitrile, benzene) to react with the $PFe^{III}O_2^{2-}$ complex to form the active species reacting with alkanes. In this case, the system need not have a nucleophilic participation of S^- in the fifth position of the prophyrin ring. Groves et al.[57] used a similar approach for the activation of porphyrin manganese complex. Benzoyl chloride was used in this case and the reaction observed was epoxidation of an olefin.

Thus, the active center selectively hydroxylating alkanes can be constructed from the dioxygen complex. Now the problem is to turn the system into the catalytic cycle, i.e., to prepare a full model of cytochrome *P*-450 monooxygenase.

Recently the first success in constructing such models was reported (see Chapter VI). Mansuy et al.[58,] and Fontecave and Mansuy[59] have described a catalytic biphasic H_2O-C_6H_6 system that had Mn^{III} (tetraphenylporphyrin)chloride and phase transfer agent as catalysts, and ascorbate as a reducing agent. The system activates dioxygen leading to selective epoxidation of olefins and oxidation of alkanes mainly to ketones. The formation of free radicals and their subsequent reaction with O_2 is possible in this system. An electrocatalytic system is described in ref. 60. It included an iron porphyrin complex immobilized on the graphite cathode immersed in acetonitrile solution containing acetic anhydride. Catalytic oxidation of cyclohexane to cyclohexanol and cyclohexanone was observed when an electric current was switched on. The observed selectivity, in particular, the partial retention of configuration in hydroxylation of *cis*-dimethylcyclohexane, indicates the nonradical mechanism of the reaction.

Another electrochemical model is based on manganese, porphryrin, and benzoic anhydride as an acylating reagent.[61] The system was used for olefin epoxidation.

Chemical models of cytochrome *P*-450 were also recently reported. In one of them[62] zinc amalgam was used as an electron donor. The iron porphyrin complex was present in acetonitrile solution together with acetic anhydride and methyl viologen (MV^{2+}), which transferred electrons from the amalgam to the iron porphyrin. Catalytic oxidation of cyclohexane was observed in this rather complex system. Using $^{18}O_2$ it was shown that one of the oxygen atoms of the molecule is found in the produced cyclohexanol, and the other one goes to acetate formed from acetic anhydride. The properties of this system, which mimics cytochrome *P*-450 monooxygenase, are close to those of the enzyme (see Table III).

Another catalytic system oxidizing alkanes was reported[63] with manganese porphyrin complex as a catalyst, with metallic zinc as the electron donor.

A molecular system as constructed by Groves et al[64] was organized with the participation of vesicles obtained from phospholipid molecules. Specially prepared manganese porphyrin complexes are incorporated into the vesicles and ascorbic acid is used as an electron donor. The system functions in water solution, the hydroxylation process taking part in the hydrophobic surrounding inside the vesicle. This model is approaching a natural enzymatic system.

Table III. Retention of Configuration, Regioselectivity, and Kinetic Isotope Effect (KIE) in Model and Native Systems

System	Retention of configuration (%)	Regioselectivity per C-H bond tert:sec:prim	KIE for hydroxylation of anisole/ CD_3-anisole
Zn/Hg - $MVCl_2$ - PFe - O_2 - Ac_2O	60	2.9:1	7
$PFeO_2^-$ - Ac_2O	60-70	3.3:1:0.08	7
Cytochrome *P*-450	100	7.5:1:0.07	7-8

Thus, systems modeling cytochrome *P*-450 are being successfully constructed. Tms as we have seen are not sufficiently active toward methane and its homologs. Apparently the approach to the oxidation of the most inert alkanes on the principles of biological oxidation must be conducted on the lines of nonheme iron complexes. A model system for these nonheme iron enzymes is nonaqueous Fenton reagent

$$Fe(II) + H_2O_2 \longrightarrow Fe^{IV}=O + H_2O$$

which was described by Groves et al.[65] in 1976. Stereoselective hydroxylation of C-H bonds by this system in acetonitrile was interpreted as evidence for $Fe^{IV}=O$ intermediate.

One possibility for involving $Fe^{II}O_2$ complexes is the use of α-ketoacids as effectors, similar to what happens in hydroxylation catalyzed by ketoglutarate-dependent dioxygenase. The first example of chemical models of this enzyme was recently demonstrated in our work[66] with the system Zn + MV^{2+} + O_2 in acetonitrile in the presence of iron perchlorate and pyruvic acid. Cyclohexane is oxidized to cyclohexanol and cyclohexanone in this system. Partial retention of configuration is observed in the oxidation of *cis*-1,2-dimethylcyclohexane.

Pyruvic acid is necessary and, because CO_2 is formed together with hydrocarbon oxidation products, it may be thought that the reaction proceeds according to the mechanism of biological oxidation (*vide supra*).

To increase the activity of the system and to involve simple alkanes including methane in the hydroxylation process a binuclear complex is evidently required with electrophilic coaction of the second iron atom. If this conclusion is correct we should soon be able to create a chemical model of methane monooxygenase.

V.   Conclusion

Homogeneous activation and functionalization of alkanes have been successfully achieved in the last 20 years both with alkane involvement in the coordination sphere of metal complexes and modeling of biological oxidation that seems to activate molecular oxygen rather than alkane molecules. Both lines of development have demonstrated possibilities for new selective reactions that are exceptionally interesting from the perspective of creating a new chemistry of alkanes.

It is hoped that further development in both directions will lead to new industrially important processes that in turn will encourage further research in this fascinating field of chemistry.

References

1.   Halpern, J., *Disc. Faraday Soc.* **1968,** *46,* 7.
2.   Shilov, A. E., *Activation of Saturated Hydrocarbons by Transition Metal Complexes.* D. Reidel, Dordrecht, 1984.
3.   (a) Garnett, J. L., Hodges, R. J., *J. Am. Chem. Soc.* **1967,** *89,* 4546; (b) Hodges, R. J., Garnett, J. L., *J.Phys. Chem.* **1968,** *72,* 1673.
4.   Gol'dshleger, N. F., Tyabin, M. B., Shilov, A. E., Shteinman, A. A., *Zh. Fiz. Khimi.* **1969,** *43,* 2174.
5.   (a) Gol'dshleger, N. F., Es'kova, V. V., Shilov, A. E., Shteinman, A. A., *Zh. Fiz. Khim.* **1972,** *46,* 1358; (b) Es'kova, V. V., Shilov, A. E., Shteinman, A. A., *Kinet. Katal.* **1972,** *13,* 534.
6.   Webster, D. E., *Adv. Organomet. Chem.* **1977,** *15,* 147.
7.   (a) Tretyakov, V. P., Rudakov, E. S., Bogdanov, A. V., Zimtseva, G. P., Kozhevina, L. I., *Dokl. Akad. Nauk SSSR* **1979,** *249,* 878; (b) Rudakov, E. S., Tretyakov, V. P., Galenin, A. A., Zimtseva, G. P., *Dopv. Akad. Nauk URSR* **1977,** ser. khim. *B,* 148.
8.   Kushch, L. A., Lavrushko, V. V., Misharin, Yu. S., Moravsky, A. P., Shilov, A. E., *Nouv. J. Chim.* **1983,** *7,* 729.
9.   Shilov, A. E., Shul'pin, G. B., *Usp. Khim.* **1987,** *56,* 754.
10.   Parshall, G. W., *Homogeneous Catalysis,* p. 179. John Wiley, New York, 1980.
11.   Crabtree, R. H., *Chem. Rev.* **1985,** *85,* 245.
12.   Langford, C. H., Gray, H. B., *Ligand Substitution Processes.* W. A. Benjamin, New York, 1965.
13.   Saillard, J. Y., Hoffmann, R., *J. Am. Chem. Soc.* **1984,** *106,* 2006.
14.   Shilov, A. E., Shteinman, A. A., *Coord. Chem.* **1977,** *24,* 97.
15.   Vinogradova, S. M., Shestakov, A. F., *Khim. Fiz.* **1984,** *3,* 371.
16.   Belluco, U., Michelin, R. A., Uguaglioti, P., Grociani, B. *J. Organomet. Chem.* **1983,** *250,* 565.
17.   Crabtree, R. H., Mihelcic, J. M., Quirk, J. M., *J. Am. Chem. Soc.* **1979,** *101,* 7738.

18. Baudry, D., Ephritikhine, M., Felkin, H., *J. Chem. Soc., Chem. Commun.* **1980**, 1243.

19. (a) Janowicz, A. H., Bergman, R. G., *J. Am. Chem. Soc.* **1982,** *104,* 352; **1983,** *105,* 2929; (b) Bergman, R. G., *Science* **1984,** *223,* 902.

20. (a) Hoyano, J. K., Graham, W. A. G., *J. Am. Chem. Soc.* **1982,** *104,* 3723; (b) Hoyano, J. K., McMaster, A. D., Graham, W. A. G., *J. Am. Chem. Soc.* **1983,** *105,* 7190.

21. Grigoryan, E. H., *Uspekhi Chim.* **1984,** *53,* 347.

22. Grigoryan, E. H., Dyachkovskii, F. S., Mullagaliev, I. R., *Dokl. Akad. Nauk SSSR* **1975,** *724,* 859.

23. Grigoryan, E. H., Dyachkovskii, F. S., Zhuk, S. Ya., Vyshinskaya, L. I., *Kinet. Katal* . **1978,** *19,* 1036.

24. Stepovik, L. P., Shilova, A. K., Shilov, A. E., *Dokl. Akad. Nauk SSSR* **1963,** *128,* 148.

25. Enikolopyan, N. S., Gyulumyan, Kh. R., Grigoryan, E. H., *Dokl. Akad. Nauk. SSSR* **1979,** *249,* 1980.

26. Grigoryan, E. H., Gyulumyan, Kh. R., Gutovaya, E. I., Enikolopyan, N. S., Ter-Kazarova, M. A., *Dokl. Akad. Nauk SSSR* **1981,** *257,*364.

27. Noskova, N. F., Sokolskii, D. V., Izteleuova, M. B., Gafarova, N. A., *Dokl. Akad. Nauk. SSSR* **1982,** *262,* 113.

28. Li, G. N., Zhang, L. F., Ding, F. J., Jin, H. L., Xiu, W. X., *Abstr. 5th Int. Symp. Homogen. Catal.,* Japan, 1986, p. 107.

29. Enikolopyan, N. C., Menchikova, G. N., Grigoryan, E. H., *Dokl. Akad. Nauk SSSR* **1986,** *291,* 11.

30. Sakakura, T., Tanaka, M., *Chem. Lett.* **1987,** 249.

31. Sakakura, T., Tanaka, M., *J. Chem. Soc., Chem., Commun.* **1987,** 758.

32. Rarfinkel, D., *Arch. Biochem. Biophys.* **1958,** *77,* 493.

33. Dalton, H., *Adv. Appl. Microbiol.* **1980,** *26,* 71.

34. Abbott, M. T., Udenfriend, S., in *Molecular Mechanisms of Oxygen Activation* O. Hayaishi, ed. Academic Press, New York, 1974.

35. (a) Sato, R., Omura, T. N. Y., eds., *Cytochrome P-450.* Academic Press, New York, 1978; (b) Guengerich, F. P., MacDonald, T.L., *Acc. Chem. Res.* **1984,** *17,* 9.

36. (a) Poulos, T. L., Finzel, B. C., Gunsalus, I. C., Wagner, G. C., Kraut, J., *J. Biol. Chem.* **1985,** *260,* 16122; (b) Poulos, T. L., Finzel, B. C., Howard, A. J., *Biochemistry* **1986,** *25,* 5314.

37. Groves, J. T., Haushalter, R. C., Nakamura, M., Nemo, T. E., Evans, B. J., *J. Am. Chem. Soc.,* **1981,** *103,* 2844.

38. Lichtenberger, F., Nastainczyk, W., Ullrich, V., *Biochem. Biophys. Res. Commun.* **1976,** *70,* 939.

39. Rudakov, E. S., *Reactions of Alkanes with Oxidants, Metal Complexes, and Radicals in Solutions* (Russian). Naukova Dumka, Kiev, 1985.

40. Hamilton, G. A., *J. Am. Chem. Soc.* **1964,** *86,* 3391.
     Hamilton, G. A., *Adv. Enzymol.* **1969,** *32,* 55.

41. White, R. E., Miller, J. P., Fanreau, L. V., Bhattacharya, A., *J. Am. Chem. Soc.* **1986,** *108,* 6024.

42. Woodland, M. P., Patel, D. S., Cammack, R., Dalton, H., *Biochim. Biophys. Acta* **1986,** *873,* 237.

43. Reem, R. C., Solomon, E. I., *J. Am. Chem. Soc.* **1987,** *109,* 1216.

44. Groves, J. T., Nemo, T. E., Myers, R. C., *J. Am. Chem. Soc.* **1979,** *101,* 1032.

45.   (a) Hill, C. L., Schardt, B. C., *J. Am. Chem. Soc.* **1980**, *102*, 6374; (b) Smegal, J. A., Schardt, B. C., Hill, C. L., *J. Am. Chem. Soc.* **1983**, *105*, 3510; (c) Smegal, J. A., Hill, C. L., *J. Am. Chem. Soc.* **1983**, *105*, 2920, 3515.

46.   Groves, J. T., Kruper, W. J., Jr., Haushalter, R. C., *J. Am. Chem. Soc.* **1980**, *102*, 6375.

47.   Groves, J. T., Haushalter, R. C., Nakamura, M., Nemo, T. E., Evans, B. J., *J. Am. Chem. Soc.* **1981**, *103*, 2844.

48.   Mansuy, D., Bartoli, J. F., Momenteau, M., *Tetrahedron Lett.* **1982**, *23*, 2781.

49.   Groves, J. T., Nemo, T. E., *J. Am. Chem. Soc.* **1983**, *105*, 5786.

50.   Khenkin, A. M., Semeikin, A. S., Koyfman, O. I., Shilov, A. E., Shteinman, A. A., *Tetrahedron Lett.* **1985**, *26*, 24247.

51.   Cook, B. R., Reinert, T. J., Suslick, K. S., *J. Am. Chem. Soc.* **1986**, *108*, 7281.

52.   Belova, V. S., Khenkin, A. M., Shilov, A. E., *Kinet. Katal.* **1987**, *28*, 1011.

53.   Collman, J. P., Gagné, R. R., Reed, C. A., Halbert, T. R., Lang., G., Robinson, W. T., *J. Am. Chem. Soc.* **1975**, *97*, 1427.

54.   Budyka, M. F., Khenkin, A. M., Shteinman, A. A., *Biochem. Biophys. Res. Commun..* **1981**, *101*, 615.

55.   Sligar, S. G., Kennedy, K. A., Pearson, D. C., *Proc. Natl. Acad. Sci. U.S.A.* **1980**, *77*, 1240.

56.   (a) Khenkin, A. M., Shteinman, A. A., *Izv. Akad. Nauk SSSR, ser. khim.* **1982**, 1668; (b) Khenkin, A. M., Shteinman, A. A., *Oxidation Commun.* **1983**, *4*, 433; (c) Khenkin, A. M., Shteinman, A. A., *J. Chem. Soc. Chem. Comm.* **1984**, 1219.

57.   Groves, J. T., Watanabe, J., McMurry, T. J., *J. Am. Chem. Soc.* **1983**, *105*, 4484.

58.   Mansuy, D., Fontecave, M., Bartoli, J. F., *J. Chem. Soc., Chem. Commun.* **1983**, 253.

59.   Fontecave, M., Mansuy, D., *Tetrahedron* **1984**, *40*, 4297.

60.   (a) Khenkin, A. M., Khenkina, T. V., Shilov, A. E., *Proc. 4th ISHC, V.2*, Leningrad, 1984, p. 46; (b) Khenkin, A. M., Shilov, A. E., *React. Kin. Cat. Lett*; **1987**, *33*, 125.

61.   Murray, R. W., Creager, S. E., Raybuck, S. A., *J. Am. Chem. Soc.* **1986**, *108*, 4225.

62.   (a) Khenkin, A. M., *Izv. Akad. Nauk SSSR, ser. khim.* **1986**, 2329; (b) Karasevich, E. I., Khenkin, A. M., Shilov, A. E., *J. Chem. Soc., Chem. Commun.* **1987**, 731.

63.   Battioni, P., Bartoli, J. F., Leduc, P., Fontecave, M., Mansuy, D., *J. Chem. Soc., Chem. Commun.* **1987**, 791.

64.   Groves, J. T., et al, *J. Am. Chem. Soc.*, in press.

65.   Groves, J. T., Van Der Puy, M., *J. Am. Chem. Soc.* **1976**, *98*, 5290.

66.   Belova, V. S., Khenkin, A. M., Shilov, A. E., *Kinet. Katal.* in press.

Chapter II

ELECTROPHILIC CHEMISTRY OF ALKANES

George A. Olah, Omar Farooq, and G. K. Surya Prakash

Donald P. and Katherine B. Loker Hydrocarbon Research Institute
Department of Chemistry
University of Southern California
Los Angeles, California 90089

I.   Introduction

Before the turn of the century, saturated hydrocarbons (paraffins) played only a minor role in industrial chemistry. They were mainly used as a source of paraffin wax as well as for heating and lighting oils. Aromatic compounds such as benzene, toluene, phenol, and naphthalene obtained from destructive distillation of coal were the main source of organic materials used in the preparation of dyestuffs, pharmaceutical products, etc. Calcium carbide-based acetylene was the key starting material for the emerging synthetic organic industry. It was the ever increasing demand for gasoline after the first world war that initiated study of isomerization and cracking reactions of petroleum fractions. After the second world war, rapid economic expansion necessitated increasingly abundant and cheap sources for chemicals and as a result the industry switched to petroleum-based ethylene as the main source of chemical raw material. One of the major difficulties that had to be overcome is the low reactivity of some of the major components of the petroleum. The lower boiling components (up to 250°C) are primarily straight-chain saturated hydrocarbons or paraffins, which, as their name indicates (*parum affinis*: too little affinity), have very little reactivity. Consequently the lower paraffins were cracked to give olefins (mainly ethylene, propylene, and butylenes). The straight-chain liquid hydrocarbons also have very low octane numbers, which makes them less desirable as gasoline components. To transform these paraffins into useful components for gasoline and other chemical applications, they have to undergo diverse reactions such as isomerization, cracking, or alkylation. These reactions, which are used on a large scale in industrial processes, necessitate acidic catalysts (at temperature around 100°C) or noble metal catalysts (at higher temperature, 200-500°C) capable of activating the strong covalent C-H or C-C bonds.[1] Since the early 1960s, superacids[2] are known to react with saturated hydrocarbons, even at temperatures much below 0°C. This discovery initiated extensive studies devoted to electrophilic reactions and conversions of saturated hydrocarbons, including the parent methane.

II. C-H (and C-C Bond) Protolysis and Hydrogen-Deuterium Exchange

The fundamental step in the acid-catalyzed conversion processes is the formation of the intermediate carbocations. Whereas all studies involving isomerization, cracking, and alkylation reactions under acidic conditions indicate that a trivalent carbocation (carbenium ion) is the key intermediate, the mode of formation of this reactive species from the neutral hydrocarbon remained controversial for many years.

$$\text{R-H} \xrightarrow{\text{Acid}} \text{R}^+ \rightleftharpoons \begin{array}{l} \longrightarrow \text{Isomerization} \\ \longrightarrow \text{Cracking} \\ \longrightarrow \text{Alkylation-homologation} \end{array}$$

In 1946, Bloch et al.[3] observed that *n*-butane would isomerize to isobutane under the influence of pure aluminum chloride only in the presence of HCl. They proposed that the ionization step takes place through initial protolysis of the alkane as evidenced by the formation of minor amounts of hydrogen in the initial stage of the reaction.

$$n\text{-C}_4\text{H}_{10} + \text{HCl} \xrightarrow{\text{AlCl}_3} sec\text{-C}_4\text{H}_9^+ \text{ AlCl}_4^- + \text{H}_2$$

The first direct evidence of protonation of alkanes under superacid conditions has been reported independently by Olah and Lukas[4] as well as by Hogeveen and co-workers.[5]

When *n*-butane or isobutane was reacted with $\text{HSO}_3\text{F:SbF}_5$ (Magic Acid), *tert*-butyl cation was formed exclusively as evidenced by a sharp singlet at 4.5 ppm (from TMS) in the ^{1}H-NMR spectrum. In excess Magic Acid, the stability of the ion is remarkable and the NMR spectrum of the solution remains unchanged even after having been heated to 110°C.

$$(CH_3)_3CH \xrightarrow[\text{room temperature}]{\text{HSO}_3\text{F:SbF}_5} (CH_3)_3C^+ \; SbF_5FSO_3^- \; + \; H_2 \xleftarrow{\text{HSO}_3\text{F:SbF}_5} n\text{-}C_4H_{10}$$

It was shown[6] that the *tert*-butyl cation undergoes degenerate carbon scrambling at higher temperatures. A low limit of $E_a \sim 30$ kcal/mol was estimated for the scrambling process, which could correspond to the energy difference between *tert*-butyl and primary isobutyl cation (the latter being partially delocalized, "protonated cyclopropane") involved in the isomerization process.

n-Pentane and isopentane are ionized under the same conditions to the *tert*-amyl cation. *n*-Hexane and the branched C_6 isomer ionize in the same way to yield a mixture of the three tertiary hexyl ions as shown by their ^{1}H-NMR spectra.

Both methylcyclopentane and cyclohexane were found to give the methylcyclopentyl ion, which is stable at low temperature, in excess superacid.[7] When alkanes with seven or more carbon atoms were used, cleavage was observed with formation of the stable *tert*-butyl cation. Even paraffin wax and polyethylene ultimately gave the *tert*-butyl ion after complex fragmentation and ionization processes.

In compounds containing only primary hydrogen atoms such as neopentane and 2,2,3,3-tetramethylbutane a carbon-carbon bond is broken rather than a carbon-hydrogen bond.[8]

$$CH_3 - \overset{\overset{\displaystyle CH_3}{|}}{\underset{\underset{\displaystyle CH_3}{|}}{C}} - CH_3 \xrightarrow{H^+} CH_4 + (CH_3)_3\,C^+$$

Hogeveen and co-workers suggested a linear transition state for the protolytic ionization of hydrocarbons. This, however, may be the case only in sterically hindered systems. Results of the protolytic reactions of hydrocarbons in superacidic media were interpreted by Olah as indication for the general electrophile reactivity of covalent C-H and C-C single bonds of alkanes and cycloalkanes. The reactivity is the result of the σ-donor ability of a shared electron pair (of σ-bond) via two-electron, three-center bond formation. The transition state of the reactions consequently are of three-center bound pentacoordinate carbonium ion nature. A strong indication for the mode of protolytic attack was obtained from deuterium-hydrogen exchange studies. Monodeuteromethane was reported to undergo C-D exchange without detectable side reactions in the HF:SbF$_5$ system.[9] d$_{12}$-Neopentane when treated with Magic Acid was also reported[10] to undergo H-D exchange before cleavage.

$$R-\underset{R}{\overset{R}{\underset{|}{\overset{|}{C}}}}-H \xrightarrow{\ H^+\ } \left[R-\underset{R}{\overset{R}{\underset{|}{\overset{|}{C}}}} \cdots \overset{H}{\underset{H}{\big\langle}} \right]^+ \longrightarrow (R_3)C^+ + H_2$$

$$CH_3D + HF:SbF_5 \longrightarrow \left[H_3C \cdots \overset{H}{\underset{D}{\big\langle}} \right]^+ \longrightarrow CH_4 + DF:SbF_5$$

Based on the demonstration of H-D exchange of molecular hydrogen (and deuterium) in superacid solutions,[11a] Olah suggested that these reactions go through trigonal isotopomeric H$_3^+$ ions in accordance with theoretical calculations and recent IR studies.[11b]

$$\underset{D \qquad\quad H}{\overset{H}{\triangle}}^+ \ , \quad \underset{D \qquad\quad D}{\overset{H}{\triangle}}^+$$

Consequently, the reverse reaction of protolytic ionization of hydrocarbons to carbenium ions, i.e., the reduction of carbenium ion by molecular hydrogen,[12,13] can be considered as alkylation of H$_2$ by the electrophilic carbenium ion through the pentacoordinate carbonium ion. Indeed Hogeveen has experimentally proved this point by reacting stable alkyl cations in superacids with molecular hydrogen.

$$R_3C^+ + \underset{H}{\overset{H}{\underset{|}{|}}} \longrightarrow \left[R_3C \cdots \overset{H}{\underset{H}{\big\langle}} \right]^+ \longrightarrow R_3CH + H^+$$

Further evidence for the pentacoordinate carbonium ion mechanism of alkane protolysis was obtained in the H-D exchange reaction observed with isobutane. When isobutane is treated with deuterated superacids (DSO$_3$F:SbF$_5$ or DF:SbF$_5$) at low temperature

(-78°C) and atmospheric pressure, the initial hydrogen-deuterium exchange is observed only at the tertiary carbon. Ionization yields only deuterium-free *tert*-butyl cation and HD.[14] Recovered isobutane from the reaction mixture shows at low temperature only methine hydrogen-deuterium exchange. This result is best explained as proceeding through a two-electron, three-center bound pentacoordinate carbonium ion.

$$(CH_3)_3CH \xrightarrow{D^+} \left[(CH_3)_3C \cdots \begin{matrix} H \\ D \end{matrix} \right]^+ \longrightarrow (CH_3)_3CD$$

$$\downarrow \text{-HD}$$

$$(CH_3)_3C^+$$

The H-D exchange in isobutane in superacid media is fundamentally different from the H-D exchange observed by Otvos et al.[15] in D_2SO_4, who found the eventual exchange of all the nine methyl hydrogens but not of the methine hydrogen.

$$CH_3-\underset{\underset{CH_3}{|}}{\overset{\overset{CH_3}{|}}{C}}-H \xrightarrow[\text{Excess}]{D_2SO_4} (CH_3)_3C^+ \xrightarrow{-H^+} \underset{CH_3}{\overset{CH_2}{\underset{}{C}}}{\overset{||}{C}}CH_3$$

$$\downarrow D^+$$

$$\underset{CH_3}{\overset{CH_3}{C^+-CH_2D}} \xleftarrow[+D^+]{-H^+}$$

$$(CD_3)_3C^+ \longleftarrow \longleftarrow$$

$$\downarrow (CH_3)_3CH$$

$$(CD_3)_3CH$$

Otvos et al.[15] suggested that under the reaction conditions a small amount of *tert*-butyl cation is formed in an oxidative step which than deprotonates to isobutylene. The reversible protonation (deuteration) of isobutylene was responsible for the H-D exchange of the methyl hydrogens, whereas the tertiary hydrogen is involved only in intermolecular hydride transfer with excess unlabeled isobutane (at the CH position). Under superacidic conditions, in which no olefin formation occurs, the reversible isobutylene protonation cannot be involved in the exchange reaction. On the other hand, a kinetic study of hydrogen-deuterium exchange in deuteroisobutane[16] showed that the exchange of the tertiary hydrogen was appreciably faster than that of methyl hydrogens.

The nucleophilic nature of the alkanes is also shown by the influence of the acidity level on their solubility. Torck and co-workers[17] have investigated the variation of the composition of the catalytic phase as a function of SbF_5 concentration in isomerization of pentane in $HF-SbF_5$.

The total amount of hydrocarbons increases from 1.6 to 14.6% in weight when the SbF_5 concentration varies from 0 to 6.8 M. The amount of carbenium ions increases linearly with the SbF_5 concentration, and solubility of the hydrocarbon itself reaches a maximum for 5 M SbF_5. As a result of the apparent decrease in solubility of the hydrocarbon protolysis as well as the change in the composition of the acid, SbF_6 anions are transformed to Sb_2F_{11} anions.

$$RH + H^+Sb_2F_{11}^- \longrightarrow R^+Sb_2F_{11}^- + H_2$$

One of the difficulties in understanding the carbocationic nature of acid-catalyzed transformations of alkanes via the hydride abstraction mechanism was that no stoichiometric amount of hydrogen gas evolution was observed from the reaction mixture. For this reason, a complementary mechanism was proposed involving direct hydride abstraction by the Lewis acid.[18]

$$RH + 2SbF_5 \longrightarrow R^+SbF_6^- + SbF_3 + HF$$

Olah has pointed out that if SbF_5 would abstract H^-, it would need to form an SbF_5H^- ion involving an extremely weak Sb-H bond compared to the strong C-H bond being broken.[14] Calculations based on thermodynamics also show that the direct oxidation of alkanes by SbF_5 is not feasible. Hydrogen is generally assumed to be partially consumed in the reduction of one of the superacid components.

$$2HSO_3F + H_2 \longrightarrow SO_2 + H_3O^+ + HF + SO_3F^- \qquad \Delta H = -33 \text{ kcal/mol}$$

$$SbF_5 + H_2 \longrightarrow SbF_3 + 2HF \qquad \Delta H = -49 \text{ kcal/mol}$$

The direct reduction of SbF_5 in the absence of hydrocarbons by molecular hydrogen necessitates, however, more forcing conditions (50 atm, higher temperature) that suggest that the protolytic ionization of alkanes proceeds probably via solvation of protonated alkane by SbF_5 and concurrent ionization-reduction.[14]

$$\left[\begin{array}{c} \underset{R}{\overset{R}{R-\overset{|}{\underset{|}{C}}}} \overset{+}{\cdots} \diagdown_{\diagup} \overset{H\cdots F}{\underset{H\cdots F}{}} \diagdown SbF_3 \\ SbF_6^{-} \end{array} \right] \longrightarrow (R_3)C^+ + 2HF + SbF_3$$
$$SbF_6^{-}$$

In studies involving solid acid-catalyzed hydrocarbon cracking reactions using HZSM-5 zeolite, Haag and Dessau[19] were able to account nearly quantitatively for the H_2 formed in the protolytic ionization step of the reaction. This is a consequence of the solid acid zeolite catalyst not being reduced by hydrogen.

It must also be pointed out, however, that initiation of acid-catalyzed alkane transformations under oxidative conditions (chemical or electrochemical) can also involve radical cations or radical paths leading to the initial carbenium ions. In the context of our present discussion we shall only briefly elaborate on this interesting chemistry and limit our treatment mainly to protolytic reactions.

Under strongly acidic conditions C-H bond protolysis is not the only pathway by which hydrocarbons are heterolytically cleaved. Carbon-carbon bonds can also be cleaved by C-C protolysis[20,21] involving pentacoordinate intermediates.

$$\diagup_{\diagdown} \!\! C \!\! - \!\! C \!\! \diagdown^{\diagup} + H^+ \longrightarrow \left[\diagup_{\diagdown} \!\! C \overset{H}{\cdots} C \!\! \diagdown^{\diagup} \right]^{+} \longrightarrow \diagdown_{\underset{|}{\diagup}} \!\! C \!\! + \quad + \quad \diagdown \!\! C \!\! - \!\! H$$

Similarly, many carbon-heteroatom bonds are also cleaved[22,23] under strong acid catalysis involving pentacoordinate carbon intermediates.

III. Alkane Conversion by Electrochemical Oxidation in Superacids

In 1973, Fleischman and co-workers[24] showed that the anodic oxidation potential of several alkanes in HSO_3F was dependent on the proton donor ability of the medium. This acidity dependence shows that there is a rapid protonation equilibrium before the electron transfer step and it is the protonated alkane that undergoes oxidation.

$$RH \xrightarrow{HSO_3F} RH_2^+ \xrightarrow[Pt\ anode]{} R^+ \underset{\diagdown}{\overset{\diagup}{\Longrightarrow}} \begin{array}{l} \overset{RH}{\longrightarrow} Oligomers \\ \overset{CO}{\longrightarrow} Acids \\ \underset{CH_3COO^-}{\longrightarrow} Esters \end{array}$$

More recently, the electrochemical oxidation of lower alkanes in the HF solvent system has been investigated by Devynck and co-workers over the entire pH range.[25] Classical and cyclic voltammetry show that the oxidation process depends largely on the acidity level. Isopentane (2-methylbutane, M2BH), for example, undergoes two-electron oxidation in HF:SbF$_5$ and HF:TaF$_5$ solutions.[26]

$$M2BH - 2e^- \longrightarrow M2B^+ + H^+$$

In the higher acidity region, the intensity-potential curve shows two peaks (at 0.9 and 1.7 V, respectively, vs the Ag/Ag+ system). The first peak corresponds to the oxidation of the protonated alkane and the second to the oxidation of the alkane itself.

The chemical oxidation process in the acidic solution (Reaction 1) can be considered as a sum of two electrochemical reactions (Reactions 2 and 3).

$$RH + H^+ \; \rightleftharpoons \; R^+ + H_2 \tag{1}$$

$$2H + 2e^- \; \rightleftharpoons \; H_2 \; (E^0 \text{ for } H^+/H_2) \tag{2}$$

$$RH \; \rightleftharpoons \; R^+ + H^+ + 2e^- \; (E^0 \text{ for } H^+/H_2) \tag{3}$$

The oxidation of the alkane (M2BH) by H+ gives the carbocation only at pH values below 5.7. In the stronger acids it is the protonated alkane that is oxidized. At pH values higher than 5.7, oxidation of isopentane gives the alkane radical, which dimerizes or is oxidized in a pH-independent process

$$M2BH \xrightarrow{-e^-} M2BH^{\bullet+} \longrightarrow M2B^\bullet + H^+$$

$$2M2B^\bullet \longrightarrow (M2B)_2$$

$$M2B^\bullet \xrightarrow{-e^-} M2B^+$$

Commeyras and co-workers[27] have also electrochemically oxidized alkanes (anodic oxidation). However, the reaction results in condensation and cracking of alkanes. The results are in agreement with the alkane behavior in superacid media and indicate the ease of oxidation of tertiary alkanes. However, high acidity levels are necessary for the oxidation of alkanes possessing only primary C-H bonds.

Once the alkane has been partly converted into the corresponding carbenium ion, then typical rearrangement, fragmentation, hydrogen transfer, as well as alkylation reactions will occur.

IV. Isomerization of Alkanes

Acid-catalyzed isomerization of saturated hydrocarbons was first reported in 1933 by Nenitzescu and Dragan.[28]   They found that when *n*-hexane was refluxed with aluminum chloride, it was converted into its branched isomers. This reaction is of major economic importance as the straight-chain C_5-C_8 alkanes are the main constituents of gasoline obtained by refining of the crude oil. Because the branched alkanes have a considerably higher octane number than their linear counterparts, the combustion properties of gasoline can be substantially improved by isomerization.

The isomerization of *n*-butane to isobutane is of substantial importance because isobutane reacts under mild acidic conditions with olefins to give highly branched hydrocarbons in the gasoline range. A variety of useful products can be obtained from isobutane: isobutylene, *tert*-butyl alcohol, methyl *tert*-butyl ether, and *tert*-butyl hydroperoxide.[25] A number of methods involving solution as well as solid acid catalysts[26] have been developed to achieve isomerization of *n*-butane as well as other linear higher alkanes to branched isomers.

A substantial number of investigations have been devoted to this isomerization reaction and a number of reviews are available.[29-31] The isomerization is an equilibrium reaction that can be catalyzed by various strong acids. In the industrial processes, aluminum chloride and chlorinated alumina are the most widely used catalysts. Whereas these catalysts become active only at temperatures above 80-100°C, superacids are capable of isomerizing alkanes at room temperature and below. The advantage is that lower temperatures thermodynamically favor the most branched isomers.

The electron donor character of C-H and C-C single bonds that leads to pentacoordinated carbonium ions explains the mechanism of acid catalyzed isomerization of *n*-butane as shown in Scheme 1. Carbon-carbon bond protolysis, however, can also take place giving methane, ethane, and propane.

Related alkanes such as pentanes, hexanes, and heptanes isomerize by similar pathways with increasing tendency toward cracking (i.e., C-C bond cleavage).

During the isomerization process of pentanes, hexanes, and heptanes, cracking of the hydrocarbon is an undesirable side reaction.

The discovery that cracking can be substantially suppressed by hydrogen gas under pressure was of significant importance. In our present-day understanding, the effect of hydrogen is to quench

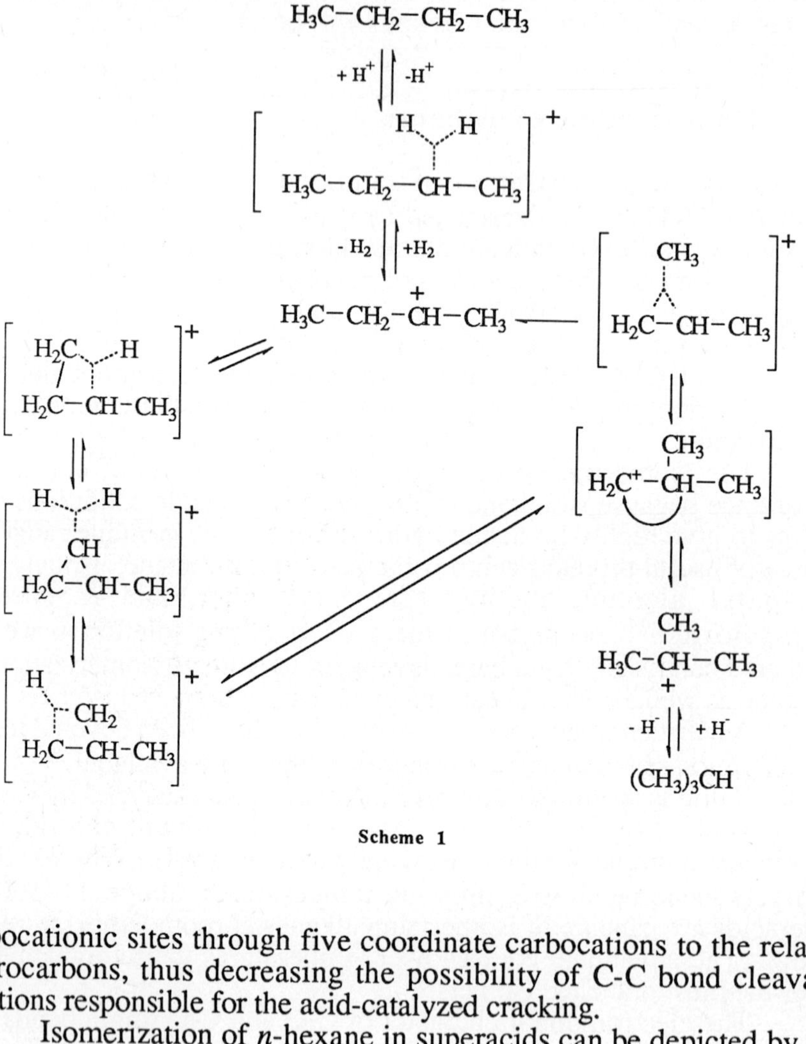

Scheme 1

carbocationic sites through five coordinate carbocations to the related hydrocarbons, thus decreasing the possibility of C-C bond cleavage reactions responsible for the acid-catalyzed cracking.

    Isomerization of *n*-hexane in superacids can be depicted by the following three steps:

STEP 1. FORMATION OF THE CARBENIUM ION:

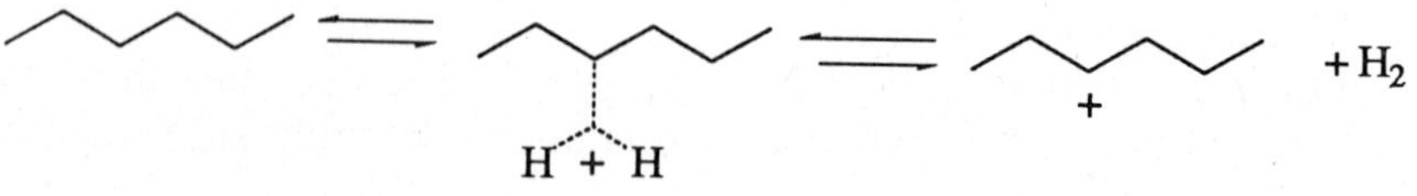

STEP 2. ISOMERIZATION OF THE CARBENIUM ION VIA HYDRIDE SHIFTS, ALKIDE SHIFTS, AND PROTONATED CYCLOPROPANE (FOR THE BRANCHING STEP):

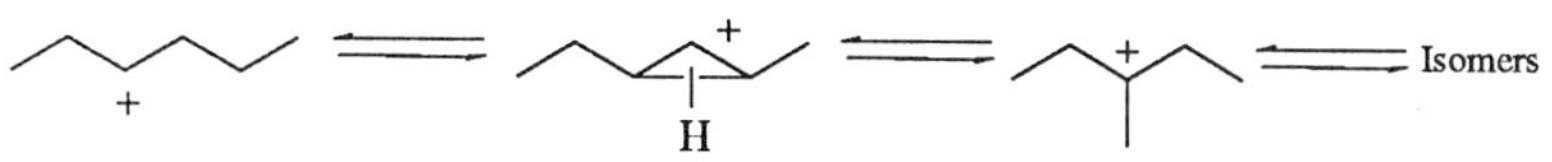

STEP 3. HYDRIDE TRANSFER FROM THE ALKANE TO THE INCIPIENT CARBENIUM ION:

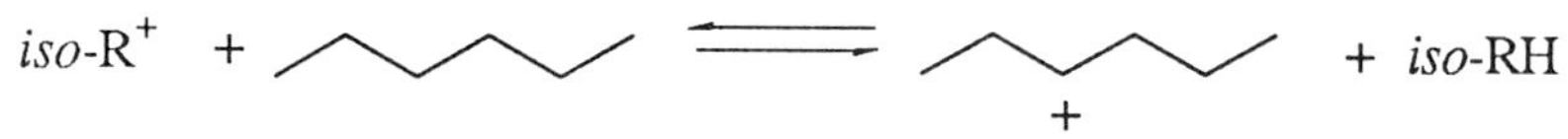

Whereas Step 1 is stoichiometric, Steps 2 and 3 form a catalytic cycle involving the continuous generation of carbenium ions via hydride transfer from a new hydrocarbon molecule (Step 3) and isomerization of the corresponding carbenium ion (Step 2). This catalytic cycle is controlled by two kinetic and two thermodynamic parameters that can help orient the isomer distribution depending on the reaction conditions. Step 2 is kinetically controlled by the relative rates of hydrogen shifts, alkyl shifts, and protonated cyclopropane formation and it is thermodynamically controlled by the relative stabilities of the secondary and tertiary ions. Step 3, however, is kinetically controlled by the hydride transfer from excess of the starting hydrocarbon and by the relative thermodynamic stability of the various hydrocarbon isomers.

For these reasons, the outcome of the reaction will be very different depending on which thermodynamic or kinetic factor will be favored. In the presence of excess hydrocarbon in equlibrium with the catalytic phase and long contact times, the thermodynamic hydrocarbon isomer distribution is attained. However, in the presence of a large excess of acid, the product will reflect the thermodynamic stability of the intermediate carbenium ions (which, of course, is different from that of hydrocarbons) if rapid hydride transfer-quenching can be achieved. Torck and co-workers[17,32] have shown that the limiting step, in the isomerization of *n*-hexane and *n*-pentane with the HF:SbF$_5$ acid catalyst, is the hydride transfer with sufficient contact in a batch reactor, as indicated by the thermodynamic isomer distribution of C$_6$ isomers.

The isomerization of *n*-pentane in superacids of the type R$_F$SO$_3$H:SbF$_5$ (R$_F$ = C$_n$F$_{2n+1}$) has been investigated by Commeyras and co-workers.[33] The influence of parameters such as acidity, hydrocarbon concentration, nature of the perfluoroalkyl group, total pressure, hydrogen pressure, temperature, and agitation has been studied.

In weaker superacids such as neat CF$_3$SO$_3$H, alkanes that have no tertiary hydrogen are isomerized only very slowly, as the acid itself is not strong enough to hydride abstract to form the initial carbocation. This lack of reactivity can be overcome by introducing initiator carbenium ions in the medium to start the catalytic process. For this purpose, alkenes may be added, which are directly converted into their corresponding carbenium ions by protonation, or alternatively the alkane may be electrochemically oxidized (anodic oxidation). Both methods are useful to initiate isomerization and cracking reactions. The latter method has been studied by Commeyras and co-workers.[27]

$$ RH \xrightarrow{-e} RH^{+} \bullet \xrightarrow{-H^{+}} R^{\bullet} \xrightarrow{-e} R^{+} $$

Olah[34] has developed a process wherein natural gas liquids containing saturated straight-chain hydrocarbons can be conveniently upgraded to highly branched hydrocarbons (gasoline upgrading) using $HF:BF_3$ catalyst. The addition of a small amount of olefins, preferably butenes, helped the reaction rate. This can be readily explained by the formation of alkyl fluorides (HF addition to olefins), whereby an equilibrium concentration of cations is maintained in the system during the upgrading reaction. The gasoline upgrading process is also improved in the presence of hydrogen gas, which helps to suppress side reactions such as cracking and disproportionation and minimize the amount of hydrocarbon products entering the catalyst phase of the reaction mixture. The advantage of the above method is that the catalysts $HF:BF_3$ (being gases at ambient temperatures) can easily be recovered and recycled.

Olah has also found CF_3SO_3H and related superacids as efficient catalysts for the isomerization of *n*-butane to isobutane.[35]

The difficulties encountered in handling liquid superacids and the need for product separation from catalyst in batch processes have stimulated research in the isomerization of alkanes over solid superacids. The isomerization of 2-methyl- and 3-methylpentane and 2,3-dimethylbutane, using SbF_5-intercalated graphite as a catalyst, has been studied in a continuous flow system.[36]

The isomerization of cycloalkanes with acid catalysts is also well known. Over SbF_5-intercalated graphite it can be achieved at room temperature without the usual ring opening and cracking reactions that occur at higher temperatures and lower acidities.[37] In the presence of excess hydrocarbon after several hours, the thermodynamic equilibrium is reached for the isomers. Interconversion between cyclohexane and methylcyclopentane yields the thermodynamic equilibrium mixture. It should be mentioned again, however, that the thermodynamic ratio for the neutral hydrocarbon isomerization is very different as compared with the isomerization of the corresponding ions. The large energy difference (> 10 kcal) between the secondary cyclohexyl cation and the tertiary methylcyclopentyl ion means that in solution chemistry in the presence of excess superacid only the latter can be observed.[7]

Whereas the cyclohexane-methylcyclopentane isomerization involves initial formation of the cyclohexyl (methylcyclopentyl) cation (i.e., via protolysis of a C-H bond) it should be mentioned that in the acid-catalyzed isomerization of cyclohexane up to 10% hexanes are also formed and this is indicative of C-C bond protolysis.

The potential of other solid superacid catalysts, such as Lewis acid-treated metal oxides for the skeletal isomerization of hydrocarbons, has been studied in a number of cases. The reaction of butane with $SbF_5:SiO_2:TiO_2$ gave the highest conversion forming C_3, *iso*-C_4, *iso*-C_5, and traces of higher alkanes. $TiO_2:SbF_5$, on the other hand, gave the highest selectivity for skeletal isomerization of butane. With $SbF_5:Al_2O_3$, however, the conversions were very low.[38,39]

Similarly, isomerization of pentane and 2-methylbutane over a number of SbF_5-treated metal oxides has been investigated. The $TiO_2:ZrO_2:SbF_5$ system was the most reactive and, at the maximum conversion, the selectivity for skeletal isomerization was found to be ~100%.[40]

A comparison of the reactivity of SbF_5-treated metal oxides with that of HSO_3F-treated catalysts showed that the former is by far the better catalyst for reaction of alkanes at room temperature, although the HSO_3F-treated catalyst showed some potential for isomerization of n-butane.[41]

$SbF_5:SiO_2:Al_2O_3$ has been used to isomerize a series of alkanes at or below room temperature. Methylcyclopentane, cyclohexane, propane, butane, 2-methylpropane, and pentane all reacted at room temperature, whereas methane, ethane and 2,2-dimethylpropane could not be activated.[41]

The isomerization of a large number of C_{10} hydrocarbons under strongly acidic conditions gives the unusually stable isomer adamantane. The first such isomerization was reported by Schleyer in 1957.[42a] During a study of the facile aluminum chloride-catalyzed *endo:exo* isomerization of tetrahydrodicyclopentadiene, difficulty was encountered with a highly crystalline material that often clogged distillation heads. This crystalline material was found to be adamantane. Adamantane can be prepared from a variety of C_{10} precursors and involves a series of hydride and alkyde shifts. The mechanism of the reaction has been reviewed in detail.[42b] Fluoroantimonic acid ($HF:SbF_5$) very effectively isomerizes tetrahydrodicyclopentadiene into adamantane.[42c] The work has been extended to other conjugate superacids,[43] wherein adamantane is obtained in quantitative yields.

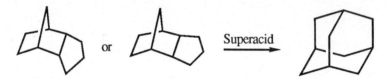

More recently[44] isomerization of polycyclic hydrocarbons $C_{4n+6}H_{4n+12}$ ($n = 1, 2, 3$) was achieved to adamantane, diamantane, and triamantane,

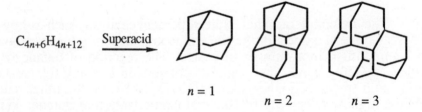

respectively, using new generation acid systems such as $B(OSO_2CF_3)_3$, $B(OSO_2CF_3)_3/CF_3SO_3H$, and SbF_5/CF_3SO_3H. The yields of these cage compounds substantially improved upon addition of catalytic amount 1-bromo (chloro, fluoro)-adamantane as a source of 1-adamantyl cation. The reactions were further promoted by sonication (ultrasound treatment).

A good example of superacid-catalyzed isomerization is the fascinating preparation of 1, 16-dimethyldodecahedrane from the seco-dimethyldodecahedrene.[45]

A novel approach to highly symmetrical pentagonal dodecahedrane was provided by Prinzbach, Schleyer, and Maier, which involved gas phase isomerization of [1.1.1.1] pagodane over 0.1% Pt/Al_2O_3 under an atmosphere of H_2 at 360°C. Dodecahedrane was obtained in 8% yield. The isomerization could involve a dehydrogenation-isomerization-hydrogenation path.

V. Acid-Catalyzed Cleavage (Cracking) Reactions of Alkanes (β-Cleavage vs C-C Bond Protolysis)

The reduction in molecular weight of various fractions of crude oil is an important operation in petroleum chemistry. The process is called cracking. Catalytic cracking is usually achieved by passing the hydrocarbons over a metallic or acidic catalyst, such as crystalline zeolites at about 400-600°C. The molecular weight reduction involves carbocationic intermediates and the mechanism is based on the β-scission of carbenium ions.

The main goal of catalytic cracking is to upgrade higher boiling oils, which, through this process, yield lower hydrocarbons in the gasoline range.[46,47]

Historically, the first cracking catalyst used was aluminum trichloride. With the development of heterogeneous solid and supported acid catalysts, the use of $AlCl_3$ was superceded, since its activity was primarily as the result of its ability to bring about acid-catalyzed cleavage reactions.

The development of highly acidic superacid catalysts in the 1960s again focussed attention on acid-catalyzed cracking reactions. $HSO_3F:SbF_5$, trade named Magic Acid, derived its name from its remarkable ability to cleave higher-molecular-weight hydrocarbons, such as paraffin wax, to lower-molecular-weight components, preferentially C_4 and other branched isomers.

Cracking

As a model for cracking of alkanes, the reaction of 2-methylpentane (MP) over SbF_5-intercalated graphite has been studied in a flow system, the hydrocarbon being diluted in a hydrogen stream.[36] A careful study of the product distribution vs time on stream showed that propane was the initial cracking product, whereas isobutane and isopentane (as major cracking products) appeared only later.

This result can be explained only by the β-scission of the trivalent 3-methyl-4-pentenyl ion as the initial step in the cracking process. Based on this and on the product distribution vs time profile, a general scheme for the isomerization and cracking process of the methylpentanes has been proposed.[36]

The propene that is formed in the β-scission step never appears as a reaction product because it is alkylated immediately under the superacidic condition by a C_6^+ carbenium ion, forming a C_9^+ carbocation that is easily cracked to form a C_4^+ or C_5^+ ion and the corresponding C_4 or C_5 alkene. The alkenes are further alkylated by a C_6^+ carbenium ion in a cyclic process of alkylation and cracking reactions. The C_4^+ or C_5^+ ions also give the corresponding alkanes (isobutane and isopentane) by hydride transfer from the starting methylpentane. This scheme, which occurs under superacidic conditions, is at variance with the scheme that was proposed for the cracking of C_6 alkanes under less acidic conditions.[27]

$$C_{12}^+ \longrightarrow C_{12-n}^+ + C_n \text{ (olefin)} \quad 8 < n < 3$$

Under superacidic conditions, it is known that the deprotonation equilibria lie too far to the left ($K = 10^{-16}$ for isobutane[16]) to make this pathway plausible. On the other hand, among the C_6 isomers, 2-methylpentane is by far the easiest to cleave by β-scission. The 2-methyl-2-pentenium ion is the only species that does not give a primary cation by this process. For this reason, this ion is the key intermediate in the isomerization-cracking reaction of C_6 alkanes.

Under superacidic conditions, β-scission is not the only pathway by which hydrocarbons are cleaved. The C-C bond can also be cleaved by protolysis.

$$\begin{array}{c} \diagdown \\ -C-C\diagup \\ \diagup \quad \diagdown \end{array} + H^+ \longrightarrow \left[\begin{array}{c} H \\ \diagdown \quad \overset{+}{} \diagup \\ -C \cdots C\diagdown \\ \diagup \quad \diagdown \end{array} \right] \longrightarrow \begin{array}{c} \diagup \\ C+ \\ \diagdown \end{array} + \begin{array}{c} H-C- \\ | \end{array}$$

The protolysis under superacid conditions has been studied independently by Olah[48] and Brouwer and Hogeveen.[49] The carbon-carbon cleavage in neopentane yielding methane and the *tert*-butyl ion occurs by a mechanism different from the usual β-scission of carbenium ions:

$$\begin{array}{c} CH_3 \\ | \\ CH_3-C-CH_3 \\ | \\ CH_3 \end{array} + H^+ \longrightarrow \left[\begin{array}{c} CH_3 \\ | \quad H \\ CH_3-C\cdots\mathopen{<} \\ | \quad CH_3 \\ CH_3 \end{array} \right]^+ \longrightarrow (CH_3)_3C^+ + CH_4$$

The protolysis occurs upon the direct protonation of the C–C bond providing evidence for the σ-basicity of hydrocarbons. Under slightly different conditions, protolysis of a C–H bond occurs yielding rearranged *tert*-pentyl ion (*tert*-amyl cation).

$$(CH_3)_3C - CH_3 + H^+ \longrightarrow \left[(CH_3)_3C - CH_2 \cdots \mathopen{<}\begin{array}{c} H \\ H \end{array} \right]^+ \longrightarrow$$

$$\begin{array}{c} CH_3 \\ | \\ CH_3 - \underset{+}{C} - CH_2CH_3 \end{array} + H_2$$

In cycloalkanes, the C–C bond cleavage leads to ring opening.[7]

This reaction is much faster than the carbon-carbon bond cleavage in neopentane, inspite of the initial formation of secondary carbenium ions. Norbornane is also cleaved in a fast reaction yielding substituted cyclopentyl ions. Thus, protonation of alkanes induces cleavage of the molecule by two competitive ways: (1) protolysis of a C-H bond followed by β-scission of the carbenium ions and (2) direct protolysis of a C-C bond yielding a lower-molecular-weight alkane and a lower-molecular-weight carbenium ion. This reaction, which is of economic importance in the upgrading of higher boiling petroleum fractions to gasoline, has also been shown to be applicable to coal depolymerization and hydroliquefaction processes.[50] The cleavage of selected model compounds representing coal structural units in the presence of HF, BF$_3$, and under hydrogen pressure has been studied by Olah and co-workers.[50] Bituminous coal (Illinois No. 6) could be pyridine solubilized to the extent of 90% by treating it with superacidic HF:BF$_3$ catalyst in the presence of hydrogen gas at 105°C for 4 hr. Under somewhat more elevated temperatures (150-170°C), cyclohexane extractibility of up to 22% and distillability of up to 28% are achieved. Addition of hydrogen donor solvents such as isopentane has been shown to improve the efficiency of coal conversion to cyclohexane-soluble products. The initial "depolymerization" of coal involves various protolytic cleavage reactions involving those of C-C bonds. The ionic hydrogenation affected by superacid systems is not sensitive to impurities in coal, a distinct advantage over conventional hydrogenation catalysts.

VI. Alkylation of Alkanes and Oligocondensation of Lower Alkanes

The alkylation of alkanes by olefins, from a mechanistic point of view, must be considered as the alkylation of the carbenium ion formed by the protonation of the olefin. The well-known acid-catalyzed isobutane-isobutylene reaction demonstrates the mechanism.

As is apparent in the last step, isobutane is not alkylated but transfers a hydride to the C$_8$+ carbocation, forming *tert*-butyl cation which is then used up in the middle step as alkylating agent. The direct alkylation of isobutane by an incipient *tert*-butyl cation would yield 2,2,3,3-tetramethylbutane, which indeed was observed in small amounts in the reaction of the *tert*-butyl cation with isobutane under stable ion conditions at low temperatures (*vide infra*).

The alkylating ability of methyl and ethyl fluoride-antimony pentafluoride complexes has been investigated by Olah and his group[51,52] showing the extraordinary reactivity of these systems. Self-condensation was observed as well as alkane alkylation. When $CH_3F:SbF_5$ was reacted with excess of CH_3F at 0°C at first only an exchanging complex was observed in the ^1H-NMR spectrum. After 0.5 hr, the starting material was converted into the *tert*-butyl cation. Similar reactions were observed with the ethyl fluoride-antimony pentafluoride

$$FCH_3 + CH_3F \rightarrow SbF_5 \longrightarrow \left[FCH_2 \cdots \overset{H}{\underset{CH_3}{\text{C}}} \right]^+ SbF_6^- \longrightarrow FCH_2-CH_3 + H^+$$

$$FCH_2CH_3 + CH_3CH_2F \rightarrow SbF_5 \rightarrow\rightarrow FCH_2CH_2CH_2CH_3 \rightarrow\rightarrow (CH_3)_3C^+$$

complex. When the complex was treated with isobutane or isopentane direct alkylation products were observed.

$$CH_3 - \overset{\overset{\displaystyle CH_3}{|}}{\underset{\underset{\displaystyle CH_3}{|}}{C}} - H + CH_3CH_2F \rightarrow SbF_5 \longrightarrow (CH_3)_3 C CH_2 CH_3$$

To improve the understanding of these alkane alkylation reactions, Olah and his group carried out experiments involving the alkylation of the lower alkanes by stable carbenium ions under controlled superacidic stable ion conditions.[48,53,54]

$$R - H + R'^+ \longrightarrow \left[R \cdots \overset{H}{\underset{}{}} \cdots R' \right]^+ \longrightarrow R - R' + H^+$$

The σ-donor ability of the C-C and C-H bonds in alkanes was demonstrated from a variety of examples. The order of reactivity of single bonds was found to be tertiary C-H > C-C > secondary C-H >> primary C-H, although various specific factors such as steric hindrance can influence the relative reactivities.

Typical alkylation reactions are those of propane, isobutane, and *n*-butane by *tert*-butyl or *sec*-butyl ion. These systems are somewhat interconvertible by competing hydride transfer and rearrangement of the carbenium ions. The reactions were carried out using alkylcarbenium hexafluoroantimonate salts prepared from the corresponding halides and antimony pentafluoride in sulfuryl chloride fluoride solution and treating them in the same solvent with alkanes. The reagents were mixed at -78ºC, warmed up to -20ºC, and quenched with ice water before analysis. The intermolecular hydride transfer between tertiary and secondary carbenium ions and alkanes is generally much faster than the alkylation reaction. Consequently, the alkylation products are also those derived from the new alkanes and carbenium ions formed in the hydride transfer reaction.

Propylation of propane by the isopropyl cation, for example, gives a significant amount (26% of the C_6 fraction) of the primary alkylation product:

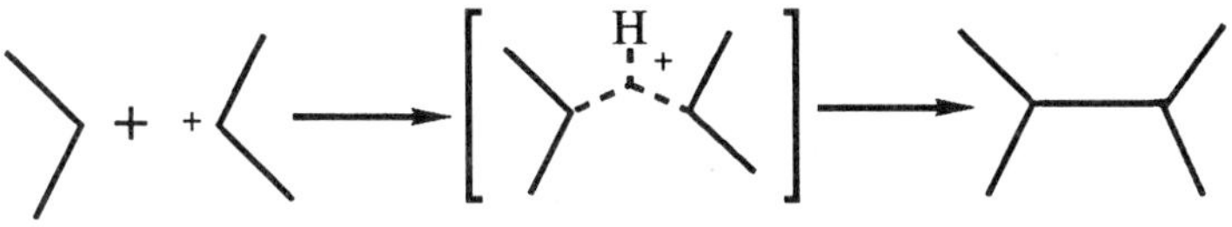

The C_6 isomer distribution, 2-methylpentane (28%), 3-methylpentane (14%), and *n*-hexane (32%), is very far from thermodynamic equilibrium, and the presence of these isomers indicates that not only isopropyl cation but also *n*-propyl cation are involved as intermediates (as shown by $^{13}C_2$-$^{13}C_1$ scrambling in the stable ion).[55]

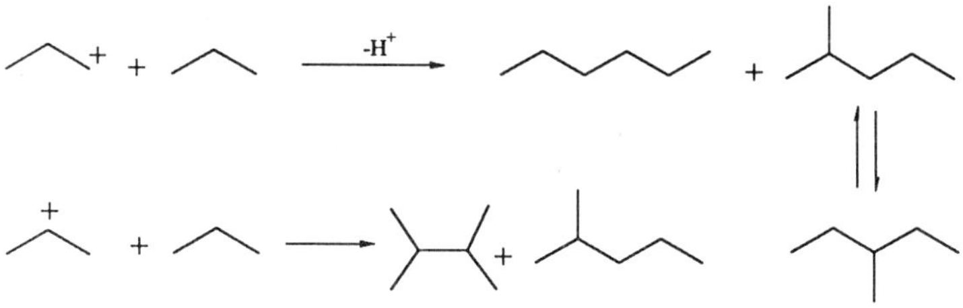

The strong competition between alkylation and hydride transfer appears in the alkylation reaction of propane by butyl cations, or butanes by propyl cation. The amount of C_7 alkylation products is rather low.

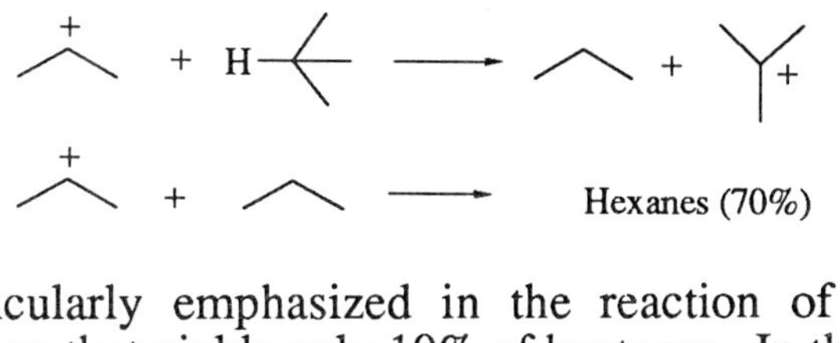

This is particularly emphasized in the reaction of propane by the *tert*-butyl cation that yields only 10% of heptanes. In the interaction of propyl cation with isobutane the main reaction is hydride transfer from isobutane to the propyl ion followed by alkylation of the propane by the propyl ions.

Even the alkylation of isobutane by the *tert*-butyl cation inspite of the highly unfavorable steric interaction has been demonstrated[48] by the formation of small amounts of 2,2,3,3-tetramethylbutane. This result also indicates that the related five-coordinate carbocationic transition state (or high lying intermediate) of the degenerate isobutane-*tert*-butyl cation hydride transfer reaction is not entirely linear, inspite of the highly crowded nature of the system.

Table I. Isomeric Octane Compositions Obtained in Typical Alkylations of Butanes with Butyl Cations

Octane	$(CH_3)_3CH$ $(CH_3)_3{}^+C$	$CH_3CH_2CH_2CH_3$ $(CH_3)_3{}^+C$	$(CH_3)_3CH$ $CH_3{}^+CHCH_2CH_3$	$(CH_3)_3CH$ $CH_3{}^+CHCH_2CH_3$
2,2,4-Trimethyl-pentane	18.0	4.0	3.8	8.5
2,2-Dimethylhexane			0.4	
2,2,3,3-Tetramethyl-butane	1-2		Trace	1-2
2,5-Dimethylhexane	43.0	0.6	1.6	29.0
2,4-Dimethylhexane	7.6	Trace		6.6
2,2,3-Trimethyl-pentane	3.0	73.6	40.6	3.2
3,3-Dimethylhexane			12.3	7.1
2,3,4-Trimethyl-pentane	1.5	7.2	15.5	6.2
2,3,3-Trimethyl-pentane	3.6		3.8	8.8
2,3-Dimethylhexane	4.2	6.9		12.8
2-Methylheptane		Trace	10.3	6.7
3-Methylheptane	19.3	7.6	6.8	9.5
n-Octane	0.2		4.8	Trace

2,2,3,3-Tetramethylbutane was not formed when *n*-butane and *sec*-butyl cations were reacted. The isomer distribution of the octane isomers for typical butyl cation-butane alkylations is shown in Table I.

The superacid [CF_3SO_3H or $CF_3SO_3H/B(O_3SCF_3)_3$]-catalyzed alkylation of adamantane with lower olefins (ethene, propene, and butenes) was also investigated.[56] Alkyladamantanes obtained show that the reaction occurs by two pathways: (1) adamantylation of olefins by admantyl cation formed through hydride abstraction from adamantane by alkyl cations (generated by the protonation of the olefins) and (2) direct σ-alkylation of adamantane by the alkyl cations via insertion into the bridgehead C-H bond adamantane through a pentacoordinate carbonium ion.

Adamantylation of Olefins

$$\text{olefin} \quad + \quad H^+ \; \rightleftharpoons \; R^+$$

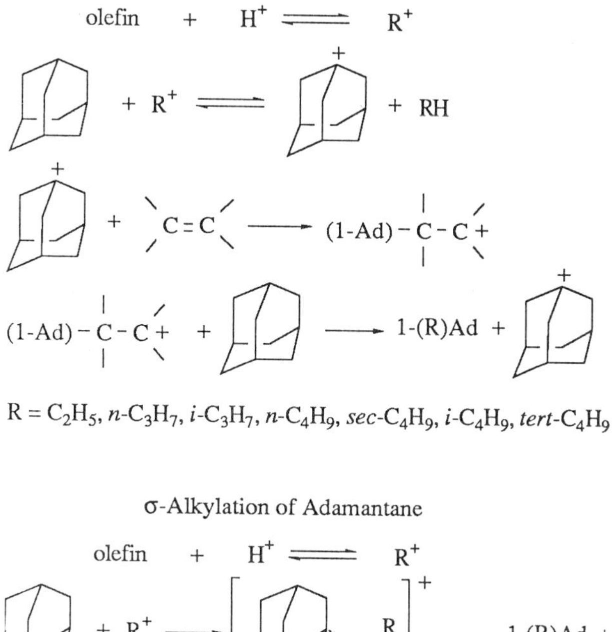

$$R = C_2H_5, \; n\text{-}C_3H_7, \; i\text{-}C_3H_7, \; n\text{-}C_4H_9, \; sec\text{-}C_4H_9, \; i\text{-}C_4H_9, \; tert\text{-}C_4H_9$$

σ-Alkylation of Adamantane

$$\text{olefin} \quad + \quad H^+ \; \rightleftharpoons \; R^+$$

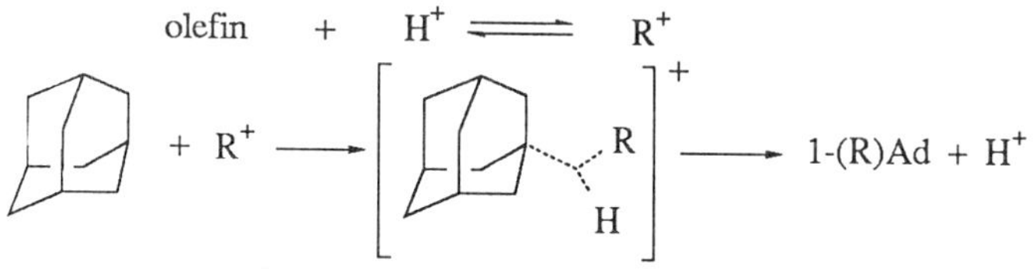

In order to gain understanding of the mechanism of the formation of alkyladamantanes, the reaction of adamantane with butenes is significant.

Reaction with 2-butenes gave mostly 1-*n*-butyladamantane, 1-*sec*-butyladamantane, and 1-isobutyladamantane. Occasionally trace amounts of 1-*tert*-butyladamantane are also formed. Isobutylene (2-methylpropane), however, consistently gave relatively good yield of 1-*tert*-butyladamantane along with other isomeric 1-butyladamantanes.

1-Butene gave only the isomeric butyladamantanes with only a trace amount of *tert*-butyladamantanes.

The formation of isomeric butyladamantanes, with the exception of *tert*-butyladamantane, is through adamantylation of olefins and carbocationic isomerization.

The formation of *tert*-butyladamantane, in the studied butylation reactions, can, however, not be explained by this path. Since in control experiments attempted acid-catalyzed isomerization of isomeric 1-butyladamantanes did not give even trace amounts of 1-*tert*-butyladmantane, the tertiary isomer must be formed in the direct σ-*tert*-butylation of adamantane by *tert*-butyl cation through a pentacoordinate carbonium ion. The same intermediate is involved in the concomitant formation of 1-adamantyl cation via intermolecular hydrogen transfer (the major reaction). The formation of even low yields of *tert*-butyladamantane in the reaction is a clear indication that the pentacoordinate carbocation does not attain a linear geometry >C---H---C< (which could result only in hydrogen transfer), inspite of unfavorable steric interactions. This reaction is similar to the earlier discussed reaction between *tert*-butyl cation and isobutane to form 2,2,3,3-tetramethylbutane.

An alternate pathway for the formation of *tert*-butyladamantane through hydride abstraction of an intermediate 1-adamantylalkyl cation would necessitate involvement of an energetic "primary" cation or highly distorted "prontonated cyclopropane," which is not likely under the reaction conditions.

The observation of 1-*tert*-butyladmantane in the superacid-catalyzed reactions of adamantane with butenes provides unequivocal evidence for the σ-alkylation of adamantane by the *tert*-butyl cation. As this involves an unfavorable sterically crowded tertiary-tertiary interaction, it is reasonable to suggest that similar σ-alkylation can also be involved in less strained interactions with secondary and primary alkyl systems. Although superacid-catalyzed alkylation of adamantane with olefins predominantly occurs via adamantylation of olefins, competing direct σ-alkylation of adamantane can also occur. As the adamantane cage allows attack of the alkyl group only from the front side, the reported studies provide significant new insight into the mechanism of electrophilic reactions at saturated hydrocarbons and the nature of their carbocationic intermediates.

The protolyic oxidative condensation of methane in Magic Acid solution at 60°C is evidenced by the formation of higher alkyl cations

$$
H-\underset{\underset{H}{|}}{\overset{\overset{H}{|}}{C}}-H \xrightarrow{H^+} \left[H-\underset{\underset{H}{|}}{\overset{\overset{H}{|}}{C}}\cdots\underset{H}{\overset{H}{<}} \right]^+ \longrightarrow CH_3^+ + H_2
$$

$$
CH_3^+ + CH_4 \longrightarrow \left[H-\underset{\underset{H}{|}}{\overset{\overset{H}{|}}{C}}\cdots\overset{H\ \ H}{\cdots}\underset{\underset{H}{|}}{\overset{\overset{H}{|}}{C}}-H \right]^+ \longrightarrow C_2H_5^+ + H_2
$$

such as *tert*-butyl and *tert*-hexyl cations.[10] It is not necessary to assume a complete cleavage of $[CH_5]^+$ to a free, energetically unfavorable methyl cation. The carbon-carbon bond formation can indeed be visualized through C-H bond of methane reacting with the developing methyl cation.

$$
H-\underset{\underset{H}{|}}{\overset{\overset{H}{|}}{C}}-H + \left[H-\underset{\underset{H}{|}}{\overset{\overset{H}{|}}{C}}\cdots\underset{H}{\overset{H}{<}} \right]^+ \longrightarrow \left[H-\underset{\underset{H}{|}}{\overset{\overset{H}{|}}{C}}\cdots\overset{H\ \ H}{\cdots}\underset{\underset{H}{|}}{\overset{\overset{H}{|}}{C}}-H \right]^+ + H_2
$$

Combining two methane molecules to ethane and hydrogen is endothermic by some 16 kcal/mol. Any condensation of methane to ethane and subsequently to higher hydrocarbons must thus overcome unfavorable thermodynamics. This can be achieved in condensation processes of an oxidative nature, in which hydrogen is removed by the oxidant.

$$2 CH_4 \longrightarrow C_2H_6 + H_2$$

In initial studies, the SbF_5 or FSO_3H component of the superacid system itself acted as oxidant. The oxidative condensation of methane was subsequently further studied in more detail.[57] It was found that with added suitable oxidants such as halogens, oxygen, sulfur, or selenium, the superacid-catalyzed condensation of methane is feasible.

$$CH_4 \xrightarrow[\text{superacid catalyst}]{\text{halogens, } O_2, S_x, Se} \text{hydrocarbons}$$

Significant practical problems, however, remain in carrying out the condensation effectively. Conversion was so far achieved only in low yields. As a result of the easy cleavage of longer chain alkanes by the same superacids C_3-C_6 products predominate.

A further approach found useful was the substitution of natural gas instead of pure methane in the condensation reaction.[57] When natural gas is dehydrogenated, the C_2-C_4 alkanes it contains are converted to olefins. The resulting methane-olefin mixture can then without separation be passed through a superacid catalyst resulting in exothermic alkylative condensation.

$$CH_4 + RCH = CH_2 \longrightarrow CH_3CHRCH_3$$

Alkylation of methane by olefins under superacid catalysis was demonstrated both in solution chemistry under stable ion conditions[58] and in heterogeneous gas phase alkylations over solid catalysts using a flow system.[59] Not only propylene and butylenes but also ethylene could be used as alkylating agents.

The alkylation of methane, ethane, propane, and *n*-butane by ethyl cation generated via protonation of ethylene in superacid media has been studied by Siskin,[60] Sommer et al.,[61] and Olah et al.[62] The difficulty lies in generating in a controlled way a very energetic primary carbenium ion in the presence of excess methane and at the same time avoiding oligocondensation of ethylene itself. Siskin carried out the reaction of a methane-ethylene (86:14) gas mixture through a 10:1 $HF:TaF_5$ solution under pressure with strong mixing. Along with the

$$CH_2 = CH_2 \xrightarrow{HF/TaF_5} CH_3 - CH_2^+ \xrightarrow{CH_4} \left[CH_3CH_2 \overset{H}{\cdots} CH_3 \right]^+$$

$$\searrow -H^+$$

$$CH_3CH_2CH_3$$

recovered reaction products 60% of C_3 was found (propane and propylene). Propylene is formed when propane, which is a substantially better hydride donor, reacts with the ethyl cation:

$$CH_3CH_2CH_3 + CH_3-CH_2^+ \longrightarrow CH_3-CH_3 + CH_3-\overset{+}{C}H-CH_3$$

$$\searrow {}_{-H^+}$$

$$CH_3-CH=CH_2$$

Propane as a degradation product of polyethylene was ruled out because ethylene alone under the same conditions does not give any propane. Under similar conditions but under hydrogen pressure, polyethylene reacts quantitatively to form C_3 to C_6 alkanes, 85% of which are isobutane and isopentane. These results further substantiate the direct alkane-alkylation reaction and the intermediacy of the pentacoordinate carbonium ion. Siskin also found that when ethylene was allowed to react with ethane in a flow system, *n*-butane was obtained as the sole product, indicating that the ethyl cation is alkylating the primary C-H bond through a five-coordinate carbonium ion.

$$CH_2=CH_2 \xrightarrow{H^+} CH_3CH_2^+ \xrightarrow{CH_3-CH_3} \left[CH_3CH_2 \cdots \overset{\overset{\displaystyle H}{\vdots}}{C} \cdots CH_2CH_3 \right]^+$$

$$\downarrow {}_{-H^+}$$

$$CH_3CH_2CH_2CH_3$$

If the ethyl cation would have reacted with excess ethylene, primary 1-butyl cation would have been formed that irreversibly would rearrange to the more stable *sec*-butyl and subsequently *tert*-butyl cation giving isobutane as the end product.

The yield of the alkene-alkane alkylation in a homogeneous HF:SbF$_5$ system depending on the alkene:alkane ratio has been investigated by Sommer and co-workers in a batch system with short reaction times.[61] The results support direct alkylation of methane, ethane, and propane by the ethyl cation and the product distribution depends on the alkene:alkane ratio.

Inspite of the unfavorable experimental conditions in a batch system for kinetic controlled reactions, a selectivity of 80% in *n*-butane was achieved through ethylation of ethane. The results show, however, that to succeed in the direct alkylation the following conditions have to be met:

    1. The olefin should be totally converted to the reactive cation (incomplete protonation favors the polymerization and cracking processes); this means the use of a large excess of acid and good mixing.

    2. The alkylation product must be removed from the reaction mixture before it transfers hydride to the reactive cation, in which case subsequent alkene formation takes place.

    3. Hydride transfer from the substrate should not be easy; for this reason the reaction produces the best yield and selectivity when methane and ethane are used.

    More recently, the direct ethylation of methane with ethylene has been investigated by Olah and his group[62] using excess ^{13}C-labeled methane (99.9% ^{13}C) over solid superacid catalysts such as TaF_5:AlF_3, TaF_5, and SbF_5:graphite. Product analyses by gas chromatography-mass spectrometry (GC-MS) are given in Table II.

Table II. Ethylation of $^{13}CH_4$ with C_2H_4

$^{13}CH_4$:C_2H_4	Catalyst	Products normalized (%)[a,b]				Label content of C_3 fraction (%)	
		C_2H_6	C_3H_8	i-C_4H_{10}	C_2H_5F	$^{13}CC_2H_8$	C_3H_8
98.7:1.3	TaF_5:AlF_3	51.9	9.9	38.2		31	69
99.1:0.9	TaF_5		15.5	3.0	81.5	91	9
99.1:0.9	SbF_5:graphite	64.1	31.5		4.4	96	4

[a]All values reported are in mole percentage.
[b]Excluding methane.

    These results show a high selectivity in a mono-labeled propane $^{13}CC_2H_8$ that can arise only from direct electrophilic attack of the ethyl cation on methane via pentacoordinate carbonium ion.

$$CH_2 = CH_2 \xrightarrow{H^+} CH_3 - CH_2^+ \xrightarrow{^{13}CH_4} \left[CH_3 - CH_2 \overset{\overset{\displaystyle H}{|}}{\diagdown} \,^{13}CH_3 \right]^+$$

$$\xrightarrow{-H^+} CH_3CH_2{}^{13}CH_3$$

An increase of the alkene:alkane ratio results in a significant decrease in single-labeled propane. Ethylene polymerization-cracking and hydride transfer become the main reactions. This labeling experiment carried out under conditions in which side reactions were negligible is indeed unequivocal proof for the direct alkylation of an alkane by a very reactive carbenium ion.

Polycondensation of alkanes over $HSO_3F:SbF_5$ has also been achieved by Roberts and Calihan.[63] Several low-molecular-weight alkanes such as methane, ethane, propane, butane, and isobutane were polymerized to highly branched oily oligomers with a molecular weight range of around 700. These reactions again follow the same initial protolysis of the C-H or C-C bond, which results in a very reactive carbenium ion. Similarly, Roberts and Calihan[64] were also able to polycondense methane with small amount of olefins such as ethylene, propylene, butadiene, and styrene to yield oily polymethylene oligomer with a molecular weight ranging from 100 to 700.

VII. Carboxylation

Alkanecarboxylic acids are readily prepared from alkenes and carbon monoxide or formic acid in strongly acidic solutions.[65] The reaction between carbenium ions generated from alkenes and carbon monoxide affording oxocarbenium ions (acyl cations) is the key step in the well-known Koch-Haaf reaction used for the general preparation of carboxylic acids.[65,66] Sulfuric acid was generally used in the reactions. Subsequent investigations[67] found that superacidic $HF:BF_3$ or CF_3SO_3H[68] is more efficient to effect this process. The volatility of the former catalyst system, however, necessitates high-pressure conditions.

The use of superacidic activation of alkanes to their related carbocations allowed the preparation of alkanecarboxylic acids from alkanes themselves with CO.

The formation of C_6 or C_7 carboxylic acids along with some ketones was reported in the reaction of isopentane, methylcyclopentane, and cyclohexane with CO in $HF:SbF_5$ at ambient temperatures and atmospheric pressures.[69]

Recently, Yoneda et al. have found that other alkanes can also be carboxylated with CO in an $HF:SbF_5$ superacid system.[70,71] Tertiary carbenium ions formed by protolysis of C-H bonds of branched alkanes in $HF:SbF_5$ undergo skeletal isomerization and disproportionation prior to reacting with CO in the same acid system to form carboxylic acids after hydrolysis. The results are shown in Table III.

When using tertiary C_5 or C_6 alkanes a considerable amount of secondary carboxylic acids[72] is produced by the reaction of CO with secondary alkyl carbenium ions. Such cations are formed as transient intermediates by skeletal isomerization of the intially formed tertiary cations. Carboxylic acids with number of carbon atoms lower than the starting alkanes are formed from the fragment alkyl cations generated by the protolysis of C-C bonds of the straight-chain alkanes. Intermolecular hydride shift between the fragment cations and the starting alkanes gives rise to carboxylic acids with alkyl groups of the same number of carbon atoms as the starting alkanes. For alkanes with more than C_7 atoms, β-scission occurs exclusively to produce C_4, C_5, and C_6 carboxylic acids.

Table III. Superacid Catalyzed Reaction of Alkanes with CO[a]

Substrate	Total yield[b]	C_3 acid	C_4 acid	C_5 acids tert-	sec-	C_6 acids tert-	sec-	C_7 acids tert-	sec-	C_8 acids
2-Methyl butane	55	Some	Some	4		95		Trace		-
				67	33	38	62			
n-Pentane	53	3	26	14		57		Trace		-
				27	73	39	61			
2-Methyl pentane	61	-	Some	3		3		94		-
				67	33	40	60	29	71	
n-Hexane	69	2	10	18		5		65		Some
				13	87	39	61	33	67	
2,2-Dimethyl butane	57	-	Some	10		15		75		-
				80	20	40	60	35	65	
n-Heptane	80	Some	47	48		5		Some		Some
				90	10	38	62			
2,2,4-Tri-methyl pentane	102	-	-	100		-		-		-
				100	0					
n-Octane	90	Some	6	87		7		Some		Some
				90	10	38	62			
n-Nonane	61	Some	34	31		33		2		Some
				87	13	38	62			

Product distribution (%)

[a]Reaction time, 1 hr. Reaction temperature 30°C. Alkane 20 mmol, SbF_5/alkane = 2 mol ratio, HF/SbF_5 = 5 mol ratio.
[b]Based on alkane used.

VIII. Formylation

Whereas electrophilic formylation of aromatics with CO was studied under both Gatterman-Koch conditions and with superacid catalysis[73-78] in some detail, electrophilic formylation of saturated aliphatics remained virtually unrecognized.

Reactions of acyclic hydrocarbons of various skeletal structures with CO in superacid media were recently studied by Yoneda et al.[70-72] as discussed in Sec. VII. Products obtained were isomeric carboxylic acids with a lower number of carbon atoms than the starting alkanes. Formation of carboxylic acids was accounted by the reactions of parent, isomerized, and fragmented alkyl cations with CO to form the oxocarbenium ion intermediates (Koch-Haaf reaction) followed by their quenching with water. No formylated products in these reactions have been identified.

Olah et al. recently reported the superacid-catalyzed reaction of polycyclic cage hydrocarbon adamantane with CO.[79] 1-Adamantanecarboxaldehyde was obtained in varying yields (Table IV) in different superacids along with 1-adamantanecarboxylic acid and 1-adamantanol (the products of the reaction of 1-adamantyl cation). The mechanism of formation of 1-adamantanecarboxaldehyde has been investigated.

Table IV. Percentage Yield of 1-Adamantanecarboxaldehyde in Different Superacids[a]

Acid systems	AdH:acid	Solvent	Yield 1-Ad-CHO (%)
CF_3SO_3H	1:12	Freon-113	0.2
	1:12	-	9.1[b]
$CF_3SO_3H + B(OSO_2CF_3)_3$	1:3	Freon-113	3.4
1:1	1:3	-	14.5[b]
$CF_3SO_3H + SbF_5$	1:3	Freon-113	8.2
1:1	1:3	-	21.0[b]

[a]All reactions were carried out from -78°C to room temperature.
[b]Yields are isolated.

Formation of aldehyde product from 1-admantanoyl cation via hydride abstraction (Scheme 2) from adamantane is only a minor pathway.

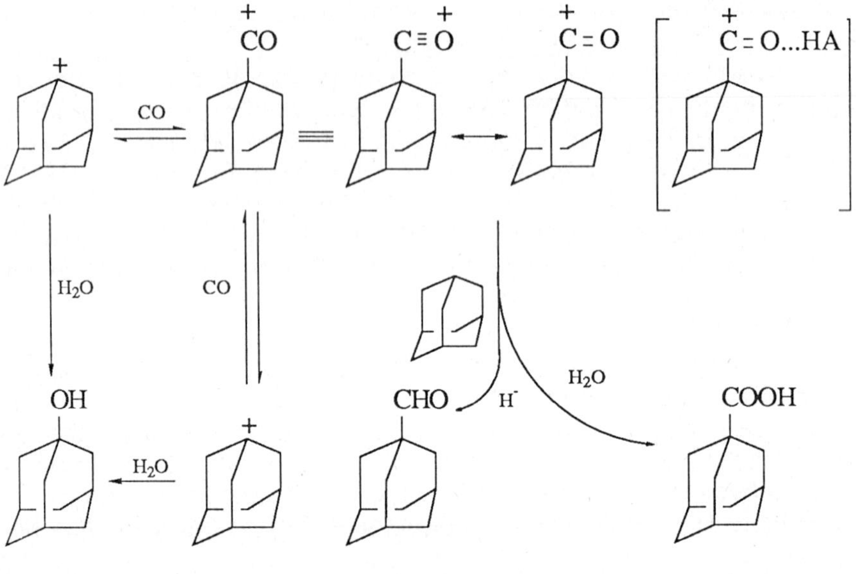

Scheme 2

The major pathway by which 1-adamantanecarboxaldehyde is formed is by direct σ-insertion of the intermediate formyl cation into the C-H bond of adamantane at the bridgehead position.

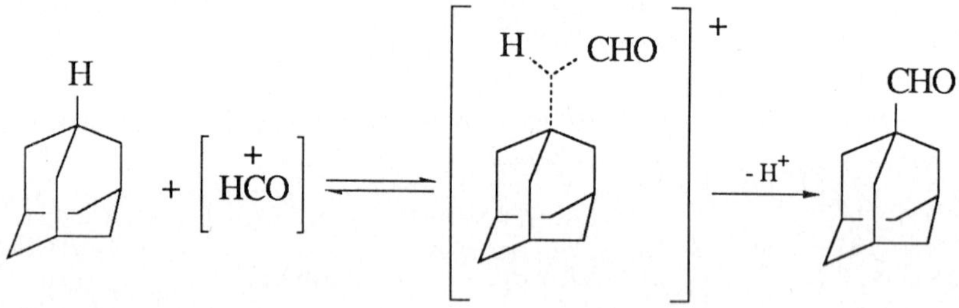

Whereas the formyl cation could not be directly observed by NMR spectroscopy,[78,79] its intermediacy has been well established in aromatic formylation reactions.[73-78] In order to account for the failure to observe the formyl cation it was suggested that CO is protonated in acid media to generate protosolvated formyl cation,[80,81] a very reactive electrophile. Protosolvation of the carbonyl oxygen allows facile deprotonation of the methine proton, thus resulting in rapid exchange via involvement of the isoformyl cation.

Insertion of protosolvated formyl cation into the C-H σ-bond has further been substantiated by carrying out the reaction of 1,3,5,7-tetradeuteroadamantane with CO in superacid media. 3,5,7-Trideutero-1-adamantanecarboxaldehyde-H and 3,5,7-trideutero-1-adamantanecarboxaldehyde-D were obtained in the ratio 94:6.

IX. Oxyfunctionalization

Functionalization of aliphatic hydrocarbons into their oxygenated compounds is of substantial interest. In connection with the preparation and studies of a great variety of carbocations in different superacid systems,[22] electrophilic oxygenation of alkanes with ozone and hydrogen peroxide was investigated by Olah and co-workers in superacid media under typical electrophilic conditions.[82] The electrophiles used were protonated ozone $(O_3H)^+$ or the hydrogen peroxonium ion $H_3O_2^+$. The reactions are depicted as taking place via initial electrophilic attack by the electrophiles on the σ-bonds of alkanes through a pentacoordinated carbonium ion transition state followed by proton elimination to give the desired product.

Table V. Products of the Reaction of Branched-Chain Alkanes with H_2O_2 in $HSO_3F{:}SbF_5{:}SO_2ClF$ Solution

Alkane	Alkane (mmol)	H_2O_2 (mmol)	Temperature (°C)	Major Products
	2	2	-78 → -20	$(CH_3)_2C=O^+CH_3$
	2	4	-78 → -20	$(CH_3)_2C=O^+CH_3$
C | C·C·C | H	2	6	-78 → -20	$(CH_3)_2C=O^+CH_3$
	2	6	+20	$(CH_3)_2C=O^+CH_3$ (trace), DAP[a] (25%), CH_3OH (50%), $CH_3CO{-}O{-}CH_3$ (25%)
C | C·C·C·C | H	2	3	-78 → -20	$(CH_3)_2C=O^+C_2H_5$ $(C_2H_5(CH_3)_2C^+$
	2	6	-78 → -20	$(CH_3)_2C=O^+C_2H_5$
C C | | C·C·C·C | | C H	2	4	-78	$(CH_3)_2C=O^+CH_3$ (50%), $(CH_3)_2C=O^+H$ (50%)
	2	6	-40	$(CH_3)_2C=O^+CH_3$ (50%), DAP (50%)

[a]DAP, Dimeric acetone peroxide.

When hydrogen peroxide is protonated in superacid media hydrogen peroxonium ion ($H_3O_2^+$) is formed. Certain peroxonium salts have been well-characterized and even isolated as stable salts.[83,84] Hydrogen peroxonium ion is considered as the incipient ^+OH ion, a strong electrophile for electrophilic hydroxylation at single (σ) bonds of alkanes. The reactions are thus similar to those described previously for electrophilic protonation (protolysis), alkylation, and formylation.

Superacid-catalyzed electrophilic hydroxylation of branched alkanes was carried out using $HSO_3F:SbF_5:SO_2ClF$ with various ratios of alkane and hydrogen peroxide at different temperatures.[85] Some of the results are summarized in Table V. Protonated hydrogen peroxide inserts into the C-H bond of alkanes. The mechanism is illustrated with isobutane.

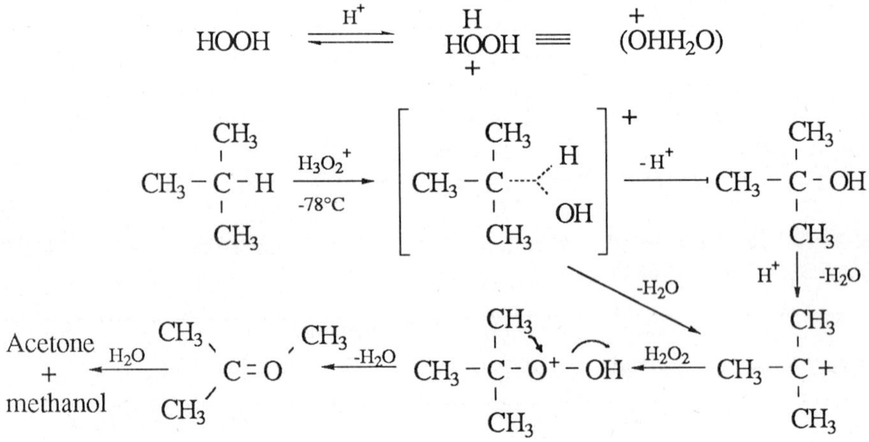

The intermediate pentacoordinate hydroxycarbonium ion transition state, in addition to giving the insertion product, gives *tert*-butyl cation (by elimination of water), which is responsible for the formation of dimethylmethylcarboxonium ion through the intermediate *tert*-butyl hydroperoxide. Hydrolysis of the carboxonium ion gives acetone and methanol. When the reaction was carried out by passing isobutane at room temperature through a solution of Magic Acid and excess hydrogen peroxide, in addition to the insertion product, a number of other products were also formed that were rationalized as arising from hydrolysis of the carboxonium ion and Baeyer-Villiger oxidation of acetone. The mechanism was further substantiated by independent treatment of alkane and Baeyer-Villiger oxidation of several ketones with hydrogen peroxide-Magic Acid systems.[86] In the superacid-catalyzed hydroxylation of straight-chain alkanes such as ethane, propane, and butane with hydrogen peroxide related oxygenated products were identified.[85]

In Olah's studies on superacid-catalyzed oxyfunctionalization of methane it was found that hydrogen peroxide in superacidic media gives methyl alcohol with very high (>95%) selectivity.[86] Electrophilic ^+OH insertion by protonated hydrogen peroxide in the C-H bonds of methane is the indicated reaction path. The reaction is limited, however, by the use of hydrogen peroxide to the liquid phase. In the superacidic medium the methyl alcohol formed is immediately protonated to methyloxonium ion $(CH_3OH_2^+)$ and thus prevented from further oxidation. Similarly, the superacid-catalyzed oxygenation of methane with ozone was studied and gives predominantly formaldehyde.[87] The reaction is best understood as electrophilic insertion of ^+O_3H ion into the methane C-H bonds leading to a hydrotrioxide, which then eliminates hydrogen peroxide giving protonated formaldehyde. The competing pathway forms protonated methyl alcohol with O_2 elimination, but this reaction is only a minor one.

$$CH_4 \xrightarrow[\text{super acid}]{H_2O_2 \text{ or } O_3} CH_3OH$$

Electrophilic oxygenation of methane to methyl alcohol under superacidic conditions proves the high selectivity of electrophilic substitution contrasted with nonselective radical oxidation. However, as indicated, hydrogen peroxide chemistry is limited to the liquid phase and the desirable goal of achieving selective heterogeneous catalytic oxidation of methane to methyl alcohol remains elusive. Consequently, in continued studies combining selective electrophilic halogenation of methane to methyl halides with subsequent hydrolysis to methyl alcohol was developed (*vide infra*).

Ozone, which is a resonance hybrid of a series of canonical structures, can react as a 1,3-dipole, an electrophile, or a nucleophile.[88] Whereas the electrophilic nature of ozone has been established in its reactions with alkenes, alkynes, arenes, amines, sulfides, phosphines, etc.,[89-93] the nucleophilic properties of ozone have been less well studied.[94]

$$\overset{+}{\underset{O \diagdown O}{O}} \longleftrightarrow \overset{+}{\underset{O = O \diagdown O}{}} \longleftrightarrow \overset{+ \diagup O \diagdown}{\underset{O O}{}}{}^{-} \longleftrightarrow {}^{-}\overset{\diagup O \diagdown}{\underset{O O}{}}{}^{+}$$

In the reaction of ozone with carbenium ions[87] the nucleophilic attack of ozone has been inferred through the formation of intermediate trioxide ion leading to dialkylakylcarboxonium ion similar to the acid-catalyzed rearrangement of hydroperoxides to carboxonium ions (Hock reaction).

$$R^2-\underset{\underset{R^3}{|}}{\overset{\overset{R^1}{|}}{C}}X \xrightarrow{SbF_5 \ SO_2ClF} R^2-\underset{\underset{R^3}{|}}{\overset{\overset{R^1}{|}}{C^+}} \xrightarrow{O_3} \left[R^2-\underset{\underset{R^3}{|}}{\overset{\overset{R^1}{|}}{C}}\text{-O-O-O} \right]^+$$

$$X = F, Cl, OH$$

$$\xrightarrow{-O_2} \left[R^2-\underset{\underset{R^3}{|}}{\overset{\overset{R^1}{|}}{C}}\text{-O}^+ \right]^+ \longrightarrow \underset{R^3}{\overset{R^2}{\diagdown}}C=\overset{+}{O}\diagup R^1$$

When a stream of oxygen containing 5% ozone was passed through a solution of isobutane in $HSO_3F:SbF_5:SO_2ClF$ solution at -78°C, dimethylmethylcarboxonium ion (CH_3CO^+) was identified by ^1H and ^{13}C-NMR spectroscopy.[87,95] Similar reactions of isopentane, 2,3-dimethylbutane, and 2,2,3-trimethylbutane resulted in the formation of related carboxonium ions as the major product (Table VI).

Table VI. Products of the Reaction of Branched Alkanes with Ozone in Magic Acid-SO_2ClF at -78°C (Ref. 87)

$\underset{H}{\overset{C}{\underset{|}{\overset{|}{C-C-C}}}}$	$(CH_3)_2C=O^+CH_3$
$\underset{H}{\overset{C}{\underset{|}{\overset{|}{C-C-C-C}}}}$	$(CH_3)_2C=O^+C_2H_5$
$\underset{H\ H}{\overset{C\ C}{\underset{|\ \ |}{\overset{|\ \ |}{C-C-C-C}}}}$	$(CH_3)_2C=O^+CH(CH_3)_2$ (60%) $(CH_3)_2C=O^+H$ (40%)
$\underset{C\ H}{\overset{C\ C}{\underset{|\ \ |}{\overset{|\ \ |}{C-C-C-C}}}}$	$(CH_3)_2C=O^+CH_3$ (50%) $(CH_3)_2C=O^+H$ (50%)

The superacid-catalyzed reaction of ozone with alkanes is considered to proceed via two mechanistic pathways as illustrated in Scheme 3.

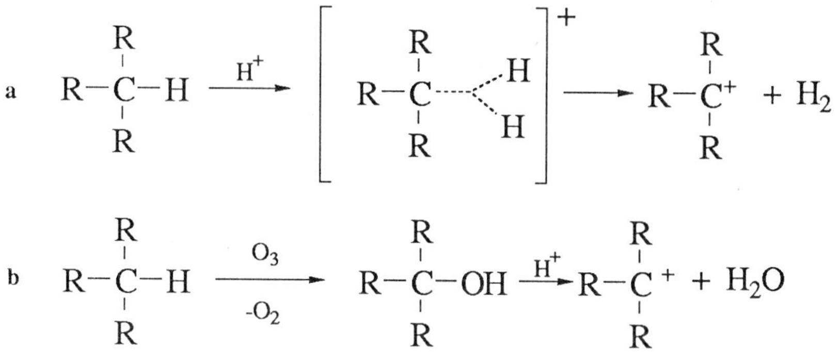

$$
\text{a}\quad R{-}\underset{\underset{R}{|}}{\overset{\overset{R}{|}}{C}}{-}H \xrightarrow{H^+} \left[R{-}\underset{\underset{R}{|}}{\overset{\overset{R}{|}}{C}}\cdots\!\!<^{\cdot\cdot H}_{H} \right]^+ \longrightarrow R{-}\underset{\underset{R}{|}}{\overset{\overset{R}{|}}{C^+}} + H_2
$$

$$
\text{b}\quad R{-}\underset{\underset{R}{|}}{\overset{\overset{R}{|}}{C}}{-}H \xrightarrow[-O_2]{O_3} R{-}\underset{\underset{R}{|}}{\overset{\overset{R}{|}}{C}}{-}OH \xrightarrow{H^+} R{-}\underset{\underset{R}{|}}{\overset{\overset{R}{|}}{C}}{}^+ + H_2O
$$

Scheme 3

Formation of the intermediate alkylcarbenium ion is the key step in both mechanisms, which subsequently will undergo nucleophilic reaction with ozone as discussed previously. Reactions of ozone with alkanes giving ketones and alcohols as involved in mechanism (b) have been reported in several instances.[96-98] The products obtained from isobutane and isoalkanes (Table VI) are in accordance with intermediate oxenium ion formation-rearrangement.

The rate of formation of the dimethylmethylcarboxonium ion from isobutane is considerably faster than that of the *tert*-butyl cation from isobutane in the absence of ozone under the same reaction conditions.[87] A solution of isobutane in excess Magic Acid:SO$_2$ClF

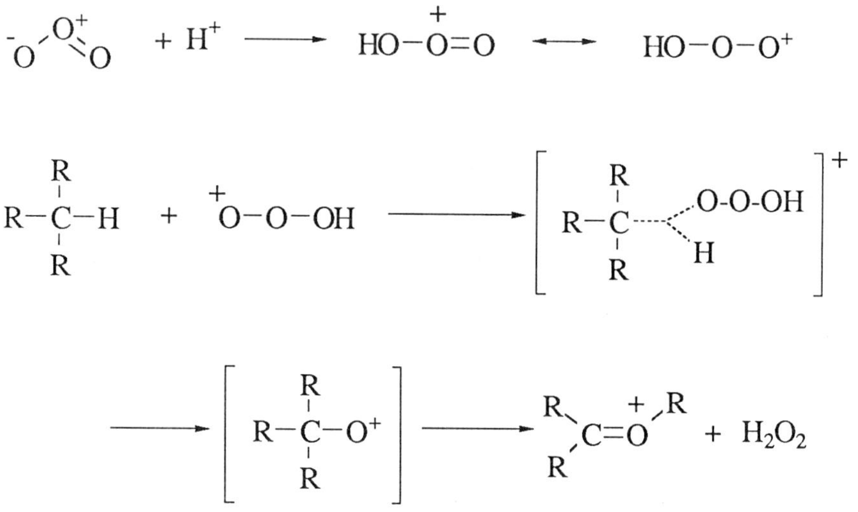

$$
{}^-O{-}\overset{+}{O}{\diagdown}_O + H^+ \longrightarrow HO{-}\overset{+}{O}{=}O \longleftrightarrow HO{-}O{-}O^+
$$

$$
R{-}\underset{\underset{R}{|}}{\overset{\overset{R}{|}}{C}}{-}H + \overset{+}{O}{-}O{-}OH \longrightarrow \left[R{-}\underset{\underset{R}{|}}{\overset{\overset{R}{|}}{C}}\cdots\!\!<^{O\text{-}O\text{-}OH}_{H} \right]^+
$$

$$
\longrightarrow \left[R{-}\underset{\underset{R}{|}}{\overset{\overset{R}{|}}{C}}{-}O^+ \right] \longrightarrow \underset{R}{\overset{R}{\diagdown}}C{=}\overset{+}{O}{\diagup}^R + H_2O_2
$$

solution gives only a trace amount of *tert*-butyl cation after standing for 5 hr at -78°C. *tert*-Butyl alcohol, on the other hand, in the same acid system gives quantitative yield of the corresponding cation. In the presence of ozone, under the same conditions it gives dimethylmethylcarboxonium ion. Isobutane does not give any oxidation product in the absence of Magic Acid under low temperature ozonization conditions. No experimental evidence for the intermediacy of the carbenium ions could be obtained.[87] The most probable path postulated for these reactions is the electrophilic attack by protonated ozone on the alkanes resulting in the formation of a pentacoordinated transition state from which the involved carboxonium is formed. Since ozone has a strong 1,3-dipole or at least a strong polarizability,[88] it is expected to be readily protonated in superacid media. Attempts to directly observe protonated ozone O_3H^+ by ^{1}H-NMR spectroscopy[87] were however inconclusive because of probable fast hydrogen exchange with the acid system.

Superacid catalyzed reaction of ozone with straight-chain alkanes has also been investigated at low temperature conditions. Magic Acid-catalyzed ozonization of ethane is shown in Scheme 4. Reaction of methane under similar condition was also investigated and discussed previously. Reactions of cycloalkanes have similarly been studied.

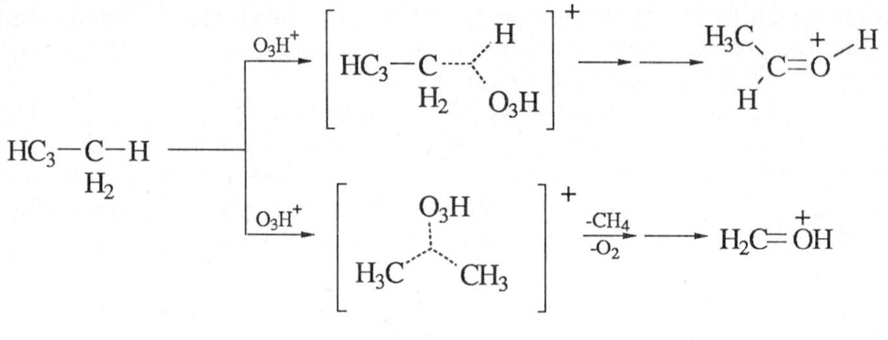

Scheme 4

Superacid-catalyzed electrophilic oxygenation of functionalized hydrocarbons has also been achieved.[82] Oxidation of alcohols, ketones, and aldehydes was carried out using protonated ozone. These reactions are illustrated in Scheme 5. In the case of carbonyl compounds, the C-H bond located farther than the γ-position seems to react with ozone in the presence of Magic Acid. The strong electron-withdrawing effect of the protonated carbonyl group is sufficient to inhibit reaction of the C-H bonds at α-, β-, and γ-positions. Superacid ($HF:SbF_5$)-catalyzed oxidation with ozone has also been carried out with certain keto steroids[99] bearing various substituents such as carbonyl, hydroxy, and acetoxy groups.

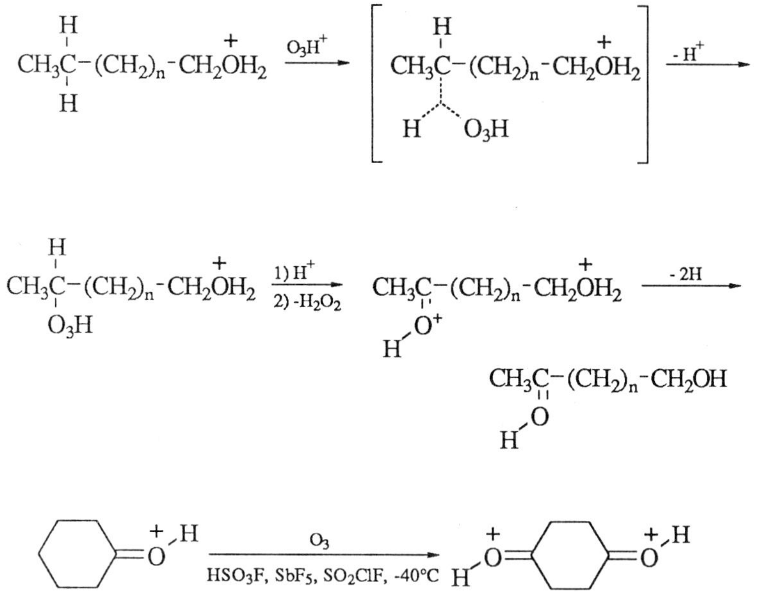

$$\text{CH}_3\overset{\overset{\displaystyle H}{|}}{\underset{\underset{\displaystyle H}{|}}{\text{C}}}-(\text{CH}_2)_n-\overset{+}{\text{CH}_2\text{OH}_2} \xrightarrow{\text{O}_3\text{H}^+} \left[\text{CH}_3\overset{\overset{\displaystyle H}{|}}{\text{C}}-(\text{CH}_2)_n-\overset{+}{\text{CH}_2\text{OH}_2} \right] \xrightarrow{-\text{H}^+}$$

$$\text{CH}_3\overset{\overset{\displaystyle H}{|}}{\underset{\underset{\displaystyle \text{O}_3\text{H}}{|}}{\text{C}}}-(\text{CH}_2)_n-\overset{+}{\text{CH}_2\text{OH}_2} \xrightarrow[2)\,-\text{H}_2\text{O}_2]{1)\,\text{H}^+} \text{CH}_3\overset{+}{\underset{\underset{\displaystyle H^{\diagdown}\text{O}^+}{|}}{\text{C}}}-(\text{CH}_2)_n-\overset{+}{\text{CH}_2\text{OH}_2} \xrightarrow{-\,2\text{H}}$$

$$\text{CH}_3\overset{}{\underset{\underset{\displaystyle H^{\diagdown}\text{O}}{||}}{\text{C}}}-(\text{CH}_2)_n-\text{CH}_2\text{OH}$$

Scheme 5

X. Nitration and Nitrosation

Nitronium ion is capable of nitrating not only the aromatic systems but also the aliphatic hydrocarbons. Electrophilic nitration of alkanes and cycloalkanes was carried out with $NO_2^+PF_6^-$, $NO_2^+BF_4^-$, and $NO_2^+SbF_6^-$ salts in CH_2Cl_2-tetramethylenesulfone or HSO_3F solution.[100a] Some representative reactions of nitronium ion with various alkanes, cycloalkanes, and polycyclic alkanes are shown in Table VII. The nitration takes place on both C-C and C-H bonds involving two-electron, three-center (2e-3c) bonded five-coordinated carbocations as depicted below with adamantane, a cage polycycloalkane.

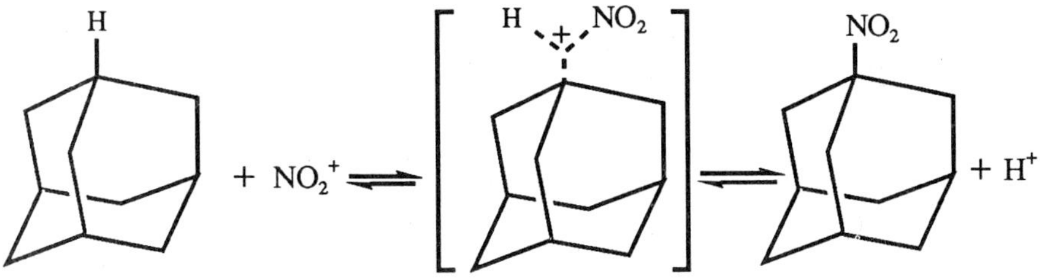

Table VII. Nitration and Nitrolysis of Alkanes and Cycloalkanes with $NO_2{}^+PF_6{}^-$

Hydrocarbon	Nitroalkane products and their mole ratio
Methane	CH_3NO_2
Ethane	$CH_3NO_2 > CH_3CH_2NO_2$, 2.9:1
Propane	$CH_3NO_2 > CH_3CH_2NO_2 > 2NO_2C_3H_7 > 1NO_2C_3H_7$, 2.8:1:0.5 :0.1
Isobutane	$tert\text{-}NO_2C_4H_9 > CH_3NO_2$, 3:1
n-Butane	$CH_3NO_2 > CH_3CH_2NO_2 > 2NO_2C_4H_9 \sim 1NO_2C_4H_9$, 5:4:1.5:1
Neopentane	$CH_3NO_2 > tert\text{-}C_4H_9NO_2$, 3.3:1
Cyclohexane	Nitrocyclohexane
Adamantane	1-Nitroadamantane > 2-nitroadamantane, 17.5:1

Formation of 1-fluoroadamantane and 1-adamantanol as byproducts in the reaction indicates that the pentacoordinate carbocation can also cleave to the 1-admantyl cation. These results also show the nonlinear nature of the ionic intermediate.

Nitrosonium ion (NO^+) is an excellent hydride-abstracting agent. Cumene undergoes hydride abstraction to provide a cumyl cation, which further reacts to give condensation products.[100b] The hydride-abstracting ability of NO^+ has been exploited in many organic transformations such as the Ritter reaction and ionic fluorination.[100c]

XI. Halogenation

Halogenation of saturated aliphatic hydrocarbons is usually carried out by free radical processes.[101] Superacid-catalyzed ionic halogenation of alkanes has been reported by Olah and his co-workers.[102] Chlorination and chlorolysis of alkanes in the presence of SbF_5, $AlCl_3$, and $AgSbF_6$ catalysts were carried out. As a representative the reaction of methane with $Cl_2:SbF_5$ system is shown:

$$\begin{array}{c} H \\ | \\ H-C-H \\ | \\ H \end{array} + \begin{array}{c} \overset{\delta+}{} \quad \overset{\delta-}{} \\ Cl-Cl \rightarrow SbF_5 \\ \text{or "Cl}^{+}\text{"} \end{array} \longrightarrow$$

No methylene chloride or chloroform was observed in the reaction. Under the stable ion conditions used, dimethylchloronium ion formation also occurs. However, this is a reversible process and helps to minimize competing alkylation of methane to ethane (and higher homologs), which becomes more predominant when methyl fluoride is formed via halogen exchange.

$$CH_4 \xrightarrow[\text{-78°C}]{SbF_5\text{-}Cl_2\text{-}SO_2ClF} \left[\begin{array}{c} H \\ | \\ H-C\cdots \overset{Cl}{\underset{H}{\diagdown}} \\ | \\ H \end{array}\right]^{+} \xrightarrow{\text{-H}^+} CH_3Cl \underset{\overrightarrow{}}{\overset{CH_3Cl}{\rightleftharpoons}} CH_3Cl^{+}CH_3$$

$Ag^{+}SbF_6^{-}$ induced electrophilic bromination of alkanes[103] has also been carried out as shown:

$$Br_2 + AgSbF_6 \rightleftharpoons \overset{\delta+}{} \overset{\delta-}{} \overset{+}{} \overset{-}{} \\ Br-Br \rightarrow AgSbF_6$$

$$R_3C-H + \overset{+}{} \overset{-}{} Br-Br \rightarrow \overset{+}{} \overset{-}{} AgSbF_6 \longrightarrow \left[\begin{array}{c} H_{\diagdown} \\ \quad \diagup\cdots Br \\ R_3C \end{array}\right]^{+} SbF_6^{-} + AgBr$$

$$\downarrow$$

$$R_3C-Br \quad + \quad H^{+}SbF_6^{-}$$

Even electrophilic fluorination of alkanes has been reported. F_2 and fluoroxytrifluoromethane have been used to fluorinate tertiary centers in steroids and adamantanes by Barton and co-workers.[104] The electrophilic nature of the reaction involving polarized but not cationic fluorine species[105] has been invoked. Gal and Rozen[106] have carried out direct electrophilic fluorination of hydrocarbons in the presence of chloroform. F_2 appears to be strongly polarized in chloroform (hydrogen bonding with acidic proton in $CHCl_3$). However, so far no positively charged fluorine species (fluoronium ions) are known in solution chemistry.

Table VIII. Chlorination and Bromination of Methane over Supported Acid Catalysts

Catalyst	CH_4/Cl_2	Reaction temperature (°C)	GHSV (ml/g/hr)	Convers (%)	Product (%) CH_3Cl	CH_2Cl_2
10% FeO_xCl_y/Al_2O_3	1:2	250	100	16	88	12
20% $TaOF_3/Al_2O_3$	1:2	235	50	14	82	6[a]
	1:2	235	1400	15	93	7
	2:1	235	1200	13	96	4
20% $NbOF_3/Al_2O_3$	1:3	250	50	10	90	10
10% $ZrOF_2/Al_2O_3$	1:4	270	100	34	96	4
Nafion-H	1:4	185	100	18	88	12
20% GaO_xCl_y/Al_2O_3	1:2	250	100	26	90	10
20% TaF_5-Nafion-H	1:2	200	100	11	97	3
25% SbF_5-graphite	1:2	180	100	7	98	2

Catalyst	CH_4/Br_2	Reaction temperature (°C)	GHSV (ml/g/hr)	Convers (%)	Product(%) CH_3Br	CH_2Br_2
20% $SbOF_3/Al_2O_3$	5:1	200	100	20	99	
20% $TaOF_3/Al_2O_3$	15:1	250	50	14	99	

Supported Platinum Metal-Catalyzed Halogenation of Methane

Catalyst	CH_4/X_2 (X = Cl,Br)	Reaction temperature (°C)	GHSV (ml/g/hr)	Convers (%)	Product (%) CH_3Cl	CH_2Cl_2
		Chlorination				
0.5% Pt/Al_2O_3	1:3	100	600	11[a]	~100	
	1:3	150	600	16[a]	92	8
	1:3	200	600	32[a]	92	8
	2:1	250	300	23[b]	98	<2
	3:1	250	300	36[b]	99	1
5% $Pd/BaSO_4$	2:1	200	600	30[b]	99	1
		Bromination				
0.5% Pt/Al_2O_3	2:1	200	300	8[b]	99	Trace

[a]Based on methane.
[b]Based on chlorine (bromine).

In extending the electrophilic hydrogen (chlorination and bromination) of methane to catalytic heterogeneous gas phase reactions Olah et al. recently found[107] that methane can be chlorinated or brominated over various solid acid or supported platinum group metal catalysts (the latter, the heterogeneous analog of Shilov's solution chemistry) to methyl halides with high selectivity under relatively mild conditions. Table VIII summarizes some of the results.

Both reactions mechanistically are electrophilic insertion reactions into the methane C-H bonds. In the platinum insertion reaction subsequent chlorolysis of the surface bound methylplatinum chloride complex regenerates the catalyst and gives methyl chloride.

$$X_2 + \text{catalyst} \rightleftharpoons \left[X^+\right]\left[\text{catalyst-}X^-\right] \quad (X = Cl, Br)$$

$$\left[X^+\right]\left[\text{catalyst-}X^-\right] + CH_4 \rightleftharpoons \left[CH_3\cdots\begin{array}{c}H\\X\end{array}\right]^+ \left[\text{catalyst-}X^-\right]$$

$$CH_3-X + \text{catalyst} + HX$$

$$CH_4 + PtCl_2 \rightleftharpoons \begin{array}{c}CH_3 \\ \diagdown \\ H \end{array}Pt\begin{array}{c}Cl \\ \diagup \\ Cl\end{array} \xrightarrow{-H^+} \left[CH_3-Pt\begin{array}{c}Cl\\ \diagup \\ Cl\end{array}\right]^- \underset{Cl^-}{\rightleftharpoons}$$

$$CH_3Pt^{II}Cl \rightleftharpoons \begin{array}{c}Cl \\ \diagdown \\ Cl\end{array}Pt^{IV}\begin{array}{c}CH_3 \\ \diagup \\ Cl\end{array} \longrightarrow CH_3Cl + Pt^{II}Cl_2$$

Concerning the electrophilic halogenation of methane it should also be pointed out that the singlet-triplet energy difference of positive halogens (as illustrated for the hypothetical X^+ ions) favor the latter.[107c] Electrophilic halogenations thus may be more complex and can involve radical ions even under conditions in which conventional radical chain halogenation is basically absent.

Combining the halogenation of methane with catalytic, preferentially gas phase hydrolysis, methyl alcohol (and dimethyl ether) can be obtained in high selectivity.[107a,b]

$$CH_4 \xrightarrow[X_2]{\text{acid cat.}} HX + CH_3X \xrightarrow[\text{cat.}]{H_2O} CH_3OH \ (CH_3OCH_3)$$

$$\text{Pt or Pd} \searrow \quad -HX$$

$$X = Cl, Br \qquad [CH_3PtX] \xrightarrow{X_2} CH_3X$$

Hydrogen halides (particularly hydrogen bromide) can be oxidatively recycled, making the process catalytic in used halogen. The hydrolysis of methyl chloride with caustic was first carried out by Berthelot in the 1830s.[108] This first preparation of methyl alcohol, however, was never utilized in a practical way. Seemingly no attempt was made until our work to extend the hydrolysis of methyl halides to catalytic gas phase conditions. Olah et al.[107a,b] carried out an extensive study of such reactions and found that methyl halides hydrolyze over alumina to methyl alcohol-dimethyl ether in yields up to 25% per pass with gaseous space velocities of up to 1500 and high turnovers.

They also found that solid acid catalysts, such as Nafion-H, are also capable of catalyzing the hydrolysis of methyl halides and yielding under the reaction temperatures of 150-170°C dimethyl ether.[109b]

$$2 \; CH_3X \quad \xrightarrow[\text{Nafion-H}]{H_2O} \quad CH_3OCH_3 + 2 \; HX$$

Halogenation of methane followed by hydrolysis consequently is a suitable way to obtain methyl alcohol (and dimethyl ether).

$$CH_4 \quad \xrightarrow[\text{cat.}]{X_2} \quad CH_3X + HCl$$

$$\searrow \begin{array}{c} X_2, H_2O \\ \text{cat.} \end{array} \qquad H_2O \Big| \; \text{cat.} \qquad\qquad (X = Cl, Br)$$

$$CH_3OH \quad [(CH_3)_2O]$$

It was also possible to show that halogens (particularly bromine) together with steam can be reacted with methane over acidic catalysts in a single step producing methyl alcohol (dimethyl ether) although conversion of methane so far was only modest.[109b]

$$CH_4 + Br_2 + H_2O \longrightarrow CH_3OH + 2 \; HBr$$

$$2 \; HBr + 1/2 \; O_2 \longrightarrow H_2O + Br_2$$

-----

overall reaction: $CH_4 + 1/2 \; O_2 \longrightarrow CH_3OH$

The feasible selective catalytic preparation of methyl halides and methyl alcohol from methane also allows subsequent condensation to ethylene and higher hydrocarbons. This can be accomplished by the Mobil Corp. ZSM-5 zeolite catalysts but also over non-shape-selective acidic-basic bifunctional catalysts, such as WO_3/Al_2O_3.

Whereas the Mobil process starts with syn gas-based methyl alcohol, Olah's studies were an extension of the previously discussed electrophilic functionalization of methane and did not involve any zeolite type catalysts. It was found that bifunctional acid-basic catalysts such as tungsten oxide on alumina or related supported transition metal oxides or oxyfluorides such as tantalum or zirconium oxyfluoride are capable of condensing methyl chloride, methyl alcohol (dimethyl ether), and methyl mercaptan (dimethyl sulfide) primarily to ethylene (and propylene).[110]

$$2\ CH_3X \longrightarrow CH_2{=}CH_2 + 2\ HX$$
$$X = halogen$$

$$2\ CH_3XH \longrightarrow CH_2{=}CH_2 + 2\ H_2X$$
$$X = O,S$$

In these reactions, methane is the major byproduct formed probably by competing radical reactions. However, since the overall starting material is methane this represents only the need for recycling.

According to Olah's studies, the conversion of methyl alcohol over bifunctional acidic-basic catalyst after initial acid catalyzed dehydration to dimethyl ether involves oxonium ion formation catalyzed also by the acid functionality of the catalyst. This is followed by basic site-catalyzed deprotonation to a reactive surface bound oxonium ylide, which is then immediately methylated by excess methyl alcohol or dimethyl ether leading to the crucial C_1-C_2 conversion step. The ethyl, methyl oxonium ion formed subsequently eliminates ethylene. All other hydrocarbons are derived from ethylene by known oligomerization-fragmentation chemistry. Propylene is formed by the reaction of ethylene with the dimethyloxonium methylide via a cyclopropane intermediate.

$$2\ CH_3OH \xrightarrow[-H_2O]{cat.} CH_2{=}CH_2$$
$$CH_3OCH_3$$

$$2\ CH_3OH \xrightarrow{acid\ cat.} CH_3OCH_3 + H_2O$$

$$\downarrow acid\ cat.$$

$$\overset{+}{CH_3}\underset{\underset{CH_3}{|}}{O}CH_3\ \overset{-}{CH_3O}{-}cat$$

$$CH_3CH{=}CH_2 \longleftarrow \underset{CH_2}{\overset{CH_2{-}CH_2}{\diagdown\diagup}} \xleftarrow{CH_2{=}CH_2} \underset{\underset{CH_2}{\overset{-}{|}}}{CH_3\diagdown\overset{+}{O}\diagup CH_3} \xrightarrow[C_1\,{-}\,C_2]{CH_3OCH_3}$$

$$\underset{\underset{CH_2CH_3}{|}}{CH_3\diagdown\overset{+}{O}\diagup CH_3} \longrightarrow CH_2{=}CH_2 + (CH_3)_2O$$

The intermolecular nature of the C_1-C_2 transformation step was shown by experiments using mono-^{13}C-labeled dimethyl ether and analyzing the isotopic composition of the product ethylene. Intramolecular Stevens-type rearrangement under the reaction conditions was clearly ruled out.

It is not necessary to invoke a free Meerwein-type trimethyloxonium ion in the heterogeneous catalytic reaction. Lewis-type coordination complexes of dimethyl ether with the acidic catalyst sites having oxonium ion character can be involved, giving subsequently via deprotonation surface bound oxonium ylides followed by methylation and elimination of ethylene.[111,112]

$$(CH_3)_2O + cat. \longrightarrow \begin{array}{c} CH_3 \\ CH_3 \end{array}\!\!O^+\!-\!\bar{c}at. \xrightarrow{-H^+} \begin{array}{c} \bar{C}H_2 \\ CH_3 \end{array}\!\!O^+\!-\!cat.$$

$$\downarrow (CH_3)_2O \; cat.$$

$$CH_3OH + CH_2\!\!=\!\!CH_2 \longleftarrow \begin{array}{c} CH_3CH_2 \\ CH_3 \end{array}\!\!O^+\!-\!cat.$$

Whereas Olah's studies did not involve zeolite catalysts, it is probable that in the ZSM-5-catalyzed Mobil process too no direct monomolecular dehydration of methyl alcohol to methylene is involved. This is thermodynamically not feasible even when considering that surface complexation could somewhat affect the otherwise very endothermic thermodynamics.

$$CH_3OH \xrightarrow{H\text{-}zeolite} CH_3\overset{+}{O}H_2 \; zeolite^- \begin{array}{c} \overset{a}{\nearrow} [CH_2 + H_3O^+] \; zeolite^- \\ \Big\updownarrow c \\ \underset{b}{\searrow} [CH_3^+ + H_2O] \; zeolite^- \end{array}$$

$$\begin{array}{cccc} CH_3OH & \longrightarrow & CH_2 & + & H_2O \\ -200.7 & & +390.4 & & -241.8 \quad \Delta H = +349.3 \; kcal \; mol^{-1} \end{array}$$

$$\begin{array}{cccc} 2\,CH_3OH & \longrightarrow & CH_3OCH_3 & + & H_2O \\ 2(-200.7) & & +184.1 & & -241.8 \quad \Delta H = -24.5 \; kcal \; mol^{-1} \end{array}$$

Therefore, it is reasonable to suggest that in the zeolite-catalyzed process condensation also proceeds via bimolecular dehydration of methyl alcohol to dimethyl ether, which subsequently is transformed via the oxonium ylide pathway and intermolecular methylation to ethyl oxonium species (the crucial C_1-C_2 bond formation step).[111] A radical or radical ion type condensation mechanism inevitably should give also competing coupling products that are not observed. Once ethylene is formed in the system it can undergo oligomerization methylene insertion or react further with methyl alcohol giving higher hydrocarbons.

With increase in the acidity of the catalyst, methyl alcohol or dimethyl ether undergoes condensation to saturated hydrocarbons and aromatics, with no olefin byproducts. With tantalum or niobium pentafluoride-based catalyst at 300°C conversion result to gasoline range branched hydrocarbons and some aromatics (30% of the product).[112] This composition is similar to that reported with H-ZSM-5 zeolite catalyst.

The condensation of methyl chloride or bromide to ethylene proceeds by a related mechanistic path involving initial acid-catalyzed dimethylhalonium ion (or related catalyst complex) formation with subsequent proton elimination to a reactive methylhalonium methylide, which is then readily methylated by excess methyl halide. The ethylhalonium ion intermediate gives ethylene by β-elimination.

$$2\,CH_3X \longrightarrow CH_2{=}CH_2 + 2\,HX$$

$$CH_3X^+ \to {}^-cat. \longrightarrow [CH_2\overset{+}{X}-cat.] \xrightarrow{CH_3X}$$

$$CH_3CH_2\overset{+}{X}-cat. \xrightarrow{-HX} CH_2{=}CH_2$$

$$2\,CH_3X \xrightarrow{acid\ cat.} CH_3\overset{+}{X}CH_3 \quad \overset{-}{Cl}-cat.$$

$$\Big\downarrow {}^{-H^+}\ base$$

$$CH_3\overset{+}{X}\overset{-}{CH_2} \xrightarrow[cat.]{CH_3X} CH_3\overset{+}{X}CH_2CH_3 \quad \overset{-}{X}-cat.$$

$$\Big\downarrow$$

$$CH_3X + CH_2{=}CH_2$$

Similar reaction paths can be visualized for condensation of methyl mercaptan or methylamines. It is interesting to note that Corey's well-known synthetic studies[113] with the use of dimethylsulfonium methylide mentioned the need to generate the ylide at low temperatures, as otherwise decomposition gives ethylene. Indeed, this reaction is similar to that involved in the higher temperature acid-base catalyzed condensation reaction.

$$2\,CH_3SH \xrightarrow{-H_2S} CH_3SCH_3 \xrightarrow[catalyst]{} (CH_3)_2S^+\ cat^-$$

$$\Big\downarrow {}^{-H^+}$$

$$CH_2{=}CH_2 + CH_3SH \xleftarrow{-H^+} CH_3\overset{+}{\underset{CH_2CH_3}{S}}cat. \xleftarrow[\substack{CH_3SCH_3\\ or\\ CH_3SH\\ catalyst}]{} CH_3\overset{+}{\underset{CH_2CH_3}{S}}cat.$$

XII. Conclusions

Olah, in the conclusion of a review on *Carbocations and Electrophilic Reactions*[18] written some 15 years ago, stated concerning the realization of the electron donor ability of shared σ-electron pairs: "More importantly, the concept of pentacoordinated carbonium ion formation *via* electron sharing of single bonds with electrophilic reagents in three-center bond formation promises to open up a whole new important area of chemistry. Whereas the concept of tetravalency of carbon obviously is not affected, carbon penta- (or tetra-) coordination as a general phenomenon must be recognized. Trivalent carbenium ions play a major role in electrophilic reactions of π- and n-donors, whereas pentacoordinated carbonium ions play an equally important similar role in electrophilic reactions of σ-donor saturated systems. The realization of the electron donor ability of shared (bonded) electron pairs (single bonds) could one day rank equal in importance with G. N. Lewis' realization[114] of the importance of the electron donor unshared (nonbonded) electron pairs. We can now not only explain the reactivity of saturated hydrocarbons and single bonds in general in electrophilic reactions, but indeed use this understanding to explore many new areas and reactions of carbocation chemistry."

It seems that the intervening years have justified that prediction to a significant degree. The electrophilic chemistry of alkanes has rapidly expanded and has started to occupy a significant role even in the conversion of methane. As other chapters discuss the reactions of coordinatively unsaturated metal compounds with C-H bonds we have not dealt with them here, although these reactions are also typical electrophilic insertion reactions.

References

1.   Asinger, F., *Paraffins, Chemistry, and Technology*. Pergamon Press, New York, 1965.

2.   (a) For a comprehensive early review, see: Olah, G. A., *Angew. Chem. Int. Ed. Engl.* **1973**, *12*, 173 and references cited therein; (b) for reviews on superacids, see Olah, G. A., Prakash, G. K. S., Sommer, J., *Superacids*. Wiley Interscience, New York, 1985; (c) Sommer, J., Olah, G. A., *La Recherce* **1979**, *10*, 624.

3.   Bloch, H. S., Pines, H., Schmerling, L., *J. Am. Chem. Soc.* **1946**, *68*, 153.

4.   Olah, G. A., Lukas, J., *J. Am. Chem. Soc.* **1967**, *89*, 2227; Olah, G. A., Lukas, J., *Ibid* **1967**, *89*, 4739.

5.   (a) Bickel, A. F., Gaasbeek, G. J., Hogeveen, H., Oelderick, J. M., Platteuw, J. C., *Chem. Commun.* **1967**, 634; (b) Hogeveen, H., Bickel, A. F., Chem. Commun. **1967**, 635.

6.   Prakash, G. K. S., Husain, A., Olah, G. A., *Angew. Chem* **1983**, *95*, 51.

7.   Olah, G. A., Lukas, J., *J. Am. Chem. Soc.* **1968**, *90*, 933.

8.   Olah, G. A., Schlosberg, R. H., *J. Am. Chem. Soc.* **1968**, *90*, 2726.

9. Hogeveen, H., Gaasbeek, C. J., *Recl. Trav. Chim. Pays-Bas.* **1968,** *87,* 319.
10. Olah, G. A., Klopman, G., Schlosberg, R. H., *J. Am. Chem. Soc.* **1969,** *91,* 3261.
11. (a) Olah, G. A., Shen, J., Schlosberg, R. H., *J. Am. Chem. Soc.* 1970, 92, 3831; (b) Oka, T., *Phys. Lett.* **1980,** *43,* 531.
12. Hogeveen, H., Bickel, A. F., *Recl. Trav. Chim. Pays-Bas.* **1967,** *86,* 1313.
13. Pines, H., Hoffman, N. E., in *Friedel-Crafts and Related Reactions,* (G. A. Olah, ed.), Vol. II, p.1216. Wiley Interscience, New York, 1964.
14. Olah, G. A., Halpern, Y., Shen, J., Mo, Y. K., *J. Am. Chem. Soc.* **1971,** *93,* 1251.
15. Otvos, J. W., Stevenson, D. P., Wagner, C. D., Beeck, O., *J. Am. Chem. Soc.* **1951,** *73,* 5741.
16. Hoffman, H., Gaasbeek, C.J., Bickel, A. F., *Recl. Trav. Chim. Pays-Bas.* **1969,** *88,* 703.
17. Bonnifay, R., Torck, B., Hellin, J. M., *Bull. Soc. Chim. Fr.* **1977,** 808.
18. Lucas, J., Kramer, P. A., Kouwenhoven, A. P., *Recl. Trav. Chim. Pays-Bas.* **1973,** *92,* 44.
19. Haag, W. O., Dessau, R. H., *Int. Catal. Congr. West Berlin, Germany* **1984** , *II,* 105.
20. Olah, G. A., Schlosberg, R. H., *J. Am. Chem. Soc.* **1968,** *90,* 2726.
21. For a review, see Brouwer, D. M., Hogeveen, H., *Prog. Phys. Org. Chem.* **1972,** *9,* 179.
22. Olah, G. A., *Angew. Chem. Intl. Ed. Engl.* **1973,** *12,* 173.
23. (a) Colvin, E. W., *Silicon in Organic Synthesis.* Butterworths, London, 1981; (b) Weber, W. P., *Silicon Reagents in Organic Synthesis.* Springer-Verlag, Berlin, Heidelberg, New York, 1983.
24. Bertram, J., Coleman, J. P., Fleischmann, M., Pletcher, D., *J. Chem. Soc. Perk.* **1973,** *II,* 374.
25. For a review, see Fabre, P. L., Devynck, J., Tremillon, B., *Chem. Rev.* **1982,** *82,* 591.
26. Devynck, J., Fabre, P. L., Ben Hadid, A., Tremillon, B., *J. Chem. Res.,* **1982,** *82,* 591.
27. (a) Germain, A., Ortega, P., Commeyras, A., *Nouv. J. Chim.* **1979,** *3,* 415; (b) Choucroun, H., Germain, A., Brunel, D., Commeyras, A., *Nouv. J. Chim.* 1982, 7, 83; (c) Choucroun, H., Germain, A., Brunel, D., Commeyras, A., *Nouv. J. Chim.* 1982, 7, 83.
28. Nenitzescu, C. D., Dragon, A., *Ber. Dtsch. Chem. Ges.* **1933,** *66,* 1892.
29. Pines, H., Joffman, N. E., in *Friedel-Crafts and Related Reactions,* (G. A. Olah, ed.), Vol.II, p.1211. Wiley-Interscience, New York, 1964.
30. Nenitzescu, C. D., in *Carbonium Ions,* G. A. Olah, P. v. R. Schleyer, eds.), Vol. II, p.490. Wiley-Interscience, New York, 1970.
31. Asinger, F., *Paraffins,* p.695. Pergamon Press, New York, 1968.
32. Bonnifay, R., Torck, B., Hellin, J. M., *Bull Soc. Chim. Fr.* **1977,** 1057; (b) **1978,** 36.
33. Brunel, D., Itier, J., Commeyras, A., Phan Tan Luu, R., Matthieu, D., *Bull. Soc. Chim. Fr.* **1979,** *II,* 249, 257.
34. Olah, G. A., U.S. Patent 4,472,268 (1984).
35. Olah, G. A., U.S. Patent 4,508,618 (1985).

36. (a) Le Normand, F., Fajula, F., Gault, F., Sommer, J., *Nouv. J. Chim.*
     1982, *6*, 411; (b) Le Normand, F., Fajula, F., Sommer, J., *Nouv. J.
     Chim.* **1982**, *6*, 291.
37. Laali, K., Muller, M., Sommer, J., *Chem. Commun.* **1980,** 1088.
38. Olah, G. A., Kaspi, J., Bukala, J., *J. Org. Chem.* **1977**, *42*, 4187.
39. Olah, G. A., U.S. Patent 4,116,880, 1978.
40. Olah, G. A., Kaspi, J., *J. Org. Chem.* **1977**, *42*, 3046.
41. Olah, G. A., DeMember, J. R., Shen, J., *J. Am. Chem. Soc.* **1973**, *95*,
     4952.
42. (a) Schleyer, P. v. R., *J. Am. Chem. Soc.* **1957**, *79*, 3292; (b) Fort, R.
     C., Jr., in *Adamantane,the Chemistry of Diamond Molecules* (P. Gassman,
     ed.), Marcel Dekker, New York, 1976; (c) Olah, G. A., Olah, J.A.,
     Synthesis 1973, 488.
43. Olah, G. A., Farooq, O., *J. Org. Chem.* **1986**, *51*, 5410.
44. Farooq, O., Farnia, S. M. F., Stephenson, M., Olah, G. A., *J. Org. Chem.*
     1988, *53*, 2840.
45. Paquette, L., Balogh, A., *J. Am. Chem. Soc.* **1982**, *104*, 774.
46. Schuit, G. C., Hoog, H., Verhuis, J., *Recl. Trav. Chim. Pays-Bas.* **1940**,
     59, 793, British Patent 535054, 1941.
47. Pines, H., *Chem. Abst.* **1974**, *41*, 474, U.S. Patent 2405516194.6 (H.
     Pines to Universal Oil).
48. For a review, see Olah, G. A., *Carbocations and Electrophilic Reactions,*
     Verlag Chemie, Weinheim and John Wiley, New York, 1974.
49. For a review, see Brouwer, D. M., Hogeveen, H., *Progr. Phys. Org. Chem.*
     1972, *9*, 179.
50. (a) Olah, G. A., Bruce, M., Edelson, E. H., Husain, A., *Fuel* **1984**, *63*,
     1130; (b) Olah, G. A., Husain, A., *Fuel* **1984**, *63*, 1247; (c) Olah, G. A.,
     Bruce, M., Edelson, E. H., Husain, A., *Fuel* **1983**, *63*, 1432.
51. Olah, G. A., DeMember, G. A., Schlosberg, R. H., *J. Am. Chem. Soc.*
     1969, *91*, 2112.
52. Olah, G. A., DeMember, G. A., Schlosberg, R. H., Halpern, Y., *J. Am.
     Chem. Soc.* **1972**, *94*, 156.
53. Olah, G. A., Mo, Y. K., Olah, G.A., *J. Am. Chem. Soc.* **1973**, *95*, 4939.
54. Olah, G. A., DeMember, J. R., Shen, J., *J. Am. Chem. Soc.* **1973**, *95*,
     4952.
55. Olah, G. A., White, A. M., *J. Am. Chem. Soc.* **1969**, *91*, 5801.
56. Olah, G. A., Farooq, O., Krishnamurthy, V. V., Prakash, G. K. S., Laali,
     K., *J. Am. Chem. Soc.* **1985,** *107*, 7541.
57. Olah, G. A., U.S. Patent (1984): 4,443,192; 4,513,164; 4,465,893;
     4,467,130; (1985): 4,513,164.
58. (a) Olah, G. A., Olah, J. A., *J. Am. Chem. Soc.* **1971**, 93; (b) Roberts,
     D. T., Jr., Calihan, L. E., *J. Macromol. Sci. (Chem.)* **1973**, A7(8), 1629;
     1641; (c) Siskin, M., *J. Am. Chem. Soc.* **1976**, *98*, 5413; (d) Siskin, M.,
     Schlosberg, R. H., Kocsis,W. P., in *New Strong Acid Catalyzed Alkylation
     and Reduction Reactions,* L. F. Albright, R. A. Gikdsktm, eds.). No. 55,
     American Chemical Society Monographs, Washington, D.C., 1977; (e)
     Sommer, J., Muller, M., Laali, K., *Nouv. J. Chim.* **1982**, *6*, 3.
59. Olah, G. A., Felberg, J. D., Lammertsma, K., *J. Am. Chem. Soc.*
     1983,105, 6529.
60. Siskin, M., *J. Am. Chem. Soc.* **1976**, *98*, 5413.
61. Sommer, J., Muller, M., Laali, K., *Nouv. J. Chim.* **1982**, *6*, 3.

62. Olah, G. A., Felberg, J. D., Lammertsma, K., *J. Am. Chem. Soc.* **1983**, *105*, 6529.
63. Roberts, D. T., Jr., Calihan, L. E., *J. Macromol. Sci. (Chem.)* **1973**, *A7(8)*, 1629.
64. Roberts, D. T., Jr., Calihan, L. E., *J. Macromol. Sci. (Chem.)* **1973**, *A7(8)*, 1641.
65. Olah, G. A., Olah, J. A., *Friedel-Crafts and Related Reactions,* Vol.3, Part 2, p.1272. Wiley-Interscience, New York, 1964,
66. Hogeveen, H., Adv. *Phys. Org. Chem.* **1973**, *10*, 29.
67. Gresham, W. F., Tabet, G. E., U.S. Patent 2,485,237, 1946.
68. Booth, B. L., El-Fekky, T. A., *J. Chem. Soc. Perk. I* **1979**, 2441.
69. Paatz, R., Weisberger, G., *Chem. Ber.* **1967**, *100*, 984.
70. Yoneda, N., Fukuhara, T., Takahasi, Y., Suzuki, A., *Chem. Lett.* **1983**, 17.
71. Yoneda, N., Sato, H., Fukuhara, T., Takahashi, Y., Suzuki, A., *Chem. Lett.* **1983**, 19.
72. Yoneda, N., Takahashi, Y., Fukuhara, T., Suzuki, A., *Bull. Chem. Soc., Jpn.* **1986**, *59*, 2819.
73. Olah, G.A., Pelizza, F., Kobayashi, S., Olah, J.A., *J. Am. Chem. Soc.* **1976**, *98*, 296.
74. Fujiyama, S., Takagawa, M., Kajiyama, S., G. P. 2,425591, 1974.
75. Delderick, J. M., Kwantes, A., B. P. 1,123,966, 1968.
76. Gresham, W. F., Tabet, G. E., U.S. Patent 21,485,237, 1949.
77. Farooq, O., Ph.D Thesis, University of Southern California, Los Angeles, CA, 1984.
78. Olah, G. A., Laali, K., Farooq, O., *J. Org. Chem.* **1985**, *50*, 1483.
79. Farooq, O., Marcelli, M., Prakash, G. K. S., Olah, G. A., *J. Am. Chem. Soc.* **1988**, *110*, 864.
80. (a) Olah, G. A., Kuhn, S., *J. Am. Chem. Soc.* **1961**, *82*, 2380; (b) Olah, G. A., Malhotra, R., Narang, S. C., *J. Org. Chem.* **1978**, *43*, 4628.
81. Christen, H. R.,*Grundlagen des Organische Chemie,* p.242. Verlag Sauerlander-Diesterweg: Aarau-Frankfurt am Main, 1970.
82. Olah, G. A., Parker, D. G., Yoneda, N., *Angew Chem. Int. Ed. Engl.* **1978**, *17*, 909.
83. Christe, K. O., Wilson, W. W., Curtis, E. C., *Inorg. Chem.* **1976**, *18*, 2578.
84. Olah, G. A., Berrier, A. L., Prakash, G. K. S., *J. Am. Chem. Soc.* **1982**, *104*, 2373.
85. Olah, G. A., Yoneda, N., Parker, D. G., *J. Am. Chem. Soc.* **1977**, *99*, 483.
86. Olah, G. A., Parker, D. G., Yoneda, N., Pelizza, F., *J. Am. Chem. Soc.* **1976**, *98*, 2245.
87. Olah, G. A., Yoneda, N., Parker, D. G., *J. Am. Chem. Soc.* **1976**, *98*, 5261.
88. Tambarulo, R., Ghosh, S. N., Barrus, C. A., Gordy, W., *J. Chem. Phys.* **1953**, *24*, 851.
89. Bartlett, P. D., Stiles, M., *J. Am. Chem. Soc.* **1955**, *77*, 2806.
90. Wibault, J. P., Sixma, E. L. J., Kampschidt, L. W. E., Boer, H., *Recl. Trav. Chim. Pays-Bas.* **1950**, *69*, 1355.
91. Wibault, J. P., Sixma, E. L. J., *Recl. Trav. Chim. Pays-Bas.* **1951**, *71*, 76.
92. Bailey, P. S., *Chem. Rev.* **1958**, *58*, 925.

93. Homer, L., Schafer, H., Ludwig, W., *Chem. Ber.* **1958**, *91*, 75.
94. Bailey, P. S., Ward, J. W., Hornish, R. E., Potts, F. E., *Adv. Chem. Ser.* **1972**, *112*, 1.
95. Further oxidation products, i.e., acetylium ion and CO_2, were reported to be observed in a number of the reactions studied. Such secondary oxidations are not induced by ozone.
96. Hellman, T. M., Hamilton, G. A., *J. Am. Chem. Soc.* **1974**, *96*, 1530.
97. Whiting, M. C., Bolt, A. J. N., Parrish, J. H., *Adv. Chem. Ser.* **1968**, *77*, 4.
98. Williamson, D. O., Cvetanovic, R. J., *J. Am. Chem. Soc.* **1970**, *92*, 2949.
99. (a) Jacquesy, J. C., Jacquesy, R., Lamande, L., Narbonne, C., Patoiseau, J. F., Vidal, Y., *Nouv. J. Chim.* **1982**, *6*, 589; (b) Jacquesy, J. C., Patoiseau, J. F., *Tetrahedron Lett.* **1977**, 1499.
100. (a) Olah, G. A., Lin, H. C., *J. Am. Chem. Soc.* **1971**, *93*, 1259; (b) Olah, G. A., Friedman, M., *J. Am. Chem. Soc.* **1966**, *88*, 5330; (c) Olah, G. A., *Acc. Chem. Res.* **1980**, *13*, 330.
101. Poutsma, M. L., in *Methods in Free-Radical Chemistry* (E. S. Huyser, ed.), Vol.II. Marcel Dekker, New York, 1969.
102. (a) Olah, G. A., Mo, Y. K., *J. Am. Chem. Soc.* **1972**, *94*, 6864; (b) Olah, G. A., Renner, R., Schilling, P., Mo, Y. K., *J. Am. Chem. Soc* **1973**, *95*, 7686.
103. Olah, G. A., Schilling, P., *J. Am. Chem. Soc.* **1973**, *95*, 7680.
104. Alker, D., Barton, D. H. R., Hesse, R. H., James, J. L., Markwell, R. E., Pechet, M. M., Rozen, S., Takeshita, T., Toh, H. T., *Nouv. J. Chim.* **1980**, *4*, 239.
105. Christe, K. O., *J. Fluo. Chem.* **1983**, 519; Christe, K. O., *J. Fluo. Chem.* **1984**, 269.
106. Gal, C., Rozen, S., *Tetrahedron Lett.* **1984**, 449.
107. (a) Olah, G. A., Gupta, B., Farnia, M., Felberg, J. D., Ip, W. M., Husain, A., Karpeles, R., Lammertsma, K., Melhotra, A. K., Trivedi, N. J., *J. Am. Chem. Soc.* **1985**, *107*, 7097; (b) Olah, G. A., U.S. Patent 7,523,040, 1985, and corresponding foreign patents; (c) work carried out in cooperation with Prof. K. Houk.
108. Berthelot, M., *Ann. Chim.* **1858**, *52*, 97.
109. (a) Weissermel, K., Arpe, H. J., *Industrial Organic Chemistry*, p.47. Verlag-Chemie, Weinheim, 1978; (b) Olah, G. A., et al., unpublished results.
110. Olah, G. A., Doggweiler, H., Felberg, J. D., Frohlich, S., Grdina, M. J., Karpeles, R., Keumi, T., Inaba, S., Ip, W. M., Lammertsma, K., Salem, G., Tabor, D. C., *J. Am. Chem. Soc.* **1984**, *106*, 2143; Olah, G. A., U.S. Patent 4,373,109, 1983, and corresponding foreign patents.
111. Olah, G. A., Prakash, G. K. S., Ellis, R. W., Olah, G. A., J. Chem. Soc. Chem. Commun. *1986*, 9.
112. Salem, G., Ph.D. Thesis, University of Southern California, 1980.
113. Corey, E. J., Chaykovsky, M., *J. Am. Chem. Soc.* **1965**, 87, 1353.
114. Lewis, G. N., *J. Am. Chem. Soc.* **1916**, *38*, 762; *Valence and Structure of Atoms and Molecules.* Chemical Catalog Corp., New York, 1923.

Chapter III

SOME RECENT ADVANCES IN ALKANE FUNCTIONALIZATION: METAL PHOSPHINE CATALYSTS AND MERCURY PHOTOSENSITIZATION

Robert H. Crabtree

Chemistry Department
Yale University
New Haven, Connecticut 06511

Two areas of alkane functionalization catalysis in which progress has been made in the last few years[1] are discussed in this chapter. The first involves the homogeneous catalysis of a variety of alkane conversions by phosphine complexes of the later transition metals. These are believed to involve oxidative addition as the key C-H bond breaking step, and may be either thermal or photochemical. Unfortunately, a number of catalyst deactivation processes so far limit the number of times the catalyst can turn over to approximately 100.

The second area is mercury photosensitization, by which alkanes can be converted to a variety of functionalized products on a preparative scale; these reactions are exclusively photochemical, but lead to a much greater variety of products, and the catalyst does not degrade significantly with time.

I. Metal Phosphine Complexes in Alkane Dehydrogenation and Carbonylation

These systems are all believed to depend on the oxidative addition of an alkane C-H bond as the first step (Eq. 1). The intermediate alkyl can then be trapped in a number of ways depending on the circumstances and the particular metal complex involved. Some examples are shown below. In the first (Eq. 2) the alkyl decomposes by β-elimination to give a free alkene, in Eq. 3 to give a polyenyl complex, and in Eq. 4 to give an aldehyde by CO insertion. Which path is followed depends in part on the extent to which the metal can accept alkane-derived fragments into the coordination sphere of the complex. For example, the $16e^-$ "CpIrL" fragment (L = 3° phosphine), which is formed by photolysis of the corresponding dihydride, as discussed in detail by Jones in Chapter IV, has only a 2e vacant site and so can accept only the

$$>M \quad + \quad -\overset{|}{\underset{/}{C}}-H \quad \longrightarrow \quad >M\overset{H}{\underset{C-}{\underset{|}{\nwarrow}}} \qquad (1)$$

$$>M \quad + \quad \wedge_R \quad \longrightarrow \quad >M\overset{R}{\underset{H}{\diagdown}} \quad \longrightarrow$$

$$>\overset{\overset{R}{\|}}{\underset{H}{M-H}} \quad \longrightarrow \quad H_2 \quad + \quad M \quad + \quad =\diagdown^R \qquad (2)$$

$$>M \quad + \quad \bigcirc \quad \longrightarrow \quad >M\overset{\diagup}{\underset{H}{\diagdown}} \quad \longrightarrow$$

$$>\overset{\bigcirc}{\underset{H}{M-H}} \quad \xrightarrow{-H_2} \quad >M-\bigcirc \quad \longrightarrow \quad >M\overset{\diagup}{\underset{H}{\diagdown}}$$

$$>\overset{H}{\underset{H}{M}}-\bigcirc \quad \xrightarrow{-H_2} \quad >\overset{\diagup}{\underset{H}{M}}-\bigcirc \qquad (3)$$

$$\tag{4}$$

$2e$ set of ligands $(R)(H)$ as shown in Eq. 5. The reaction is therefore stoichiometric.

$$\text{CpIrH}_2\text{L} \xrightarrow[-\text{H}_2]{h\nu} \{\text{CpIrL}\} \xrightarrow{\text{R-H}} \text{CpIr}(R)(H)\text{L} \tag{5}$$

Catalysts that dehydrogenate alkanes to free alkenes all have additional labile ligands that allow the alkyl hydride intermediate to become coordinatively unsaturated and so allow the alkyl ligand to β-eliminate. An example of such a catalyst is $\text{IrH}_2(\eta^2\text{-O}_2\text{CCF}_3)\text{L}_2$. Photolysis of this complex in alkane solution gives alkene and free H_2 catalytically.[2] The carboxylate group is known to be able to open and close easily, and the addition of a variety of ligands L' readily gives $\text{IrH}_2\text{L}'(\eta^1\text{-O}_2\text{CCF}_3)\text{L}_2$. The proposed mechanism of alkane dehydrogenation (Eq. 6) involves photoextrusion of H_2 and oxidative addition of alkane to the intermediate to give an alkyl hydride $\text{IrRH}(\eta^2\text{-O}_2\text{CCF}_3)\text{L}_2$.

The acetate group in this species is proposed to open up so that the alkyl hydride attains a $16e^-$ configuration, which allows it to β-eliminate. This β-elimination gives the alkene and the original dihydride is regenerated. The metal can reenter the cycle by photoextrusion of H_2 once again, and so the system is catalytic (Eq. 6). A more detailed discussion of the intermediates shown in Eq. 6, will be given after we look at the thermal chemistry of $\text{IrH}_2(\text{O}_2\text{CCF}_3)\text{L}_2$.

Although Eq. 7 is thermodynamically uphill at 25°C, photochemical dehydrogenation reactions can proceed directly because the reaction is driven by the energy of the absorbed photons (Eq. 8). In some cases a gentle stream of inert gas improves the efficiency of the process by physically removing H_2 from the system.

$$hv \atop -H_2$$

$$RCH_2CH_3 \xrightarrow{\text{cata}} RCH{=}CH_2 + H_2 \qquad (7)$$

$$hv + RCH_2CH_3 \xrightarrow{\text{cata}} RCH{=}CH_2 + H_2 \qquad (8)$$

Catalytic thermal dehydrogenation reactions are also possible, but these require a hydrogen acceptor. This can be an alkene that takes up the H_2 removed from the alkane substrate and so drives the dehydrogenation reaction. We originally introduced *tert*-BuCH=CH$_2$ (tbe) as a hydrogen acceptor for alkane dehydrogenation,[3] but others have also been used more recently. Tbe is bulky and so does not coordinate strongly to vacant sites at the metal, yet it is very active in removing hydride ligands from the metal by hydrogenation to give *tert*-BuCH$_2$CH$_3$.

It also has a high heat of hydrogenation because the steric interference between the *tert*-Bu group and the *cis* vinyl C-H bond is relieved on hydrogenation. Equation 10 is therefore thermodynamically downhill in almost all cases under the conditions employed.

$$RCH_2CH_3 + \textit{tert-}BuCH = CH_2 \xrightarrow{\text{cata}} RCH = CH_2 + \textit{tert-}BuCH_2CH_3 \quad (10)$$

(substrate) (hydrogen-acceptor)

This reaction is catalyzed by several complexes, e.g., $[IrH_2(O_2CCF_3)(PPh_3)_2]$,[2] $ReH_7(PPh_3)_2$,[4] $RuH_4(P\{\underline{p}\text{-}FC_6H_4)_3\}_3$,[5a] and $IrH_5(P\underline{i}\text{-}Pr_3)_2$.[5]

Clooctane has proved to be the most reactive alkane of those tried, presumably because the heat of dehydrogenation is unusually large as a result of the decrease in unfavorable transannular interactions on going to the product cyclooctene. This alkane is therefore a good choice for assaying new potential catalysts for alkane dehydrogenation activity. It is one of the least reactive alkanes in stoichiometric oxidative addition, however.

Sakakura and Tanaka[6] have recently observed alkane carbonylation by photochemical means using $RhCl(CO)(PMe_3)_2$ as catalyst. This may well operate by initial photoextrusion of CO from the catalyst, oxidative addition of the alkane C-H bond, addition of CO to the metal, followed by insertion, and then reductive elimination as was shown in Eq. 4. Several hundred turnovers have been observed to date, and the system is very promising. Carbonylation of alkanes such as cyclohexane, which contain only 2° C-H bonds is much less efficient.

It may be true that in all these cases the oxidative addition step is thermodynamically unfavorable, but that the later steps in the catalytic cycles draw on a small equilibrium concentration of alkyl hydride and

28 turnovers

+   (11)

<0.6 turnovers

drive the overall process. This is not well established, however, because the strengths of M-C and M-H bonds are not known in a sufficiently wide variety of cases. Bergman et al.[7] have estimated Ir-C and Ir-H as 51 and 74 kcal/mol in $Cp^*Ir(PMe_3)H(C_6H_{11})$. On the other hand Halpern[8] has suggested that M-C and M-H bond strengths will generally be closer to 30 and 60 kcal/mol, respectively. The latter estimates may be affected by the fact that Halpern's methods of measuring M-C bond strengths work best for compounds with relatively weak M-C bonds. The M-C and M-H bond dissociation energies for $L_nM(R)(H)$ will be larger if L_nM is an unstable fragment, and this will lead to an increase in the driving force for oxidative addition in the case of an unstable fragment, such as $Cp^*Ir(PMe_3)$. Comparing the estimates of Bergman et al.[7] for the unstable $Cp^*Ir(PMe_3)$ system with those obtained for adducts of the much more stable complex $IrCl(CO)L_2$ suggests that the M-X bond strengths in a variety of oxidative addition products are systematically 15-20 kcal/mol lower in the case of adducts of $IrCl(CO)L_2$ [e.g., (values for $IrCl(CO)L_2$ are given second): Ir-H, 74.2, 60; Ir-Cl, 90.3, 71 kcal/mol].[7] Bond strengths may prove to be less easily transferable from one system to another in organometallic than in organic chemistry.

$$L_nM \ + \ R\text{-}H \longrightarrow L_nM(R)(H) \qquad\qquad (12)$$

It is notable that Cp^*IrL, the proposed intermediate in the systems of Bergman et al.[9] (L = 3° phosphine) and Hoyano and Graham[10] (L = CO), is an Ir(I) species that cannot become square planar and may be particularly unstable thermodynamically for this reason.

In the catalytic systems, the most probable intermediate leading to the C-H bond breaking is a 14e 3-coordinate T-shaped species. This would be a reasonable structure for an $\{IrL_2(\eta^1\text{-}OCOCF_3)\}$ intermediate in the $IrH_2(\eta^2-O_2CCF_3)L_2$ system, and for $IrHL_2$ in Felkin's iridum system, as well as for $\{RhClL_2\}$ in the Tanaka system. The $ReH_7(PR_3)_2$-based systems are unlikely to be directly analogous, but the 14e ReH_3L_2 has been proposed[4] as the key reactive intermediate. This preference for 14e intermediates is a reflection of the feeling that simple 16e square planar d^8 species are unlikely candidates for alkane C-H bond breaking, because, unlike CpIrL, so many stable examples are known.

A most important point about all these systems is their selectivity for initial attack at relatively unhindered C-H bonds. This is shown not only in Eq. 11 but also in the dehydrogenation of methylcyclohexane by various catalysts, such as $IrH_5(P\underline{i}-Pr_3)_2$,[5] $[IrH_2(O_2CCF_3)L_2]$,[2] and ReH_7L_2[4] (Eq. 13). Methylenecyclohexane is the major product for the iridium catalysts, and 2- and 3-methylcyclohexene are the major products in the Re case. Since 1-methylcyclohexene is the most stable product thermodynamically, the alkenes observed must be kinetic products. Initial attack at the methyl group (Ir) or the less hindered ring CH_2 groups of methylcyclohexane (Re) is therefore favored. In the case of the Ir catalysts, most of the other alkenes have been shown to be produced by subsequent isomerization of the initially formed methylenecyclohexane; the IrH_5L_2 catalysts are more active for isomerization than is the iridium trifluoroacetate, and so more of the initial product distribution is preserved in the latter case. The ReH_7L_2 catalysts are not significantly active for alkene isomerization, and so a greater part of the kinetic products survive the reaction conditions and are observed among the products.

$$\text{IrH}_5\text{L}_2, \text{IrH}_2(\text{O}_2\text{CCF}_3)\text{L}_2 \text{ or ReH}_7\text{L}_2$$

tbe, heat

| major kinetic product (Ir) | thermodynamic product (Ir) | major products (Re) | absent |

(13)

The iridium catalysts also dehydrogenate linear alkanes.

Terminal alkenes are significant products, but all the other possible alkenes are also formed. Most of these come from the subsequent isomerization of the initially formed 1-alkene. As will be seen in detail below, the rhenium catalysts dehydrogenates linear alkanes only stoichiometrically, but with very high selectivity for the 1-alkene.

The reason for this selectivity pattern appears to be the nature of the kinetic pathway for C-H oxidative addition. A Burgi-Dunitz analysis of a number of complexes containing C-H⋯M interactions led to the conclusion[11a] that the CH bond approaches the metal H-first with a large M⋯H-C angle. As the C-H bond gets closer it pivots about the H atom so as to reduce the M-H-C angle and bring the carbon atom close to the metal. The transition state, shown schematically below, appears to be late, with a short M-C distance. Substituents on the carbon atom are therefore likely to interfere with the approach to the transition state and so retard the reaction.

$$M \overset{\displaystyle H}{\underset{\displaystyle C}{\big<}}$$

The same analysis also suggested that cyclometallation (Eq. 14) might have a higher barrier than the activation of an external C-H bond such as that of the substrate because of unfavorable conformational factors in the approach of a ligand C-H bond. Jones et al.[12] have shown that for the $CpIrH_2P(iPr)_3/h\nu/CH_3CH_2CH_3$ system, there is indeed a greater kinetic barrier for ligand C-H oxidative addition than for the addition of an alkane C-H bond.

$$\overset{}{M}\overset{L}{\diagdown}\overset{(CH_2)_n}{\underset{H}{\overset{|}{C}}} \longrightarrow \overset{}{M}\overset{L}{\diagup}\overset{(CH_2)_n}{\underset{C}{\overset{}{\diagdown}}} \qquad (14)$$

Chencey and Shaw[13] showed many years ago that cyclometalation is strongly accelerated by steric crowding. This means that steric crowding should be avoided if we want to favor alkane activation over cyclometalation. Indeed, all the proposed intermediates in the successful alkane-activating systems are relatively unhindered: MXL_2 (M = Rh, Ir), $IrH_2L_2^+$ ReH_3L_2, and CpIrL.

For all the catalytic systems previously discussed, radical and carbonium ion mechanisms can be excluded on several grounds, perhaps the most convincing being the 1° > 2° > 3° selectivity for attack at different types of C-H bond. Mechanisms involving heterogeneous catalysis by metal surfaces, which might be formed by decomposition of the catalyst, have been excluded by several tests including the use of metallic mercury, which poisons platinum group metal surfaces.[3e]

A detailed study[2b] of the chemistry of $[IrH_2(O_2CCF_3)L_2]$ (**1**, L = P(p-FC_6H_4)$_3$) in thermal alkane dehydrogenation with tbe as hydrogen acceptor has produced considerable insight into the mechanism of the C-H activation reactions. The addition of CO or even of such relatively weakly binding ligands as alkenes or MeCN (L') leads to opening of the trifluoroacetate to give the mono-adduct.

$$
\text{(15)}
$$

Interestingly, the corresponding CH_3CO_2 complex is much more reluctant to undergo opening of the acetate group. Accordingly, the complex gives Reaction 15 only with CO but not with MeCN or alkenes, and it does not act as a catalyst. The reaction of the trifluoroacetate with alkenes is reversible and the K_{eq} values shown in Table I were measured. These values illustrate the low coordinating power of tbe relative to other alkenes. This is a very useful property for a hydrogen acceptor, because it must not block the active sites at the metal that it has helped to create and so exclude the substrate alkane. The strong binding of $Me_3SiCH=CH_2$ relative to tbe shows how this electrophilic system prefers a relatively electron-releasing alkene.

$[IrH_2(tbe)(\eta^1\text{-}OCOCF_3)L_2]$ slowly changes with time at low temperature to give a species for which the $^1H^-$, $^{13}C^-$, and ^{31}P-NMR data suggest an agostic alkyl hydride structure as shown in Figure 1. It is reasonable to suppose that an agostic alkyl hydride may also be formed on the oxidative addition of the alkane in the alkane dehydrogenation reaction. This result is interesting in that the agostic character of the alkyl hydride would tend to increase the thermodynamic driving force to form the alkyl hydride and so facilitate the alkane

Table I. K_{eq} Values for the Reaction of Various Alkenes with $IrH_2(O_2CCF_3)\{P(p\text{-}FC_6H_4)_3\}_2$

Alkene structure	K_{eq}
Ph (isopropenylbenzene)	0
t-Bu vinyl	0.6
cyclopentene	1.9
Ph allyl	3.2
Si (vinylsilane)	1250
cyclooctene	5500
ethylene	large

oxidative addition. The final products from $[IrH_2(tbe)(\eta^1\text{-}OCOCF_3)L_2]$ are *tert*-$BuCH_2CH_3$ and an orthometalated species $[IrH(\eta^2\text{-}OCOCF_3)PPh_3(PPh_2(o\text{-}C_6H_4)]$, which can be isolated.

The orthometalated species is an effective catalyst precursor for alkane dehydrogenation and so we think it is in equilibrium with the active catalyst. This was something of a surprise because, in planning

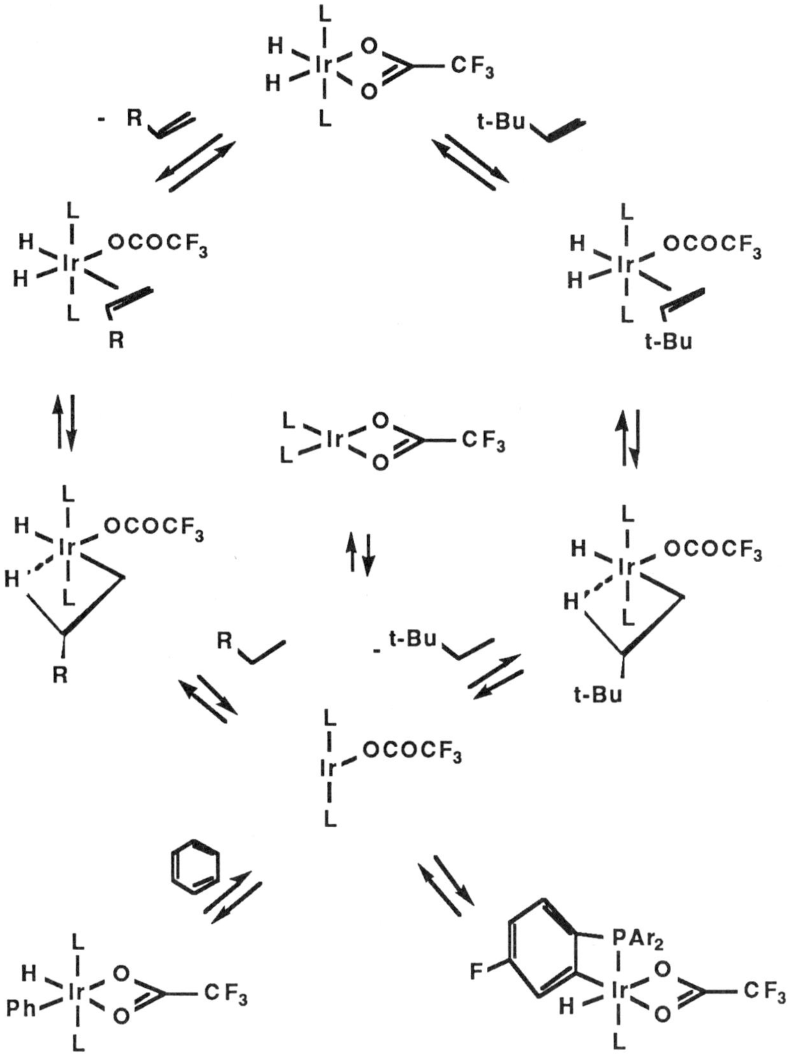

Figure 1. Thermal alkane dehydrogenation with $IrH_2(O_2CCF_3)L_2$. All but the 14e species have been isolated or detected spectroscopically.

our approach to the alkane conversion problem, we originally thought that cyclometalation would prevent a given catalyst from reacting with alkanes. Even though this particular cyclometalation is reversible, it must still lower the overall activity of the catalyst by removing a large part of the active form from the system.

In benzene as solvent, the reaction of $IrH_2(O_2CCF_3)L_2$ with tbe does not give the cyclometalation product but rather an isolable phenyl hydride, $[IrPhH(\eta^2\text{-}OCOCF_3)L_2]$, is formed instead. This is the first case in which a catalytically active C-H activation system has also given a stable C-H oxidative addition product. The reasons for this stability are clear: the Ir-Ph bond is unusually strong because of d_π-p_π interactions and the phenyl group cannot easily β-eliminate. k_H/k_D for the thermal oxidative addition of benzene is 4.5. This seems to be a reasonable value for an oxidative addition, although lower isotope effects (k_H/k_D = ~1.4) were observed for cyclohexane and benzene addition to the much more reactive intermediates $Cp^*M(PMe_3)$[9,12](M = Rh, Ir). The reaction scheme shown in Figure 1 is therefore proposed for thermal alkane dehydrogenation with this catalyst. The phenyl hydride undergoes first-order decomposition at 65°C to give an equilibrium mixture with the cyclometalation product; K_{eq} = 3.6 in favor of the phenyl hydride. Like the cyclometalation product, the phenyl hydride is therefore a satisfactory catalyst precursor for alkane dehydrogenation.

A most serious problem with all the catalysts previously described is rapid degradation under catalytic conditions. This has been studied in detail[2b] in the case of $[IrH_2(O_2CCF_3)\{P(p\text{-}C_6FH_4)_3\}_2]$, where it has been shown that the main degradation pathway is P-C bond hydrogenolysis, a process that has been reviewed by Garrou,[14a] and treated theoretically by Hoffman.[14b] The process seems to involve oxidative addition of the P-C bond to the metal, to produce a catalytically inactive cluster, as shown schematically below. ArH has also been detected from $ReH_7(PAr_3)_2$[4] and propane from $IrH_5(P\underline{i}\text{-}Pr_3)_2$- catalyzed[5b] alkane dehydrogenation reactions.

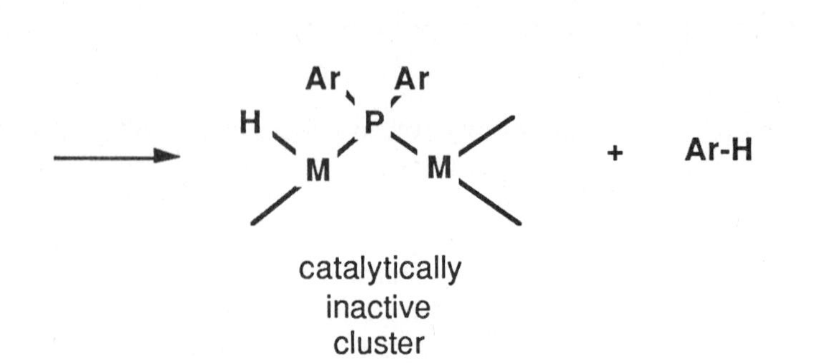

catalytically
inactive
cluster

$$+ \quad \text{Ar-H} \qquad (16)$$

The photochemical catalyst $[IrH_2(O_2CCF_3)\{P(C_6H_{11})_3\}_2]$ is much less sensitive to decomposition than the thermal catalyst, which contains $P(p\text{-}FC_6H_4)_3$. The production of cyclooctene from cyclooctane, although slow, is roughly linear with time (12 turnovers after 7d, 23 turnovers after 14d photolysis). After 7d, 70% of the catalyst can be recovered in pure form even from the dilute solution employed. It is not clear whether the greater stability is a result of moving from an aryl to an alkyl phosphine or from a thermal to a photochemical process.

It has been shown[2b] that the catalytic activity of $IrH_2(O_2CCF_3)(PAr_3)_2$ for alkane dehydrogenation falls as the amount of ArH produced by P-C bond hydrogenolysis rises. The formation of 0.5 mol of ArH per Ir is sufficient to deactivate the catalyst entirely. These and other deactivation pathways require further study and general methods for slowing or preventing these processes are badly needed. Such methods might also be applicable to other homogeneous catalysts and lead to practical applications.

Although the reaction scheme shown (Eq. 17) is stoichiometric, it is sufficiently remarkable to warrant special comment.

$$\text{ReH}_7\text{L}_2$$

$$(17)$$

In contrast to the iridium catalysts, which dehydrogenate

n-alkanes to alkenes, ReH_7P_2 gives stable diene complexes instead.[4b] All the possible isomers of the diene are in rapid equilibrium, as shown by spin saturation transfer measurements in the ^{1}H-NMR. The addition of $P(OMe)_3$ to this mixture induces the transfer of two of the hydride ligands to the diene to give the free monoene, and essentially only the terminal isomer is formed. A possible reaction pathway is shown in Eq. 17. Presumably, the terminal diene complex is the most reactive and the $P(OMe)_3$ selectively liberates the terminal C=C double bond of the diene in this species from the metal. The C=C double bond that remains coordinated is then hydrogenated. The other isomers of the diene complex must convert into the reactive isomer sufficiently rapidly so that essentially all the product is formed via the more reactive isomer.

Baker and Field[11b] have reported that $(dmpe)_2FeH_2$ gives $(dmpe)_2FeH(1\text{-}pentyl)$ on photolysis in *n*-pentane at -30°C. On warming to room temperature, further photolysis releases 1-pentene. The reaction is not catalytic because the iron complex reacts with the product alkene to give the very stable $(dmpe)_2Fe(1\text{-}pentene)$ and $(dmpe)_2FeH(CH=CHn\text{-}Pr)$. Presumably the dmpe ligand is converted to the η^1 form on irradiation, and this allows the alkyl hydride to β-eliminate.

$[IrH_2(Me_2CO)_2L_2]^+$ has four labile ligands, the two hydrides and the two acetone groups. Loss of both acetone groups from an alkyl hydride intermediate $[IrRH(Me_2CO)_2L_2]^+$ could allow double β-elimination. As might be anticipated, thermal reaction of $[IrH_2(Me_2CO)_2L_2]^+$ with cyclopentane and tbe has been shown to lead to dehydrogenation beyond the cyclopentene stage to give a cyclopentadienyl complex, $CpIrHL_2^+$, as the ultimate product.

The dehydrogenation of cyclic alkanes to arenes has been observed using $[IrH_2(Me_2CO)_2L_2]SbF_6$.[3c] At 80°C, a cyclohexadienyl complex and a benzene complex were formed from cyclohexane. At 150°C, free arene was liberated. The reaction was not catalytic because the temperatures required to liberate the arene also led to deactivation reactions, such as P-C bond hydrogenolysis. No solvent that was able to dissolve the iridium complex proved satisfactory as a reaction medium, and so the reactions had to be carried out in alkane suspension. The cationic iridium species does not seem to be at all soluble in alkanes and so the reactions may take place at the surface of the crystallites. Unusual counter ion effects were seen, the SbF_6 salt being much more effective than the BF_4 salt, for example; these are probably solid state effects. The reason for dehydrogenation of the alkane beyond the alkene stage with the cationic Ir species and with ReH_7L_2 may be the greater number of labile ligands present, which allows more binding sites at the metal for alkane-derived fragments than does $IrH_2(O_2CCF_3)L_2$.

A number of interesting catalytic reactions of methane have been observed. Catalytic H-D exchange between C_6D_6 and methane has been seen using a number of different metal phosphine catalysts. [5b, 12b, 15] As early as 1969, Goldshleger et al.[15a] reported catalytic H-D exchange in methane with $CoH_3(PPh_3)_3$. This is particularly important historically in that it was the first case in which a metal phosphine complex was shown to effect alkane activation. However, the work was not developed because attention was concentrated on the $PtCl_4^{2-}$ catalysts.[15b] Grigoryan et al[15c] have more recently shown H-D exchange in methane for a variety of Ziegler-Natta systems. In later work, Grigoryan et al.[15d] have shown that methane can be added to ethylene to give propane with $Ti(Ot-Bu)_4$-$AlEt_3$, or to acetylene to give propene with $Fe(acac)_3$-$AlEt_3$. Although yields are small in these systems, this work nevertheless points the way to alkane conversions of significant practical interest. Jones has shown that $CpReH_2(PPh_3)_2$ loses phosphine on photolysis to give a species active for the C_6D_6-CH_4 isotopic exchange reaction; 68 turnovers were observed after 3 hr.[12b] Felkin has studied thermal H-D exchange in methane using $IrH_5\{P(i-Pr)_3\}_2$ activated by *tert*-$BuCH=CH_2$ and C_6D_6 as the D source. About 40 turnovers were observed after 46 days at 80°C. The thermal Ir catalysts also gave evidence for the formation of other products. Ethane, toluene, and biphenyl were all observed in small and variable amounts from the C_6H_6-CH_4 reaction. These may have been formed by reductive elimination from an IrH_3L_2RR' species.

Several other stoichiometric systems are relevant to the discussion. Whitesides[16] showed that the bis-neopentyl platinum complex of Eq. 18 gave facile cyclometalation. They were surprised that the reverse reaction, which would of course be an alkane activation with neopentane as the substrate, did not take place because, thermodynamically, the bonds formed should compensate for the bonds lost. Presumably, it is the unfavorable entropic term that is the major factor in preventing the reverse reaction, but in addition, the substantial steric bulk of the neopentyl group may tend to reduce the Pt-neopentyl bond strength in a bis-neopentyl complex.

$$\text{(18)}$$

Green et al.[17] showed that photolysis of ReH_5L_3 (L = PMe_2Ph) liberates L rather than H_2, and in the presence of tbe as hydrogen acceptor, they found that cyclopentane can be dehydrogenated to

CpReH$_2$L$_2$ as Felkin[4] had previously observed by a thermal route from ReH$_7$L$_2$.

Direct vinyl C-H activation has been reported. Cp*IrL reacts with C$_2$H$_4$ to give both Cp*IrL(C$_2$H$_4$) and Cp*IrL(C$_2$H$_3$)H. The ethylene complex does not go on to give the vinyl hydride and so the formation of the latter was thought to occur by direct C-H oxidative addition.[18a] Faller and Felkin[18b] have reported IrH$_5$(Pi-Pr)$_2$-catalyzed H-D exchange between solvent C$_6$D$_6$ and tbe to give predominantly trans-*tert*-BuCH=CHD. Once again, direct C-H oxidative addition was proposed because insertion would have allowed free rotation about the *tert*-BuCH$_2$-CH$_2$Ir bond and so cis and trans isomers of the product *tert*-BuCH=CHD would have been observed. Baker and Field[11c] have also observed vinylic C-H activation with (dmpe)$_2$FeH$_2$, by photoextrusion of H$_2$, as mentioned above. There is a thermodynamic advantage to aryl C-H activation, because the M-aryl bond is considerably stronger (~15 - 20 kcal/mol) than a M-alkyl bond. This advantage probably carries over to the case of vinyl C-H activation; this bond should be at least as strong as M-Ph, perhaps more so, in view of the smaller size of the vinyl group. Arene C-H bond activations are very numerous.[15b]

Alkane C-C bond cleavage is a possible alternative strategy for alkane activation. So far, direct C-C bond activation by metal phosphine complexes has not yet been observed. This is probably because C-C bond cleavage is kinetically much slower than C-H cleavage, and is illustrated (for the reverse reaction) by the contrast between the much readier reductive elimination of alkyl hydride complexes compared to their dialkyl analogs. In addition, a direct oxidative addition of a C-C bond would put two bulky alkyl groups on the metal; this would lead to a reduced M-C bond strength. Nolan et al.[7] finds that the C-C oxidative addition reaction of Cp*Ir(PMe$_3$) with C$_6$H$_{11}$-C$_6$H$_{11}$ would be exothermic by 22 kcal/mol using the Ir-C$_6$H$_{11}$ bond strength data derived from Cp*Ir(PMe$_3$)C$_6$H$_{11}$(H); steric effects would probably reduce the exothermicity, however. Suggs and Cox[19a] has observed cyclometalation reactions involving C-C bond cleavage.

We have shown how the unstrained alkane, 1,1-dimethylcyclopentane can be dehydrogenated to the corresponding diene complex by [IrH$_2$(Me$_2$CO)$_2$L$_2$]SbF$_6$. This product, [(η^4-5,5-Me$_2$C$_5$H$_4$)IrL$_2$]SbF$_6$ can then undergo a reaction, first studied by Benfield and Green[19b] and by Eilbracht,[19c] in which a methyl group migrates from the diene to the metal with C-C bond cleavage to give [(η^5-MeC$_5$H$_4$)IrMeL$_2$]SbF$_6$.[3d] The driving force for this reaction is the formation of the aromatic cyclopentadienyl group. Extending the study to the 1,1-diethylcyclopentadiene analogs led to the unexpected formation of 1,2- and 1,3-diethylcyclopentadienyl iridium *hydrides*. A mechanism involving migration of an ethyl group to the metal from the diene as before, but now followed by migration of the ethyl group back to a different carbon atom of the C_5 ring was proposed. In no case have catalytic reactions involving C-C bond cleavage been seen; even at the high temperatures required ($\sim$150°C).

Oxidative addition of a C-C bond to the metal is much more rapid in strained hydrocarbons, because of the relief of strain in the transition state. The earliest report dates from 1955.[20]

The catalytic rearrangement of strained alkanes is common.[21] Cassar and Halpern[22] showed that in the $[RhCl(CO)_2]_2$-catalyzed rearrangement of quadricyclane to norbornadiene, the admission of CO led to the formation of an acyl complex that seemed to be formed by trapping the initial oxidative addition product by a migratory insertion. On the other hand, the Ag^+ catalyzed rearrangement of strained alkanes goes via carbonium ion intermediates that can be trapped by such nucleophiles as MeOH.[23]

II.   Mercury Photosensitization

We routinely carry out tests for catalyst homogeneity in our studies of catalytic alkane conversion, in order to rule out the possibility that the catalyst has partially decomposed to give the platinum metal in the form of a film on the apparatus or a colloid in the solvent. This metal might then be the active catalyst. One of the tests involves adding a drop of mercury to the system.   If the catalytic activity is authentically homogeneous, then the reaction rate should not be affected by the presence of the mercury, otherwise, the mercury poisons the metal by adsorbtion and the activity is lost. In carrying out this test in the case of the photochemical iridium catalyst for cyclooctane dehydrogenation mentioned in the previous section, Mark Burk, a graduate student in my group, noticed that large amounts of an involatile residue was formed along with the expected cyclooctene. Further study revealed that this was the result of a vapor phase mercury photosensitized reaction of cyclooctane to give bicyclooctyl; no iridium is required. The systems described in the discussion that follows contain only elemental mercury and the organic substrate. According to the literature,[24] the general principle involved in these reactions is the homolytic cleavage of an alkane C-H bond by the $^3P_1$ excited state of mercury (Hg*) formed by irradiation at 254 nm. Equations 21-24 crudely describe the sequence of events.

$$Hg + h\nu \longrightarrow Hg^* \tag{21}$$

$$Hg^* + RH \longrightarrow R\bullet + H\bullet + Hg \tag{22}$$

$$H\bullet + RH \longrightarrow R\bullet + H_2 \tag{23}$$

$$2R\bullet \longrightarrow R_2 \tag{24}$$

Steve Brown noted that oligomers tend not to be formed in these reactions even at high conversions. This is not the case in a solution phase radical reaction, because the product R_2 is intrinsically more reactive than the initial substrate, RH. The reason we do not see further conversion of the R_2 in this system proved to be that the dimer has so low a vapor pressure that essentially only the original substrate, cyclooctane, is present in the vapor phase.

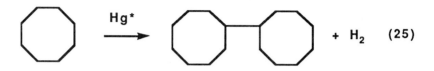

$$ \quad + \quad H_2 \qquad (25) $$

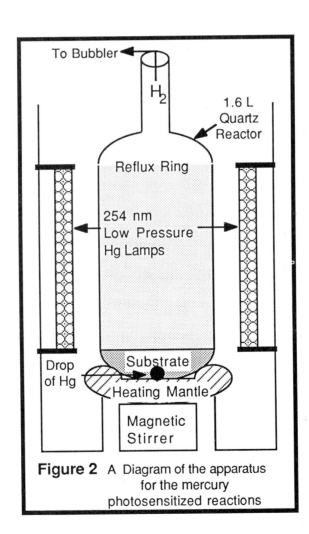

Figure 2 A Diagram of the apparatus for the mercury photosensitized reactions

This vapor selectivity effect has been very useful to us in developing our work.[25] Figure 2 shows the apparatus we developed to help take full advantage of vapor selectivity effects. The substrate and a drop of mercury are refluxed together in a quartz vessel and irradiated at 254 nm (4 x 8 W low pressure Hg lamps). In this way the product dimers are quickly returned to the liquid phase and the vapor phase is constantly replenished with substrate. The volume of hydrogen evolved allows us to monitor the progress of the reaction.

The observed selectivity is $3° > 2° > 1°$ as expected for a homolytic pathway and species with $4°$-$4°$ C-C bonds are very efficiently assembled. The product ratio can be altered by diluting the alkane with gases such as H_2. This leads to increased selectivity for the $4°$-$4°$ product. A possible explanation is the Hg* - induced homolysis of H_2 leading to H atoms which then add to alkene products of radical disproportionation, as discussed in more detail below.

A few previous reports[26] deal with alcohols, ethers, amines, and silanes. We find that reactions involving these and other saturated, but heteroatom-substituted species are preparatively useful when carried out under our conditions. The system is selective for C-H bonds α- to the heteroatom and for Si-H bonds.

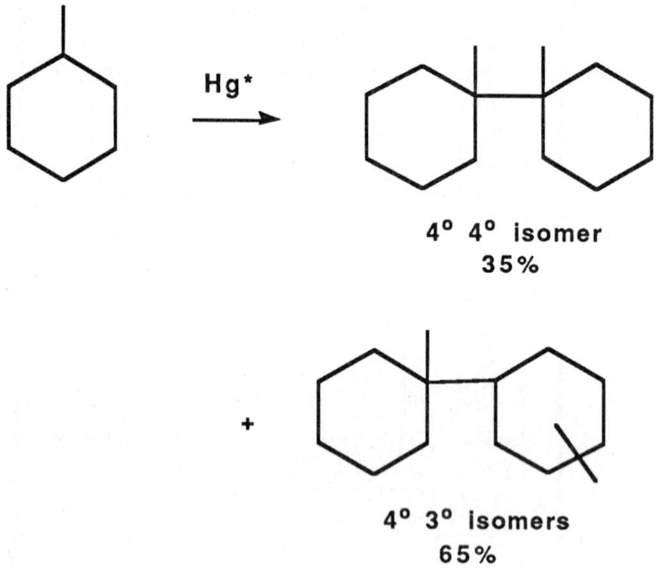

4° 4° isomer
35%

4° 3° isomers
65%

(26)

$$CH_3OH \longrightarrow HOCH_2CH_2OH \qquad (27)$$

$$tBuOCH_3 \longrightarrow tBuOCH_2CH_2OtBu \qquad (28)$$

$$(29)$$

$$\text{Et}_3\text{SiH} \longrightarrow \text{Et}_3\text{Si–SiEt}_3 \qquad (30)$$

We were very pleased to find that alkanes can be cross-dimerized with other species. Equation 31 shows the results from cyclohexane and methanol. The three products are formed in approximately statistical amounts. The polarities of the three species are so different that simple solvent extraction suffices to obtain the cross-dimer itself; the glycol can be removed with water and the bicyclohexyl with pentane. Reactions of this type constitute a rapid functionalization of an alkane and can be carried out efficiently on a large scale (tens of grams overnight with 4 to 16 8-W low-pressure

$$(31)$$

mercury lamps). The absence of significant amounts of byproducts is an unexpected property of what is generally thought to be a radical reaction. The synthetic utility of the method is shown by the fact that we have used it for the preparation of ligands and starting materials needed for unrelated purposes.

All these reactions can be pushed to high conversion (95%) without significant fall off in selectivity and in this way tens of grams of functionalized product can be obtained over 20 hr. This is vastly more efficient than any of the systems described in the first section of this chapter, in part because mercury does not degrade under the reaction conditions. Under certain circumstances more than 10^4 turnovers can be observed per Hg atom without a perceptible decrease in reaction rate.

Cross-dimerizing cyclohexane with the formaldehyde trimer gives a material that yields cyclohexanecarboxyaldehyde on hydrolysis (Eq. 32). This reaction is therefore equivalent to the one described by Sakakura and Tanaka[6] (Eq. 11), except the selectivity is different.

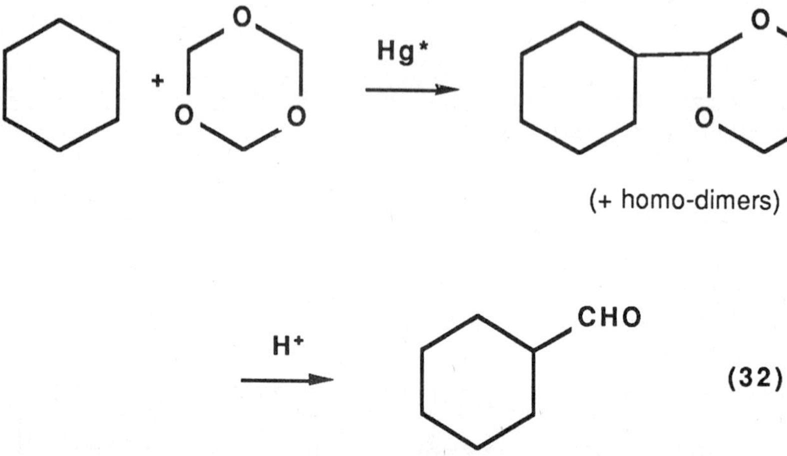

(+ homo-dimers)

(32)

Cross-dimerizations involving other functionalized species also work well. Equation 33 shows the formation of a silylethanol from the silane and ethanol. This is a remarkable transformation because the much greater strength of Si-O over Si-C bonds normally favors the formation of a alkoxysilane (e.g., R3SiOEt). The Si-H bond is much more reactive than the C-H bonds of the R3Si alkyl groups and so essentially all the product derives from Si-H cleavage. Equation 34 shows the formation of a hydroxymethyl ether from an ether and an alcohol.

$$R_3Si\text{-}H + CH_3CH_2OH \longrightarrow R_3SiCH(CH_3)OH + \text{homodimers} \qquad (33)$$

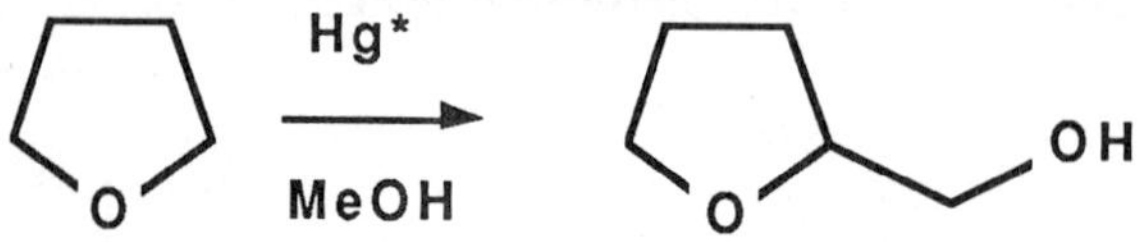

(34)

Oxidation reactions also take place under Hg photosensitized conditions. As Gray[26b] showed in 1952, CH_4-O_2 mixtures give MeOOH.

In contrast to the millimolar amounts of alkenes formed and turnovers of < 100 so far observed in the systems based on oxidative addition, we can now make tens of grams of functionalized product in 24 hr. by cross-dimerization. As a relatively unreactive atom it is not surprising that the mercury atom is not degraded in this system. At least 10^4 turnovers can be observed without a significant decrease in the rate.

This emphasizes an important advantage of the very simple Hg system over the more complicated metal phosphine catalysts such as $[IrH_2(O_2CCF_3)(PR_3)_2]$, which have many deactivation pathways open to them. On the other hand the phosphine-substituted catalysts do have the very useful property of selectively attacking the least hindered C-H bond in the alkane. This means that the selectivity is opposite to that observed in the Hg system and in all the 'classical' alkane functionalization reactions.

The problem is that no one has been able to discover a stable catalyst. In few of the known catalysts of this type, reported by Burk et al.[2], by Felkin et al.,[4,5] or by Sakakura and Tanaka[6], are the catalysts stable enough to give even 100 turnovers of product. For the moment, this severely limits their usefulness, but suggests that a careful study of their deactivation pathways might lead to the synthesis of a more stable catalyst.

The reason for the different selectivities of the two systems is believed to be the steric bulk of a catalyst such as $RhCl(CO)L_2$, compared with the small Hg atom, as well as the sterically demanding side-on transition state for oxidative addition (left),[11] compared with the end-on transition state (right) believed to operate for the Hg reaction.[16]

$$\begin{array}{c} H \\ | \diagdown \\ | M \\ | \diagup \\ C \end{array} \qquad\qquad C \text{---} H \text{---} Hg$$

Certain puzzling features of the reaction made us think that the generally accepted mechanism for the Hg photosensitization reaction, based on Eqs. 21-24 did not adequately describe the situation. In particular, we felt that the C-C bond forming step could not be the simple recombination of two radicals. Our conditions are very different from those used in the earlier studies, and we thought that this might be a factor. We were therefore tempted to postulate intermediates such as RHg·. Although we still cannot exclude such species, we now believe that the main features can be explained on the usual radical mechanism, taking account of the conditions we use.

Some of the initially puzzling factors follow. (1) The high

conversions, yields, and selectivities in our system contrast with those normally found in radical reactions. (2) The formation of only small quantities of alkenes in the products from alkanes such as *cis*-1,4-dimethylcyclohexane was unexpected, because we should see a substantial amount of **3** if R· is allowed to recombine and disproportionate in the normal way (Eq. 35). For example, the reaction of Eq. 35 runs very well even though the ratio of disproportionation to recombination rates, k_d/k_r is 7.1. Indeed, the addition of up to 10% of the corresponding alkene has an insignificant effect on the product ratios. (3) *tert*-Bu$_2$CH· is said not to dimerize,[27a] yet we find that *tert*-Bu$_2$CH$_2$ gives some *tert*-Bu$_2$CHCH*tert*-Bu$_2$ among the products (Eq. 36). This reaction is also interesting in that a 1° C-H bond has been broken to give the major product.

If **3** is not seen as a significant product in Eq. 35, either the radical mechanism is flawed, or **3** reacts further. Since **3** is an alkene, it might well be more reactive than the starting alkane and so be selectively destroyed. To see if disproportionation really is taking place, we decided to look for the alkane disproportionation product, which could not undergo an unusually rapid subsequent reaction. Normally, the alkane disproportionation product is not seen because the alkene formed in this way cannot be distinguished from the starting substrate alkane. To get around this problem, we used (Eq. 35) specifically the cis isomer of the starting alkane and in this way we can also see at least part of the alkane product of disproportionation, because both cis and trans alkanes will be formed. This allows us to estimate k_d/k_r. We find a value of 7.1, which seems reasonable. This result further suggests that the alkene, **3**, is indeed being formed as well but that it must be reacting subsequently at a much faster rate than the starting alkane, as discussed below.

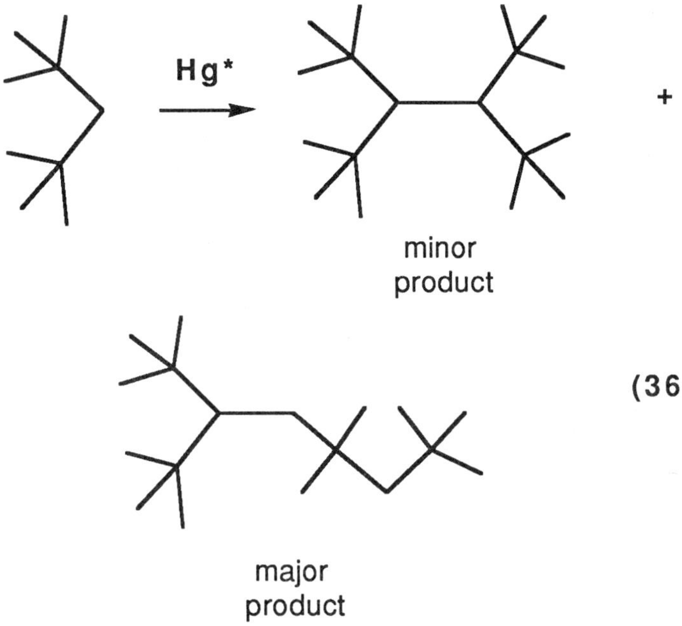

minor
product

(36)

major
product

We believe several factors are important in defining the properties of the system. Hg* is one of the most selective reagents for the homolysis of C-H bonds, 3° bonds appear to react faster than 2° by a factor of 7-10 to 1; 1° C-H bonds are less reactive still and dimers involving a 1° bond homolysis are seen only under unusual circumstances (e.g., Eq. 36, where unusually large steric effects operate). Being a relatively unreactive atom, cold Hg does not seem to interfere very much with the subsequent chemistry. Certainly no HgR_2 has ever been detected in any of these systems, and RHg· is only weakly bound.

The second important factor is that an H atom is produced along with R·. This means that when alkene is formed by disproportionation (Eq. 37) it is rapidly converted in one of the two ways shown in Eqs. 38-39. Since the reaction of Eq. 38 is faster than Eq. 39, most of the alkene goes back to R·. Earlier workers considered Eq. 38 but not Eq. 39.[24] Equation 39 gives a dimeric radical. If this subsequently reacts with H·, then the normal R_2 dimer is formed, if with R·, then trimer will be formed. We always see some trimer, and believe that this may be the main route for its formation.

The same mechanism that recycles the alkene product of

$$2 \text{ R•} \longrightarrow R_2 + \text{alkene} \qquad (37)$$

$$(38)$$

$$(39)$$

disproportionation also accounts for the anomalously high selectivity for forming C-C bonds from what were 3°, rather than 2° carbons in the reagents. The selectivity in the initial bond breaking step is reported[27b] to be ~6:1 for 3°:2° C-H bonds. The dimeric products, on the other hand, often show a 3°:2° selectivity of 7-10:1. If H• adds to the alkene formed from a 2° radical, it can emerge as a 3° radical. If H• adds to the alkene formed from a 3° radical, the product is more likely to return to the same starting radical, this 3° radical being more stable. 2° radicals are constantly being converted into 3° radicals by this mechanism and so the selectivity of the system is amplified. This explains the increased selectivity under H_2 mentioned above.

Other metal atoms have been studied in the matrix and it has been shown that many of them, when photoexcited, can give oxidative addition with alkane C-H bonds. For example the sequence shown in Eq. 40 has been studied by Billups et al.[28] and by Ozin and McCaffrey.[29] Irradiation at 300 nm causes the oxidative addition, which can be reversed by subsequent irradiation at 400 nm. Cu(^{2}P) atoms have also been shown to oxidatively add to the C-H bond of methane, but the initial adduct, believed to be MeCuH, decomposed photolytically under the conditions of its formation to give CuH and CuMe.[30]

$$\text{Fe} + \text{CH}_4 \underset{400 \text{ nm}}{\overset{300 \text{ nm}}{\rightleftarrows}} \text{H—Fe—CH}_3 \qquad (40)$$

Green et al.[31] has used metal vapor synthesis (MVS) to make isolable quantities of a number of organometallic species directly from alkanes. In each case the complex has the stoichiometry of the parent alkane. Equation 42 is reminiscent of some of the chemistry we saw in the homogeneous metal phosphine chemistry discussed earlier in this chapter.

$$C_5H_{10} \xrightarrow{\text{W, PMe}_3} CpW(PMe_3)H_5 \qquad (42)$$

In Eqs. 41 and 43, the metals bind the hydrogens in μ-H sites in the cluster. In the rhenium case, the carbon fragment from the alkane is bound as a μ-carbene in the resulting cluster. This may be a reasonable analogy for the way alkane-derived fragments are bound on metal surfaces. The coligands, benzene and trimethylphosphine, are required to stabilize the complex. It is perhaps surprising that these ligands are not activated in the products observed. Of course, other species may have escaped detection, and the observed products are no doubt thermodynamic traps, so we cannot deduce anything about the relative reactivity of arene, phosphine, and alkane C-H bonds. Photochemical processes may also occur under the MVS conditions, because the presence of a brightly glowing bead of metal near its melting point must generate a broad spectrum of radiation.

There are also many examples of alkane activation by transition metal ions in the gas phase. This work has been carried out by ion beam mass spectroscopy or ion cyclotron resonance (ICR). In these studies it is usually only the mass of the charged products that can be observed, and so the structures of the products are largely conjectural. A species said to be $M(CH_2=CH_2)^+$ might in reality be $M(=CH_2)_2^+$, for example. A feature of these reactions is the ease with which C-C as well as C-H bonds are broken. An example involving Fe^+ and isobutane is shown below.[32]

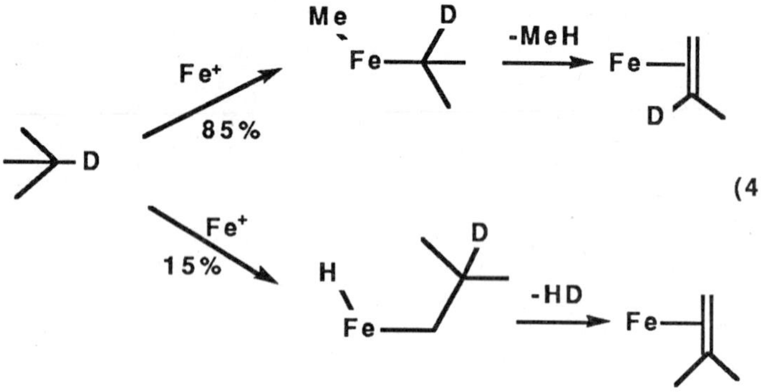

(44)

This kinetic facility for C-C bond breaking may be related to the fact that Fe^+ is sterically very compact. The thermodynamic stabilities of M-alkyl bonds in the naked $[M-R]^+$ species tend to exceed the M-H bond strength for $[M-H]^+$, in contrast to the situation for metal complexes (at least of the d-block), in which the ordering is $[M-R] < [M-H]$. The dissociation energy for $[Fe-Me]^+$ has been estimated as 69 ± 5 kcal/mol. This is much stronger than Halpern's estimate for M-C bonds in metal complexes, but is less than the estimates obtained by Marks et al.[33] for Cp_2ThMe_2 (D(Ir-Me) = 82.5 kcal/mol).

Interesting ligand effects have been observed in some of the metal ion work. For example, Co^+ activates alkane C-H bonds but Co_2^+ does not. The addition of a CO ligand, as in $Co_2(CO)^+$ restores the activity.[32]

III.   Conclusion

There are now several examples of catalytic systems based on oxidative addition of alkane C-H bonds to metal phosphine complexes. Perhaps their most important feature is the selectivity they show for unhindered C-H bonds. Several different types of catalytic alkane conversion reactions have been observed, both thermal and photochemical. The great thermodynamic stability of alkanes means that relatively few alkane conversion processes are exothermic. This means that either photochemistry has to be employed so that the energy of the photons drives the reaction, or an exothermic process has to be coupled to an endothermic alkane conversion. Reactions involving O_2 with the formation of water as the coupled reaction would be strongly exothermic, but metals capable of oxidative addition with alkanes are rarely stable to O_2. Felkin et al.[4] find that their rhenium system still operates in the presence of oxygen, although no oxygenated products are formed, so there may still be hope in this direction. An interesting area for further development involves additions of alkane C-H bonds across unsaturated species such as alkenes and alkynes, as discovered by Grigoryan.

The efficiency of all these catalysts seems to be sharply restricted by deactivation reactions. These are as yet insufficiently understood. The key goal in this area is to develop catalysts that will be resistant to deactivation, perhaps by moving to ligands other than PR_3. The work of Shilov (Chapter I) shows how simple ligands such as Cl⁻ and acetate can be incorporated into an active catalyst, so we may expect developments in this direction.

The mercury system is capable of rapidly assembling relatively complex molecules from simple starting materials, and doing so on a preparatively useful scale. Exxon Corp. has therefore taken an interest in possible commercial applications of the "MERCAT" process. The catalyst resists degradation largely because it is a simple atom of a rather unreactive element.

A more general principle that emerges from the mercury work is that unusual selectivity patterns may be achieved by physical separation of different species. It may be possible to take advantage of this principle in other cases in which the products of a reaction are much more reactive than the starting material and therefore tend not to survive the reaction. The selective oxidation of methane to methanol may be an example.

Acknowledgment

We thank the National Science Foundation, the Petroleum Research Fund, the Department of Energy, and Exxon Corp. for funding our work in this area.

References

1. R. H. Crabtree, *Chem. Rev.* **1985**, *85*, 245.
2. (a) Burk, M. J., Crabtree, R. H., McGrath, D. V.,*Chem. Commun..* **1985**, 1829; (b) Burk, M. J., Crabtree, R. H., *J. Am. Chem. Soc.* **1987**, *109*, 8025.
3. (a) Crabtree, R. H., Mihelcic, J. M., and Quirk, J. M., *J. Am. Chem. Soc.* **1979**, *101*, 7738; (b) Crabtree, R. H., Mellea, M. F., Mihelcic, J. M., Quirk, J. M., *J. Am. Chem. Soc.* **1982**, *104*, 107; (c) Burk, M. J., Crabtree, R. H., Parnell, C. P., Uriarte, R. J., *Organometallics* **1984**, *3*, 816; (d) Crabtree, R. H., Dion, R. P., Gibboni, D. J., McGrath, D. V., Holt, E. M.,*J. Am. Chem. Soc.* **1986**, *108*, 7222; (e) Anton D. R., Crabtree, R. H., *Organometallics* **1983**, *2*, 855.

4.   Baudry, D., Ephritikine, M., Felkin, H., *Chem. Commun.*. **1982**, *606*, **1983**, 788; Baudry, D., Ephritikine,M., Felkin, H., Zakrzewski, J., *Chem. Commun.*. **1982**, 1235, *Tetrahedron. Lett* **1984,** 1283.

5.   (a) Felkin, H., Fillebeen Khan, T., Gault, Y., Holmes-Smith, R., Zakrzewski, J., *Tetrahedron. Lett.* **1984**, 1279; (b) Cameron, C. J., Felkin, H., Fillebeen-Khan, T., Forrow, N. J., Guittet, E., *Chem. Commun.* **1986**, 801; (c) Felkin, H., Fillebeen-Khan, T., Holmes-Smith, R., Lin, Y., *Tetrahedron. Lett.* **1985**, *26*, 1999.

6.   (a) Sakakura, T., Tanaka, M., *Chem. Commun.* **1987**, 758; (b) *Chem. Lett.* **1988**, 263.

7.   Nolan, S. P., Hoff, C. D., Stoutland, P. O., Newman, L. J., Buchanan, J. M., Bergman, R. G., Yang G. K., Peters, K. S., *J. Am. Chem. Soc.* **1987**, *109*, 3143.

8.   Halpern, J., *Acct. Chem. Res.* **1982**, *15*, 238.

9.   Wax, M. J., Stryker, J. M., Buchanan, J. M., Kovac, C. A., Bergman, R. G., *J. Am. Chem. Soc.* **1984**, *106*, 1121.

10.   Hoyano, J. K., Graham, W. A. G., *J. Am. Chem. Soc.* **1982**, *104*, 3723.

11.   (a) Crabtree, R. H., Holt, E. M., Lavin, M. E., Morehouse, S. M., *Inorg. Chem.* **1985**, *24*, 1986; (b) Baker, M. V., Field, L. D., *J. Am. Chem. Soc.* **1987**, *109*, 2825; (c) Baker, M. V., Field, L. D., *J. Am. Chem. Soc.* **1986**, *108*, 7433.

12.   (a) Jones, W. D., Feher, F. J., *J. Am. Chem. Soc.* **1982**, *104*, 4240; **1986**, *108*, 4814; (b) Jones, W. D., Maguire, J. A., *Organometallics* **1986**, *5*, 590.

13.   Chencey, A. J., Shaw, B. L., *J. Chem. Soc.* Dalton **1972**, *754*, 860.

14.   (a) Garrou, P., *Chem. Rev.* **1985**, *85*, 171; (b) Hoffman, R., *Helv. Chim. Acta* **1984**, *67*, 1.

15.   (a) Goldshleger, N. F., Tyabin, M. B., Shilov, A. E., Shteinman, A. A., *Zh. Fiz. Khim.* **1969**, *43*, 2174; (b) Shilov, A. E., *The Activation of Saturated Hydrocarbons by Transition Metal Compounds*. Riedel, Dordrecht, 1984; (c) Grigoryan, E. A., Dyachkovskii, F. S., Mullagaliev, I. R., *Dokl. Akad. Nauk SSSR* **1975**, *224*, 859; Grigoryan, E. A., Dyachkovskii, F. S., Zhuk, S. Ya., Vyshinskaja, L. I., *Kinet. Katal.* **1978**, *19*, 1063; (d) Enikolopjan, N. S., Gyulumjan, Kh. R., Grigoryan, E. A., *Dokl. Acad. Nauk SSSR* **1979**, *249*, 1380; Grigoryan, E. A., Gyulumjan, Kh. R., Gurtovaya, E. I., Enikolopjan, N. S., Ter-Kazarova, M. A., *Dokl. Acad. Nauk SSSR* **1981**, *257*, 364.

16.   Ibers, J. A., DiCosimo, R., Whitesides, G. M., *Organometallics* **1982**, *3*, 1.

17.   Green, M. A., Huffman, J. C., Caulton, K. G., Ziolkowski, J. J., *J. Organometal. Chem.* **1981**, *218*, C39.

18.   (a) Stoutland, P. O., Bergman, R. G., *J. Am. Chem. Soc.* **1985**, *107*, 4581; (b) Faller, J. W., Felkin, H., *Organometallics* **1985**, *4*, 1488.

19.   (a) Suggs, J. W., Cox, S. D., *J. Organometal. Chem.* **1981**, *221*, 199; (b) Benfield, F. W., Green, M. L. H., Dalton, J. C. S., **1974**, 1325; (c) Eilbracht, P., *Chem. Ber.* **1980**, *113*, 542 and 1033.

20.   Tipper, C. F. H., *J. Chem. Soc.* **1955**, 2043; Adams, D. M., Chatt, J., Guy,R., Shepard, N., *J. Chem. Soc.* **1961**, 738.

21.   Bishop, K. C., *Chem. Rev.* **1976**, *76*, 461; Hogeveen, H., Volger, H. C., *J. Am. Chem. Soc.* **1967**, *89*, 2486

22.   Cassar, L., Halpern, J., *J. Chem. Soc.* D **1971**, 1082

23.   Hogeveen, H., Nusse, B. J., *Tetrahedron. Lett.* **1974**, 159.

24. Cvetanovic, R. J., *Progr. React. Kinet.* **1964**, *2*, 77.

25. Brown, S. H., Crabtree, R. H., *Chem. Commun.* **1987,** 970; *Tetrahedron. Lett.* **1987**, *28*, 5599.

26. (a) Porter, R. F., *Inorg. Chem.* **1980**, *19*, 447; Plotkin, J. S., Sneddon, L. G., *J. Am. Chem. Soc.* **1979**, *101*, 4155; Nay, M. A., Woodall, G. N.C., Strausz, O. P., Gunning, H. E., *J. Am. Chem. Soc.* **1965**, *87*, 179; Mains, G. J., *Inorg. Chem.* **1966**, *5*, 114; (b) Gray, J. A., *J.Chem.Soc.*, **1952**, 3150.

27. (a) Ingold, K. U., *J. Am. Chem. Soc.* **1974**, *96*, 2441; (b) Holroyd, R. A., Klein, G. W., *J. Phys. Chem.* **1967**, *63*, 2273.

28. Billups, W. E., Konarski, M. M., Hauge, R. H., Margrave, J. L., *J. Am. Chem. Soc.* **1982**, *102*, 7393.

29. Ozin, G. A., McCaffrey, J. G., *J. Am. Chem. Soc.* **1982**, *104*, 7351.

30. Ozin, G. A., McIntosh, D. F., Mitchell, S. A., *J. Am. Chem. Soc.* **1981**, *103*, 1574.

31. Bandy, J. A., Cloke, F. G. N., Green, M. L. H., O'Hare, D., Prout, K., *Chem. Commun.* **1984,** *240*; 1885, *355*, 356; Green, M. L. H., Parker, G., *Chem. Commun.* **1984**, 1467.

32. (a) Allison, J., Freas, R. B., Ridge, D. P., *J. Am. Chem. Soc.* **1979**, *101*, 1332; (b) Freas, R. B., Ridge, D. P., *J. Am. Chem. Soc.* **1980**, *102*, 7129; **1984**, *106*, 825.

33. Bruno, J. W., Marks, T. J., Morss, L. R., *J. Am. Chem. Soc.* **1983**, *105*, 6824.

Chapter IV

ALKANE ACTIVATION PROCESSES BY CYCLOPENTADIENYL COMPLEXES OF RHODIUM, IRIDIUM, AND RELATED SPECIES

William D. Jones

Department of Chemistry
University of Rochester
Rochester, New York 14627

This chapter presents organometallic complexes in which the first simple step of breaking the C-H bond was directly observed. This fundamental reaction was believed to be the most difficult since the subsequent reactions (which include β-elimination, insertion, and π-allyl hydride formation) had ample precedent in the organometallic chemistry of metal alkyl complexes. In 1982 Janowicz and Bergman reported the first well-characterized example of simple oxidative addition of an unactivated alkane to a homogeneous metal complex using a permethylcyclopentadienyl iridium complex. This observation generated tremendous excitement in the academic and industrial communities, as it led to the employment of alkanes in homogeneous organometallic reactions. Several discoveries of new low-valent metal complexes containing C_5Me_5 and/or PR_3 ligands that could also activate the C-H bonds of aliphatic hydrocarbons rapidly followed. This chapter presents an account of the discoveries and compares the mechanisms, thermodynamics, and selectivities of the reactive compounds.

I.   Initial Observations of Alkane Activation

The discoveries began with an interest in the hydrogenolysis reactions being studied by Bergman and co-workers. It had been established that treatment of $CpCo(PPh_3)Me_2$ with H_2 and PPh_3 induced the formation of methane and $CpCo(PPh_3)_2$ by way of a labile dihydride intermediate

$CpCo(PPh_3)H_2$. The permethylcyclopentadienyl complex $(C_5Me_5)Co(PPh_3)Me_2$ was found to undergo the same type of reaction, but now the dihydride $(C_5Me_5)Co(PPh_3)H_2$ was slightly more stable and could be observed in the reaction mixture, although it was still too unstable to be isolated. No reactions with solvent were noted.[1] The iridium analog $(C_5Me_5)Ir(PPh_3)H_2$ was synthesized in an attempt to characterize the dihydride species, and was found to be quite stable with respect to hydrogen loss. As is commonly observed with metal polyhydrides, photolysis ($\lambda \geq 275$ nm) of the complex in benzene solution at room temperature results in dihydrogen loss. Two products were observed, a hydridophenyl complex and an orthometalated complex in approximately equal amounts (Eq. 1). The benzene solvent (~12 M) apparently competes equally well with one of the ortho hydrogens of the PPh_3 ligand for the postulated reactive intermediate $[(C_5Me_5)Ir(PPh_3)]$.[2]

$$(1)$$

The complex $(C_5Me_5)Ir(PMe_3)H_2$ was examined in an effort to decrease the orthometalation reaction. In this case, irradiation in benzene solvent gave $(C_5Me_5)Ir(PMe_3)(Ph)H$ as the exclusive product (Eq. 2). There was no evidence for metalation of the PMe_3 ligand to

$$(2)$$

give $M(\eta^2\text{-}PCH_2Me_2)$ as has been found with other metal-trimethylphosphine complexes (M = Fe, Co, Ru, Os, Re, Mo, W, Ta).[3] When the irradiation was carried out in cyclohexane, oxidative addition to the solvent occurred. A 90% yield of $(C_5Me_5)Ir(PMe_3)(c\text{-}C_6H_{11})H$ was observed after 74% conversion of the dihydride.

Similarly, irradiation of the dihydride in neopentane solvent produced the neopentyl hydride adduct in 80% yield after 83% conversion. C-H bond activation by the proposed intermediate [(C$_5$Me$_5$)Ir(PMe$_3$)] must be thermodynamically downhill and kinetically favorable with respect to dimerization or formation of other products, as indicated in the free energy diagram for the activation reaction shown in Figure 1.[2] Again, no internal metalation of the phosphine was observed. The air-sensitive

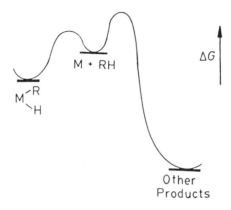

Figure 1.

cyclohexane oxidative addition product was thermally stable up to 100°C. Irradiation of the cyclohexyl hydride complex in benzene slowly generated (C$_5$Me$_5$)Ir(PMe$_3$)(Ph)H over 12 hr, indicating that the

$$(3)$$

alkyl hydride complex was only slightly photosensitive. These two properties were crucial to the discovery of alkane oxidative addition with (C$_5$Me$_5$)Ir(PMe$_3$)H$_2$, and can be contrasted with the photochemistry of Cp$_2$WH$_2$ studied earlier by Green. Photolysis of Cp$_2$WH$_2$ in alkanes

gave only tungstenocene dimers[4] (Eq. 3), in spite of the fact that
$Cp_2W(CH_3)H$ was known to be stable up to 60°C.[5] Irradiation in
mesitylene solvent gave only $Cp_2W(CH_2C_6H_3Me_2)_2$. The high
photosensitivity of the intermediate $Cp_2W(CH_2C_6H_3Me_2)H$ was
demonstrated independently to lead to the latter disubstituted product
under photolysis conditions.[6] Since the iridium alkyl hydride
complexes were also generated under photochemical conditions, they
cannot be as photosensitive as the $(C_5Me_5)Ir(PMe_3)H_2$ precursor, a key
factor that permitted observation of the initial C-H activation products
upon irradiation with this system. The C_5Me_5 ligand was shown not to
be a factor distinguishing the tungsten and iridium systems. Green and
co-workers found that irradiation of $(C_5Me_5)_2WH_2$ leads to dihydrogen
loss followed by activation of one of the methyl groups of the C_5Me_5
ligand (Eq. 4).[7]

$$\qquad\qquad\qquad\qquad\qquad\qquad\qquad\qquad\qquad\qquad\qquad\qquad (4)$$

Irradiation of the iridium dihydride in liquid propane solvent
gives two products in a 4.5:1 ratio, corresponding to activation of either
a primary or secondary C-H bond. The ^{1}H-NMR resonances of the
hydride ligand were found to be a useful probe of the position of
activation, with primary alkyl hydrides appearing at lower field than
secondary alkyl hydrides. Since the statistical preference for primary
activation is 3:1 in propane, the unsaturated iridium intermediate only
slightly prefers primary over secondary C-H bonds. Also, since the
hydrido alkyl products are fairly stable under the reaction conditions,
this ratio corresponds to the *kinetic* selectivity of $(C_5Me_5)Ir(PMe_3)$ for
the two types of C-H bond.[2]

Little selectivity is also seen upon irradiation of the dihydride in
pentane, but now activation of the C-2 hydrogens generates a pair of
diasteriomers so that a total of four different products are observed
(Eq. 5). The kinetic selectivities of this intermediate for various other
hydrocarbons will be compared with those of other complexes later in
the chapter.[2]

The hydrido alkyl complexes of the type
$(C_5Me_5)Ir(PMe_3)(R)H$ were found to be thermally labile at high
temperatures, allowing the establishment of the relative *thermodynamic*
stabilities of a series of hydrido alkyl complexes as opposed to the
kinetic selectivities of $(C_5Me_5)Ir(PMe_3)$ for various hydrocarbon C-H
bonds obtained in the photolysis studies. This observation also
indicated that the species $[(C_5Me_5)Ir(PMe_3)]$ generated in the

$$(5)$$

photochemical experiments *did not* receive its special reactivity as a result of energy present in the form of an electronically excited state.[8]

Upon warming a pentane solution of the four possible pentane activation products shown in Eq. 5 to 110°C in a sealed tube, the ^{1}H-NMR resonances for the three secondary hydrido alkyl complexes were found to decrease as the fourth resonance for the primary hydrido alkyl complex increased. This experiment suggested that the *n*-pentyl complex was thermodynamically more stable than the 2- or 3-pentyl complexes. In fact, thermolysis of the mixture of pentyl hydrides in cyclohexane solvent at 110°C resulted in the conversion of the secondary pentyl complexes into the cyclohexyl complex, whereas the primary pentyl complex remained unchanged. Higher temperatures were found to be required to labilize the 1-pentyl and cyclohexyl complexes. Reductive elimination of alkane was observed to be a first-order process.[8]

When either the *n*-pentyl hydride complex or the cyclohexyl hydride complex was heated to 140°C in a solvent mixture composed of *n*-pentane and cyclohexane, equilibration occurred over the next 50 hr producing a 1:1 ratio of the two compounds. The equilibrium constant of 10.8 corresponds to a -2.0 kcal/mol free energy favorability for the 1-pentyl complex over the cyclohexyl complex at 140°C. Similar equilibration experiments were used to provide a series of thermodynamic stabilities for various alkyl hydride complexes, and are presented later in this chapter. *The general trend found is for activation of the least substituted carbon to give the most stable complexes.* The phenyl hydride complex was not labile up to 200°C, at which temperature decomposition set in.

The Bergman intermediate [$(C_5Me_5)Ir(PMe_3)$] also activates vinylic C-H bonds. Thermolysis of $(C_5Me_5)Ir(PMe_3)(c\text{-}C_6H_{11})H$ at 130°C under 20 atm ethylene in cyclohexane solvent gives the hydrido vinyl product in 66% yield plus 34% of the ethylene complex (Eq. 6). An X-ray structure of the hydrido vinyl complex showed coplanar arrangement of the H-Ir-CH=CH$_2$ unit, although no C-H interaction was observed.[9]

$$\text{(C}_5\text{Me}_5\text{)Ir(PMe}_3\text{)(H)(c-C}_6\text{H}_{11}) + \text{C}_2\text{H}_4 \xrightarrow{130\,°C} \text{(C}_5\text{Me}_5\text{)Ir(PMe}_3\text{)(H)(C}_2\text{H}_4) + \text{(C}_5\text{Me}_5\text{)Ir(PMe}_3\text{)(C}_2\text{H}_3) \quad (6)$$

Shortly after the discovery of alkane activation by Janowicz and Bergman, Hoyano and Graham reported another $(C_5Me_5)IrL$ system that reacted with hydrocarbons. They found that irradiation of $(C_5Me_5)Ir(CO)_2$ in either cyclohexane or neopentane solvent resulted in the oxidative addition of the alkane to the metal center (Eq. 7). The

$$(C_5Me_5)Ir(CO)_2 \xrightarrow[\text{RH}]{h\nu} (C_5Me_5)Ir(CO)(R)(H) \quad (7)$$

$$R = c\text{-}C_6H_{11}, \ CH_2CMe_3$$

air-sensitive products were characterized spectroscopically in solution and by conversion to the alkyl chloro derivatives by treatment with CCl_4. The parent compound $CpIr(CO)_2$ also underwent similar photochemically induced reactions in alkane solvents,[10] and had been previously suggested by Rausch et al. to activate benzene upon irradiation.[11]

Shortly after the reports with iridium, Jones and Feher reported that heating of $(C_5Me_5)Rh(PMe_3)(C_6H_5)H$ in arene solvents leads to arene exchange reactions. Thermolysis of the phenyl hydride complex in C_6D_6 solution gave >90% C_6H_6 and the perdeutero complex $(C_5Me_5)Rh(PMe_3)(C_6D_5)D$. A mixture of the *meta* and *para* tolyl hydride products was obtained upon thermolysis in toluene. Photolysis of the dihydride $(C_5Me_5)Rh(PMe_3)H_2$ in benzene solution gave the phenyl hydride complex, just as in the iridium system studied by Bergman. Perhaps the most important discovery differentiating the rhodium system and its iridium analog (other than the obvious enhanced lability) was a series of labeling experiments demonstrating that the aryl hydride complexes were in fact undergoing a series of rapid intramolecular [1,2] rearrangements at temperatures above -15°C. This isomerization provided a valuable clue to the mechanism of arene activation, and will be discussed more fully later in this chapter.[12]

In comparison, the corresponding rhodium alkyl hydride complexes were found to be far less stable than the aryl hydride complexes. Synthesis of the methyl hydride could be carried out only at low temperature by borohydride reduction of $[(C_5Me_5)Rh(PMe_3)(CH_3)(THF)]^+$. Above -20°C, methane was formed by reductive elimination just as in $Pt(PPh_3)_2MeH$.[13] Irradiation of the dihydride $(C_5Me_5)Rh(PMe_3)H_2$ in alkane solvents at room temperature did not give rise to the formation of observable alkyl hydride complexes as in the case of iridium. The ability of the intermediate $[(C_5Me_5)Rh(PMe_3)]$ to reversibly activate arenes combined with the stability of the rhodium alkyl hydride only at low temperatures suggested that the rhodium system might activate alkanes at temperatures below -20°C. This proved to be the case, as irradiation of $(C_5Me_5)Rh(PMe_3)H_2$ in liquid propane at -50°C produced the *n*-propyl hydride product as the exclusive kinetic product of the reaction (Eq. 8).

$$
\underset{\substack{H}}{\overset{}{Rh}}\!\!-\!\!PMe_3 \quad \xrightarrow[\substack{C_3H_8 \\ -55°C}]{h\nu} \quad \underset{\substack{H_3C-CH_2}}{\overset{}{Rh}}\!\!-\!\!PMe_3 \qquad (8)
$$

The rhodium system therefore mimics the iridium system, but at an approximately 140°C lower temperature range.[10] Periana and Bergman later extended this work to include other alkanes.[14]

Methane was found to be quite reactive with these reactive intermediates, an observation that attracted interest from the methane industry. Graham and co-workers found that irradiation of the iridium dicarbonyl in perfluorohexane solvent under 8 atm of methane resulted in the production of $(C_5Me_5)Ir(CO)(CH_3)H$.[15] In contrast to Graham's results, irradiation of $(C_5Me_5)Ir(PMe_3)H_2$ in perfluoroalkane solvent under 4 atm methane gave no detectable methane activation. Irradiation in cyclooctane solvent under methane pressure produced only the hydrido cyclooctyl complex. However, thermolysis of $(C_5Me_5)Ir(PMe_3)(c\text{-}C_6H_{11})H$ at 150°C in cyclooctane under 20 atm methane for 14 hr produced the hydrido methyl product in 58% yield. The hydrido methyl complex must therefore be the thermodynamic sink under these conditions, in agreement with the general trend of increased thermodynamic stability being associated with less substitution at the activated carbon center.[6] Photochemical elimination of methane is apparently facile. There has been a report of a new precursor for the facile formation of the reactive species $[(C_5Me_5)IrL]$. Chetcuti and Hawthorne have found that thermolysis of

$(C_5Me_5)Ir(CO)(\eta^2-NCC_6H_4Cl)$ gives rise to loss of the bound nitrile ligand and oxidative addition to benzene solvent.[16] The generality of this reaction is yet to be established.

Another cyclopentadienyl analog of the iridium complex investigated by Bergman et al. was found to activate alkanes. Irradiation of the complexes $CpRe(PMe_3)_3$, $(C_5Me_5)Re(PMe_3)_2(CO)$, and $(C_5Me_5)Re(PMe_3)(CO)_2$ in alkane solvent led to the formation of the intermediates $CpRe(PMe_3)_2$, $(C_5Me_5)Re(PMe_3)_2$, and $(C_5Me_5)Re(PMe_3)(CO)$ by dissociation of either PMe_3 or CO. Activation of C-H bonds of the solvent and the formation of a variety of oxidative addition adducts occurred, although with a lower stability and different selectivity than in the iridium system.[17]

Irradiation of $CpRe(PMe_3)_3$ in benzene or cyclopropane solvent leads to the corresponding C-H addition complexes as indicated in Eq. 9. The products are very air sensitive and are stable only for a period of several hours at room temperature. In *n*-hexane solvent the *n*-hexyl hydride product is observed at -20°C, although its half-life at room temperature is only ~10 min.[17]

Irradiation in cyclohexane solvent does not lead to solvent activation, but instead produces the metallocycle in which a PMe_3 methyl group has been activated. This isolable metallocycle was found to undergo reversible reductive elimination, providing a self-contained method for generating the coordinatively unsaturated species $[CpRe(PMe_3)_2]$ at room temperature. Dissolving the metallocycle in benzene, cyclopropane, or hexane gives the corresponding alkyl or phenyl hydride complex (Eq. 9).[17]

(9)

Since cyclohexane was found to be an inert solvent for the photochemical reactions (or at least only formed a transiently stable cyclohexyl hydride complex), the irradiation of $CpRe(PMe_3)_3$ in cyclohexane under 25 atm of methane gave the corresponding methyl hydride complex in 42% yield. Similarly, irradiation in cyclohexane containing ethylene (25 atm) gave the vinyl hydride complex in 42% yield.[17]

Similar but not identical results were obtained with the permethylcyclopentadienyl complexes, except that the reactive intermediate generated in this case was $(C_5Me_5)Re(PMe_3)(CO)$. The oxidative addition adducts proved even more difficult to isolate than in the parent C_5H_5 series. Cycloalkanes and even *n*-alkanes (larger than methane) were not activated, with the $Re(\eta^2\text{-}PMe_2CH_2)$ metallocycle being formed upon irradiation in these solvents. As with $CpRe(PMe_3)_2$, the metallocycle species proved to be the most convenient precursor for the generation of $(C_5Me_5)Re(PMe_3)(CO)$ for the activation of benzene, methane, or cyclopropane.[17] More recently, the dinitrogen adduct $(C_5Me_5)Re(PMe_3)(CO)(N_2)$ has been prepared by Sutton and co-workers and shown to be an efficient photochemical precursor for the unsaturated species $[(C_5Me_5)Ir(PMe_3)]$.[18]

A related demonstration of alkane activation was made by Jones and Maguire employing a rhenium III/V rather than a I/III couple. Irradiation of $CpRe(PPh_3)_2H_2$ in a THF-d_8-RH solvent mixture resulted in catalytic H-D exchange between the THF-d_8 and the arene or alkane RH (RH = CH_4, C_2H_6, C_3H_8, etc.; Eq. 10). Similar observations were made using C_6D_6 as the deuterium source. However, photolysis

$$R\text{-}H + R'\text{-}H \xrightarrow[h\nu,\ C_6D_6]{CpRe(PPh_3)_2H_2} R\text{-}D + R'\text{-}D \qquad (10)$$

of the $CpRe(PPh_3)_2H_2$ does not result in reductive elimination of dihydrogen as in the iridium and rhodium systems, perhaps because of the trans dispositions of the two dihydride ligands.[19] Instead, phosphine is photolabilized as demonstrated by the inhibition of the H-D exchange upon addition of PPh_3 and the photochemical exchange of PMe_3 into $CpRe(PPh_3)_2H_2$. Loss of phosphine must also be involved in the H-D exchange process since the reversible elimination of H_2 and oxidative addition of alkane would not result in any net H-D exchange. One of the unique features of this system is that these alkane H-D exchange reactions can be run in dry air, as $CpRe(PPh_3)_2H_2$ is not oxygen sensitive. Hundreds or thousands of turnovers are typically observed.[20]

Very recently, Ghosh and Graham reported a pyrazolylborate analog of $(C_5Me_5)Ir(CO)_2$ that also demonstrates the ability to activate alkane C-H bonds. A methylated pyrazolylborate iridium complex has been shown to undergo an efficient photochemical reaction in hydrocarbon solvents (benzene, neopentane, or cyclohexane) to give oxidative addition adducts (Eq. 11). The related rhodium complex gives a stable phenyl hydride upon irradiation in benzene. Irradiation in

$$\text{(11)}$$

cyclohexane gives the dihydrido carbonyl product, apparently by way of β-elimination of the initially formed cyclohexyl hydride. The latter can be trapped by addition of CCl_4 immediately following photolysis (Eq. 12).[21]

Methane activation could also be achieved in a thermal reaction of the rhodium pyrazolylborate complex. Irradiation in cyclohexane solvent gives rise to the unstable cyclohexyl hydride complex, which upon treatment with methane (1 atm) in the dark produced some of the methyl hydride derivative. From the ratio of cyclohexyl and methyl hydride products, and the solubility of methane, Graham calculated an equilibrium constant of 190 for the reaction. The preference for the

$$\text{(12)}$$

unsubstituted methane C-H bond is again seen.

Some noncyclopentadienyl complexes have also been reported recently that activate alkane C-H bonds, all of which invoke binary metal(0)/trialkylphosphine reactive intermediates. Baker and Field have found that irradiation of $Fe(dmpe)_2H_2$ in pentane at low temperatures

(-90°C) gives the *n*-pentyl hydride complex *cis*-Fe(dmpe)$_2$(*n*-pentyl)H [dmpe = bis(dimethylphosphino)ethane]. The complex decomposes rapidly upon warming to -20°C (Eq. 13), ultimately forming

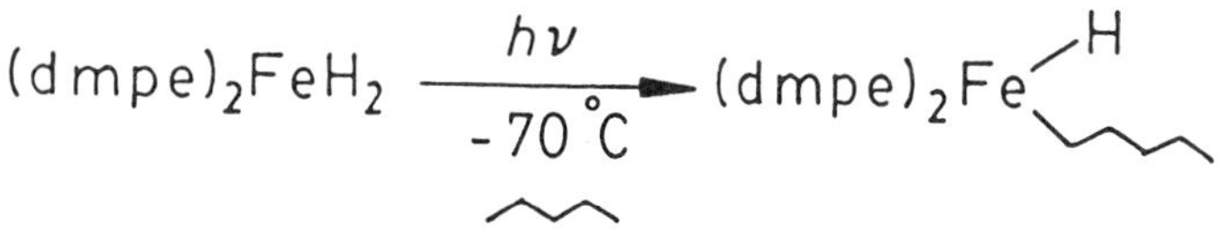

$$(\text{dmpe})_2\text{FeH}_2 \xrightarrow[-70\,°\text{C}]{h\nu} (\text{dmpe})_2\text{Fe}{<}^{\text{H}} \qquad (13)$$

Fe$_2$(dmpe)$_3$ and oligomeric species. The sp^2-hybridized C-H bonds of arenes (benzene, toluene) and alkenes (cyclopentene, ethylene) are also activated upon irradiation of the iron dihydride complex in cyclopentane solution at -80°C. As with (C$_5$Me$_5$)Ir(PMe$_3$), the products of the reaction of the iron complex with ethylene include both the vinyl hydride and the π-complex in a 10:1 ratio. The vinyl hydride complex is observed to isomerize to the π-complex upon warming to room temperature.[22]

Flood and co-workers have examined the related thermal chemistry of Os(PMe$_3$)$_4$(CH$_2$CMe$_3$)H, as summarized in Scheme 1.

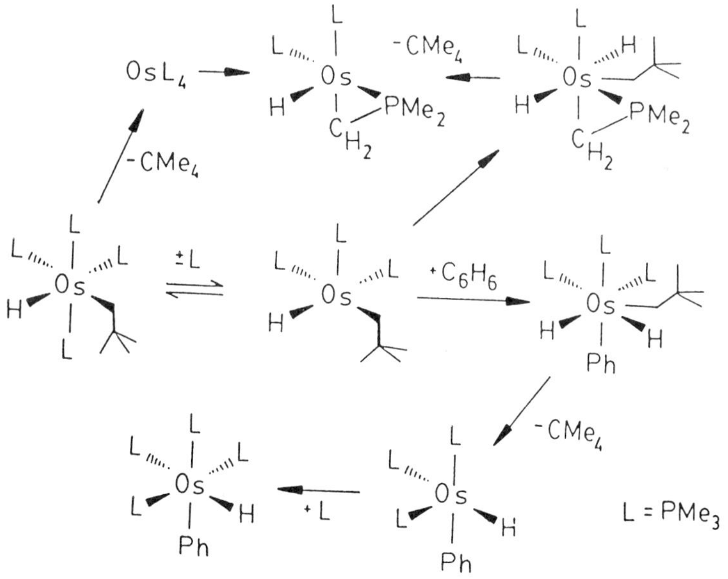

Scheme 1

This system is a little more complicated in that both PMe$_3$ dissociation and neopentane reductive elimination are found to occur at 80°C. In benzene solvent PMe$_3$ dissociation is followed by oxidative addition of benzene and reductive elimination of neopentane. This sequence of events is supported by the observation of neopentane-d_1 when the reaction is carried out in C_6D_6, the inhibition of the reaction by added PMe_3, and the rapid exchange of $P(CD_3)_3$ into the complex. The metallocycle $Os(PMe_3)_3(\eta^2-PMe_2CH_2)H$ is also formed by two independent pathways, one that is dissociative in PMe_3 and one that is not. Thermolysis of the neopentyl hydride complex in the presence of methane produces small quantities of a product whose ^{31}P-NMR spectrum is consistent with that of independently prepared $Os(PMe_3)_4(CH_3)H$; the major product is the metallocycle.[23]

In somewhat related experiments, Whitesides and co-workers have discovered a bisphosphine platinum(0) complex that oxidatively adds to alkanes under thermal conditions. Heating a solution of Pt[bis(dicyclohexylphosphino)ethane](neopentyl)H in cyclopentane solvent leads to the loss of neopentane and the formation of $Pt(L_2)$(cyclopentyl)H (25% yield), and an insoluble white precipitate. Aromatic and benzylic C-H bonds are also activated by the platinum(0) intermediate (Eq. 14).[24]

$$\text{(14)}$$

R = Me, Ph, c-pentyl, benzyl

In 1983, Kaska and co-workers made an independent report of the activation of alkane solvent by treatment of the chelate complex shown in Eq. 15 with bistrimethylsilylamide. Unfortunately, several products were obtained and the reaction was not well characterized. No complete report on this reaction has since appeared. Use of deuterated solvent gave a resonance at δ -24.0 in the ^2H-NMR spectrum, consistent with solvent activation. Kaska and co-workers also proposed intramolecular alkyl group activation in the reaction.[25]

$$\text{(15)}$$

One last set of compounds to mention are the IrL_2Cl and RhL_2Cl [L = P(*i*-Pr)$_3$] complexes studied by Werner et al. These complexes are formed *in situ* by the reaction of [(COE)$_2$MCl]$_2$ (M = Rh, Ir) with P(*i*-Pr)$_3$ (COE = cyclooctene). Heating a benzene solution of the iridium complex gives a 1:2 ratio of the dihydride and phenyl hydride complexes (Eq. 16). (The former product arises from reaction with cyclooctene.)[26] Although Werner et. al. have not reported reaction of

$$[Ir(PPr^i_3)_2Cl]_2 \xrightarrow{C_6H_6} Cl-Ir\overset{PPr^i_3}{\underset{PPr^i_3}{\mid}}\overset{\ldots H}{\underset{H}{}} + Cl-Ir\overset{PPr^i_3}{\underset{PPr^i_3}{\mid}}\overset{\ldots C_6H_5}{\underset{H}{}} \quad (16)$$

these complexes with alkanes, a recent paper by Sakakura and Tanaka[27] has shown that irradiation of the related species $RhCl(CO)(PMe_3)_2$ in pentane solvent under a CO atmosphere gives rise to alkane carbonylation products (Eq. 17). This insertion reaction must be photochemically driven since it is well known that aldehyde decarbonylation is a thermodynamically favorable process.[28]

$$\wedge\wedge + CO \xrightarrow[RhCl(CO)(PMe_3)_2]{h\nu} \wedge\wedge\wedge CHO , \overset{CHO}{\wedge\wedge} \quad (17)$$

As outlined, the general approach to alkane C-H activation by oxidative addition commonly involves a low-valent metal with good donor ligands [hydride, C_5Me_5, P(alkyl)$_3$]. Both photochemical and thermal activation of alkanes can be accomplished. The next two sections discuss the selectivities of alkane and arene activation as revealed by the systems previously described. Insights into the mechanisms of C-H bond activation will then be discussed citing experiments with these molecules.

II. Selectivities in C-H Bond Activation

In examining the competitive reactions of reactive intermediates with C-H bonds, there are two distinct types of selectivity that should be considered. The *kinetic selectivity* refers to the C-H bonds that are the first to be broken when a metal species encounters a hydrocarbon. The *thermodynamic selectivity* refers to the relative free energies of the oxidative addition products under reaction conditions, and will determine the product distribution once equilibrium has been obtained.

This distinction is important in that it is necessary to be careful in interpreting the results of a specific experiment in terms of whether the product distribution reflects the kinetic or thermodynamic selectivity for the reaction. Generally, a product distribution can be shown to be kinetic only if the observed product ratios change over time to give the thermodynamic mixture. It is not possible to distinguish kinetic and thermodynamic selectivity by simply observing a fixed distribution of products. In the following discussion these two types of selectivity will be carefully distinguished, as they are found to be different in many cases. In addition, when examining the product distribution of a competitive reaction there is selectivity not only between the activation of C-H bonds in two separate molecules, but also between the C-H bonds within a single molecule. This distinction will be important in discussing the mechanism of C-H bond activation in the successive section.

1. Arene versus Arene Selectivity

Arene-arene selectivities have been examined in many systems, since arene activation was established over 20 years ago. The best studied examples include Cp_2MH_3 (M = Nb, Ta), $Ir(PMe_3)_2H_5$, Cp_2WH_2, $Fe(dmpe)_2(naphthyl)H$, and $(C_5Me_5)Rh(PMe_3)(Ph)H$, and will be presented here briefly, citing kinetic selectivity studies first and then thermodynamic studies. The complex Cp_2NbH_3 catalyzes the exchange of deuterium in aromatic C-H bonds by way of a reductive elimination/oxidative addition pathway as shown in Scheme 2.

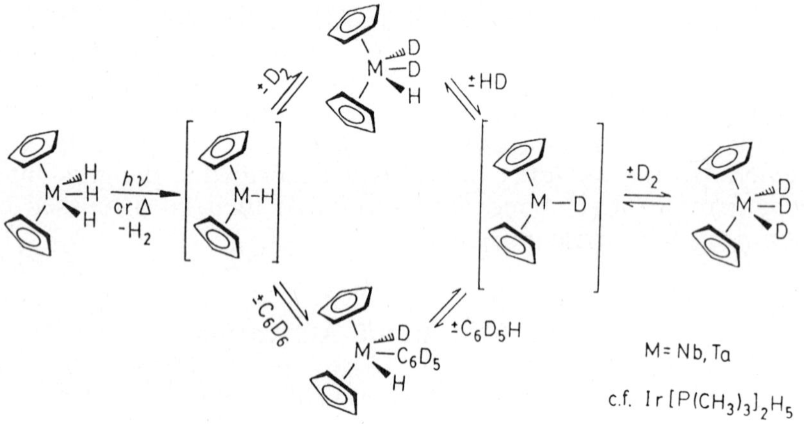

Scheme 2

The rate of exchange of deuterium into various arenes was measured by monitoring the incorporation of D_2 with time. Since the initial products of the exchange were monitored (35% exchange), the selectivities obtained represent the kinetic selectivity of the intermediate $[Cp_2NbH]$

for various arenes. The total relative amount of deuterium incorporation was found to follow the order p-$C_6H_4F_2$ > C_6H_5F > C_6H_6 > $C_6H_5OMe \cong C_6H_5Me$ > p-$C_6H_4Me_2$. The selectivity trend follows the electron deficiency of the arene, and is consistent with activation by a nucleophilic metal complex. The analogous tantalum complex showed an even smaller range of selectivity with the same arenes, as does the deuterium exchange catalyst $Ir(PMe_3)_2H_5$. Table I shows the relative rates of arene deuteration for these three complexes relative to the rate of exchange of C_6D_6. Overall, the kinetic selectivities are quite small.[28]

Table I. Arene-Arene Selectivities by Metal Complexes for D_2 Exchange

Catalyst	p-$C_6H_4F_2$	C_6H_5F	C_6H_6	C_6H_5OMe	C_6H_5Me	p-$C_6H_4Me_2$	Exchange (%)
Cp_2NbH_3	4.2	2.1	1.0	0.8	0.8	0.04	35
Cp_2TaH_3	1.1	1.0	1.0	0.8	0.9	0.3	51
$Ir(PMe_3)_2H_5$	--	4.0	1.0	0.9	0.6	--	60

The relative kinetic selectivity for the the *ortho*, *meta*, and *para* positions of the aromatic ring as a function of the substituent was also determined for the Cp_2NbH_3 catalyst (Fig. 2). The electronic directing effects of the substituents is small, with *meta* and *para* exchange occurring much faster than *ortho*. *Ortho* exchange is slow presumably on steric grounds, and accounts for the meager reactivity of *p*-xylene as well as the rapid ortho exchange in fluorobenzene.[28]

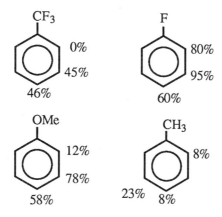

Figure 2.

The reactions of Cp_2WH_2 and $Cp_2W(CH_3)H$ with substituted arenes gives information about the kinetic selectivity of the reactive intermediate [Cp$_2$W] since the reactions were carried out at temperatures below that at which the arene in the complex Cp$_2$W(aryl)H can undergo reductive elimination. However, the photochemical reactions are known to produce secondary photolysis products, the formation of which might alter the primary product ratios. In addition, it was not reported whether the observed product ratios change on heating to the temperature at which the arene adds reversibly.

Irradiation of Cp$_2$WH$_2$ in toluene solvent proceeds with loss of H$_2$ to give 4 products. The *meta-* and *para-*tolyl hydride complexes Cp$_2$W(tolyl)H were formed in 60% yield in a 1:10 ratio, with no *ortho* activation observed. The secondary products Cp$_2$W(CH$_2$Ph)(*p*-tolyl) and Cp$_2$W(CH$_2$Ph)(*m*-tolyl) were formed in 40% yield in a 3:1 ratio. The preference for *meta* and *para* activation is again seen here.[29]

Irradiation of Cp$_2$WH$_2$ in fluorobenzene gives a 2:3 ratio of *meta-* to *para-*Cp$_2$W(C$_6$H$_4$F)H products, again with no *ortho* activation.[30] The same product ratio is obtained by thermolysis of Cp$_2$W(CH$_3$)H in fluorobenzene at 70°C, confirming that this is a kinetic product distribution.[31] Similarly, methylbenzoate gives a 3:1 ratio of *para-* to *meta-*aromatic activation products. Irradiation of Cp$_2$WH$_2$ in anisole or *o*-xylene gives only a single product in each case, Cp$_2$W(*p*-C$_6$H$_4$OMe)H or Cp$_2$W(3,4-C$_6$H$_3$Me$_2$)H, respectively.[30] The kinetic selectivity for the positional activation of toluene by (C$_5$Me$_5$)Rh(PMe$_3$) was determined by irradiation of the dihydride complex in toluene solvent at -45°C. Since interchange of the tolyl hydrido positional isomers occurs at -10°C in this complex (see Sec. III), the product mixture was quenched with bromoform in order to convert the hydrido complexes into the stable bromo derivatives (Scheme 3).

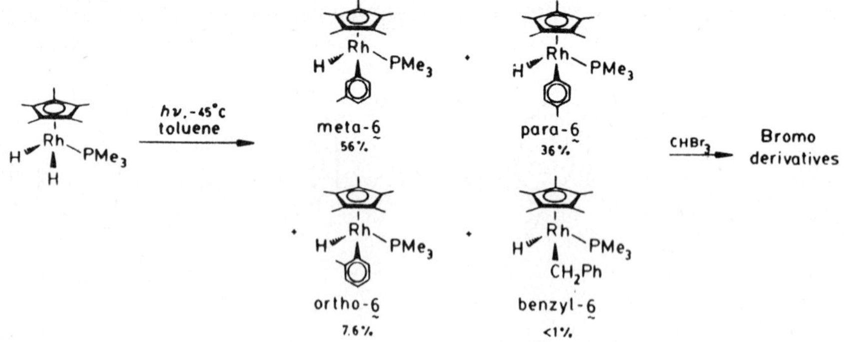

Scheme 3

The ratio of bromo isomers was taken as representing the kinetic selectivity of the 16-electron rhodium intermediate toward the four types of toluene C-H bonds.[32] *Meta* activation was observed to be about twice as favorable as *para* activation. However, ~8% *ortho* activation could also be detected and a trace of the benzylic product was observed.

The small amount of the latter species was somewhat surprising considering the weakness of the benzylic C-H bond and the fact that $(C_5Me_5)Rh(PMe_3)$ can activate both alkane and arene C-H bonds with comparable ease. This interpretation is subject to the assumption that the photolysis or bromoform quench did not alter the primary reaction product ratios, however, and may not represent the true kinetic selectivity for the benzylic C-H bond.

These experiments give information about the kinetic product distributions in arene activation. Studies with $Fe(dmpe)_2(naphthyl)H$ by Tolman et al.[33] give thermodynamic product distributions, since the reactive species $Fe(dmpe)_2$ is generated by the thermally induced reductive elimination of naphthalene in C_6D_6 solvent containing an arene. The reactions are complicated in that mixtures of the *cis*- and *trans*-$Fe(dmpe)_2(aryl)H$ complexes are present, but ^{31}P-NMR spectroscopy was used to assign the product distributions reliably.

Reaction of $Fe(dmpe)_2(aryl)H$ in fluorobenzene gives only the *ortho* activation product, $Fe(dmpe)_2(o\text{-}C_6H_4F)H$. Trifluoromethylbenzene gives *para*- and *meta*-$Fe(dmpe)_2(C_6H_4CF_3)H$ in a 1:4 ratio. *Ortho*-bis(trifluoromethyl)benzene gives only $Fe(dmpe)_2[3,4\text{-}C_6H_3(CF_3)_2]H$. All of the above complexes form quantitatively in C_6D_6 solvent, indicating a strong preference for electron-deficient arenes.[33]

With toluene, only the *meta*- and *para*-isomers $Fe(dmpe)_2(tolyl)H$ are observed in a 0.8:1.0 ratio. The equilibrium constant for formation of the tolyl hydrido complexes in C_6D_6 was 0.17, indicating a destabilizing effect of methyl substitution. Acetophenone gave 9% *para*, 52% *meta*, and 29% *methyl* activation adducts, with 10% $Fe(dmpe)_2(C_6D_5)H$ also being present at equilibrium. Benzonitrile gave 50% *ortho*, 36% *meta*, and 14% *para* activation. Pyridine showed 56% *ortho*, 34% *meta*, and 10% *para* activation.[33]

As with the kinetic activation studies with Cp_2NbH_3 and $Ir(PMe_3)_2H_5$, the activation of *meta*- and *para*-aryl C-H bonds is preferred. Exceptions to this trend occur when steric interactions do not destabilize the formation of *ortho*-aryl complexes, such as with fluorobenzene, benzonitrile, or pyridine.

The last complex to be mentioned with regard to arene-arene selectivity are derivatives of the type $(C_5Me_5)Rh(PMe_3)(aryl)H$ studied by Jones and Feher. Thermolysis of the complexes at 60°C in arene solvents permitted equilibration of both the positional isomers as well as between arenes.[32] As with the $Fe(dmpe)_2(aryl)H$ complexes, these experiments give thermodynamic product distributions since the arene is labile under the reaction conditions.

The equilibrium constants for equilibration of several substituted arenes with $(C_5Me_5)Rh(PMe_3)(C_6H_5)H$ are given in Table II. Once again, it can be seen that electron-donating substituents decrease the stability of the aryl hydrido complex, and that *ortho* activation is strongly disfavored with any substituents other than those with the least steric demand.[32]

Table II.   Equilibrium Constants at 60°C for the Arene Exchange Reaction:
$(C_5Me_5)Rh(PMe_3)(C_6H_5)H + ArH \rightleftharpoons (C_5Me_5)Rh(PMe_3)(Ar)H + C_6H_6$

ArH	K_{eq}	Positional distribution
Toluene	0.37	2:1, *para:meta*
o-Xylene	0.13	100% $3,4-C_6H_3Me_2$
m-Xylene	0.083	100% $3,5-C_6H_3Me_2$
p-Xylene	0.017	100% $2,5-C_6H_3Me_2$

2.   Alkane versus Alkane Selectivity

The $(C_5Me_5)ML_n$ species with M = Ir, Rh, or Re and L = PPh_3, PMe_3, or CO provide information about the selectivity of a metal for different types of alkane C-H bonds in an intermolecular reaction. Since the iridium complexes form alkyl hydride complexes that are stable to >100°C, irradiation of $(C_5Me_5)Ir(PMe_3)H_2$ or $(C_5Me_5)Ir(CO)_2$ in alkane solvent mixtures gives information about the kinetic selectivities of the intermediates $(C_5Me_5)Ir(PMe_3)$ and $(C_5Me_5)Ir(CO)$, respectively. As in all photochemical reactions, these observed selectivities are also subject to the assumption that the photolysis conditions do not appreciably alter the primary product compositions. The products were examined at low percentage conversions to minimize secondary photolysis reactions.

Bergman has examined a wide range of alkane substrates with the iridium and rhodium PMe_3 complexes. Irradiation of $(C_5Me_5)M(PMe_3)H_2$ in various alkane solvent mixtures (the rhodium studies being done at low temperature to ensure stability of the products) allows the generation of the alkyl hydride product mixtures shown in Table III. This table also shows the relative reactivity on a "per hydrogen" basis for a comparison of the intrinsic reactivity of various types of C-H bonds. Cyclohexane was taken as having a relative kinetic reactivity of one.[2,32,34]

The relative thermodynamic stability of the alkyl hydrido complexes was established by two independent methods. The first involves heating complexes of the type $(C_5Me_5)Ir(PMe_3)(R)H$ in a mixture of alkane solvents. The equilibrium constants obtained and the corresponding $\Delta G°$ values corresponding to these values are summarized in Table IV. A second method involves a combination of reaction calorimetry and pulsed laser photoacoustic calorimetry, and provides absolute Ir-X bond strengths. Primary alkyl products are formed exclusively over secondary products under equilibrium

Table III. Kinetic Selectivities of $(C_5Me_5)M(PMe_3)$ (M = Ir, Rh) for C-H Bonds[a]

Alkane	$k_{relative}$		k_{per} hydrogen	
	Rh	Ir	Rh	Ir
Cyclohexane	1.0	1.0	1.0	1.0
Benzene	9.8	1.8	19.5	3.5
Cyclopropane	5.2	1.1	10.4	2.1
n-Hexane (1°)	3.0	1.4	5.9	2.7
n-Hexane (2°)	0.0	0.13	0.0	0.2
Propane (1°)	1.3	0.75	2.6	1.5
Propane (2°)	0.0	0.05	0.0	0.3
Cyclopentane	1.5	0.92	1.8	1.1
Cyclooctane[b]	-	0.12	-	0.09
Cyclodecane[b]	-	0.38	-	0.23
Neopentane[b]	-	1.14	-	1.14

[a] -60°C.
[b] 0 - 10°C.

conditions, establishing the greater thermodynamic stability of primary alkyl compounds. Similarly, formation of the phenyl hydrido complex from any of the alkyl hydrido complexes is irreversible, indicating its even stronger thermodynamic stability.[35] The calorimetric experiments support these conclusions, as indicated by the bond strengths for $(C_5Me_5)Ir(PMe_3)X_2$ (X = H, Cl, Br, I, Ph, c-hexyl) given in Table V. The thermodynamic stability of the phenyl hydride can be attributed to the high Ir-Ph bond strength of 80.6 kcal/mol, which is 6 kcal/mol stronger than the Ir-H bond, and 30 kcal/mol stronger than the Ir-C_6H_{11} bond. The photoacoustic experiments (based upon the reaction of *tert*-BuO$^•$ with $(C_5Me_5)Ir(PMe_3)H_2$) provide an Ir-H bond strength of 72.9 ± 4.3 kcal/mol, in excellent agreement with the reaction calorimetric results.[36]

Several general trends can be noted from the above observations on the $(C_5Me_5)M(PMe_3)$ systems. First, there is both a kinetic and thermodynamic preference for the activation of primary over secondary C-H bonds. For the iridium system, the kinetic selectivity is small whereas a large thermodynamic difference is found. Rhodium shows a higher kinetic selectivity. The lower thermodynamic stability of secondary alkyl systems was employed in using the iridium cyclohexyl complex to activate the C-H bonds of methane under conditions of thermodynamic control. The large effect of α-substitution on carbon on the Ir-C bond strength can probably be attributed to steric interactions of the β-substituents with the metal, as discovered by Halpern and co-workers in a series of cobalt macrocycle complexes (Table VI).[37]

Table IV. Thermodynamic Selectivity of $(C_5Me_5)Ir(PMe_3)$ for C-H Bonds for the Equilibria: $(C_5Me_5)Ir(PMe_3)(R_1)H + R_2H \rightleftharpoons (C_5Me_5)Ir(PMe_3)(R_2)H + R_1H$

R_1	R_2	K_{eq} (140°C)	$\Delta G°$ (140°C) (kcal/mol)
c-Hexyl	n-Pentyl	10.8	-2.0
-CH$_2$CHMeCHMe$_2$	n-Pentyl	3.5	-1.0
c-Pentyl	-CH$_2$CHMeCHMe$_2$	1.5	-0.3
c-Hexyl	c-Pentyl	2.0	-0.6
-CH$_2$CMe$_2$CH$_2$CH$_3$	-CH$_2$CH$_2$CMe$_3$	>20	<-2.5

Table V. Bond Strength Estimates (kcal/mol) for $(C_5Me_5)Ir(PMe_3)X_2$ from Reaction Calorimetry

X	Bond strength
H	74.2
Cl	90.3
Br	76.0
I	63.8
Ph	80.6
c-C$_6$H$_{11}$	50.8

Table VI. Cobalt-Alkyl Bond Energies in the Compounds R-Co(Saloph)(pyr)

R	CH$_2$CH$_2$CH$_3$	CH(CH$_3$)$_2$	CH$_2$C(CH$_3$)$_3$	CH$_2$C$_6$H$_5$
D$_{Co-R}$ (kcal/mol)	25	20	18	22

Second, there is an *inverse* correlation of the C-H bond strength and the reactivity of the C-H bond. The strongest C-H bonds appear to be the most reactive, according to the trend shown below. This correlation is just the opposite of what might have been predicted based upon C-H bond breaking alone. In fact, it is the M-C bond strength that controls the selectivity. The trend of M-C bond strengths parallels that of H-C bond strengths, except that the differences are greater, leading to the apparent inverse correlation.

Bond strengths:

H-Ph > H-vinyl > H-CH$_3$ > H-CH$_2$R > H-CHR$_2$ > H-CR$_3$ > H-CH$_2$Ph;　M-Ph >> M-vinyl >> M-CH$_3$ >> M-CH$_2$R >> M-CHR$_2$ >> M-CR$_3$ >>M-CH$_2$Ph

This trend can be accounted for on comparison of the steric effects in the X-C bond. With X = H, the bonding of the small hydrogen 1s orbital would be expected to be far less sensitive to the geometry and substitution at carbon than with X = metal. In the latter case, the larger, more diffuse d-orbital used in bonding would enhance the relative importance of steric destabilization of the X-C bond. The aryl and vinyl groups, with substituents on the bound carbon 120° from the metal, form the strongest bonds since they are sterically unencumbered.

Third, the rhodium system shows a greater thermodynamic selectivity for the less hindered bonds than does iridium. Secondary C-H bond activation is observed only in reactions with solvent molecules that offer no choice other than secondary activation. That is, substrates such as propane or butane show exclusively primary activation products, yet competitive studies of cyclopentane vs benzene or benzene vs propane show that activation of all of the bonds (except the secondary propane bonds) is competitive. It is possible that kinetic activation of propane by rhodium gives both primary and secondary products, but that the secondary products rearrange under the reaction conditions. This possibility will be discussed more fully when the mechanism of activation is addressed in Sec. III.

The studies with CpRe complexes provide further information about alkane selectivity. The species CpRe(PMe$_3$)$_2$ and (C$_5$Me$_5$)M(PMe$_3$)(CO) were observed to give oxidative addition products in which only primary C-H bonds of linear alkanes or the secondary C-H bonds of cyclopropane had been activated. As with the rhodium complex, it is possible that secondary activation of alkanes occurs, but that the secondary hydrido alkyl complexes undergo reductive elimination under the reaction conditions giving way to the more stable primary adducts. Benzene was found to undergo irreversible oxidative addition, indicating its greater thermodynamic stability just as in the iridium complexes. Ethylene is also activated to give a vinyl hydride adduct.[38]

Another system that gives information about alkane selectivity is the H-D exchange catalyzed by CpRe(PPh$_3$)$_2$H$_2$. Irradiation of this complex in C$_6$D$_6$ solvent containing two alkanes allows the determination of the *kinetic* preference of the intermediate for the two alkanes, as indicated earlier in Eq. 10. The results of the competitive exchange experiments are summarized in Table VII and show a remarkable similarity to the kinetic selectivities seen with rhodium and iridium.

The competition between benzene and propane in THF-d$_8$ solvent shows an 8:1 preference for benzene activation. This result implies that the alkane-arene selectivity is low and that the actual

Table VII. Competitive Kinetic Selectivity of CpRe(PPh$_3$)$_2$H$_2$ on Irradiation in C$_6$D$_6$

RH/R'H	Products	Ratio
CH$_4$/C$_2$H$_6$	CH$_3$D/C$_2$H$_5$D	2:1
C$_3$H$_8$	CH$_2$DCH$_2$CH$_3$/CH$_3$CHDCH$_3$	20:1
THF	α-d$_1$-THF/β-d$_1$-THF	1.7:1
C$_6$H$_6$/C$_3$H$_8$^{a}	C$_6$H$_5$D/C$_3$H$_7$D	8:1

^{a}THF-d$_8$ solvent employed as deuterium source.

turnover numbers for C-H bond activation by CpRe(PPh$_3$)H$_2$ is quite large (~10^4), but that most of the activations result in degenerate exchange with the solvent.[20]

3. Alkane versus Arene Selectivity

The kinetic and thermodynamic selectivities for arene vs alkane C-H bonds have been examined with several of the complexes described earlier in the chapter. The early observations of arene but not alkane activation led to the widespread belief that alkane oxidative addition was thermodynamically uphill. The observations of spontaneous reductive elimination of alkane from alkyl hydride complexes synthesized at low temperatures supported this notion of unfavorable alkane activation. The work with the complexes (C$_5$Me$_5$)M(PMe$_3$) and those that followed show that these beliefs were only partially correct.

The kinetic selectivity for benzene vs cyclohexane by (C$_5$Me$_5$)Ir(PMe$_3$)H$_2$ was determined by irradiating the complex in a mixture of the two hydrocarbons at room temperature. Under these conditions, the kinetic product distribution showed a 4:1 preference for aromatic C-H bond activation, in spite of the stronger C-H bond energy (110 kcal/mol for benzene vs ~100 kcal/mol for cyclohexane). On heating this solution to the temperature at which the cyclohexane underwent reductive elimination (110°C), the mixture converted exclusively to the hydrido phenyl complex. Thermolysis of the hydrido phenyl complex up to 200°C did not result in the elimination of benzene.[2] These experiments show that the activation of alkane C-H bonds can be kinetically competitive with arene activation, but that the latter is strongly preferred thermodynamically.

Bergman also reported the kinetic selectivity of the iridium species toward *p*-xylene providing a striking example of the inverse correlation of bond strength vs reactivity. A 3.7:1 preference (statistically corrected for the number of available hydrogens) for aromatic over benzylic activation was observed, in spite of a >20 kcal/mol C-H bond energy difference![2]

Graham found a similar lack of kinetic selectivity with the complex $(C_5Me_5)Ir(CO)$. Irradiation of $(C_5Me_5)Ir(CO)_2$ in a benzene-cyclohexane solvent mixture gave a 2.5:1 ratio of the phenyl to cyclohexyl product. Irradiation in a mixture of benzene-neopentane showed a 2.0:1 ratio of the phenyl to neopentyl product. The thermodynamic selectivity of the iridium carbonyl complex has not yet been reported, but a strong preference for the hydrido aryl complex is to be expected.[8]

The competition of the rhodium species $(C_5Me_5)Rh(PMe_3)$ toward a benzene-alkane mixture also shows competitive activation of both types of C-H bonds. When the reaction is carried out at low temperature, a slight kinetic preference for aromatic activation is seen. On warming this solution above -20°C, reductive elimination of alkane was observed, resulting in the exclusive formation of the hydrido phenyl complex.[32]

With a mixture of *p*-xylene and benzene solvent at -10°C, the rhodium intermediate $(C_5Me_5)Rh(PMe_3)$ produces a 6:1 ratio of benzene to xylene aromatic activation[1]. At this temperature, however, any benzylic activation product would be expected to be unstable, so that this result provides information only about the aromatic kinetic selectivity. As previously mentioned and shown in Scheme 3, toluene showed a marked kinetic preference for aromatic activation.[32]

Few examples of alkane functionalization have been reported based upon the above reactions. The alkyl hydride products in the $(C_5Me_5)Ir(PMe_3)$ reactions were found to undergo reductive elimination of alkane upon treatment with bromine, $ZnBr_2$, or hydrogen peroxide. Reaction with bromoform did give the alkyl bromide complexes $(C_5Me_5)Ir(PMe_3)RBr$ in high yield, which could then be cleaved with $HgCl_2$ and bromine to give functionalized alkyl bromides (Eq. 18). The rhodium analog reacts directly with bromine to give alkyl bromide.[2]

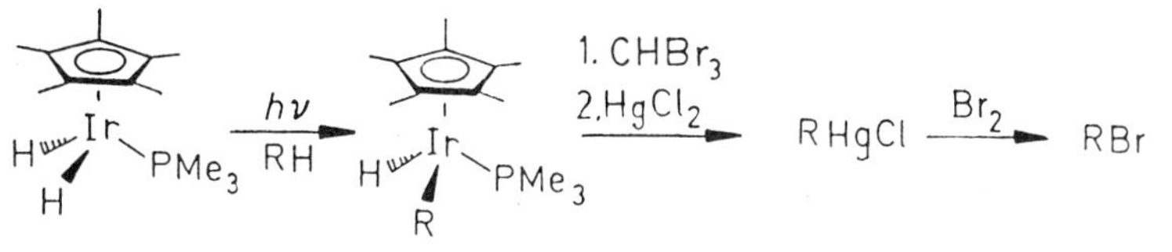

III.   The Mechanism of C-H Bond Activation

1.   Arene Activation-η^2-Arene Complexes

The mechanism of hydrocarbon C-H bond activation by nucleophilic metal complexes has been studied in some detail for several of the complexes that give observable oxidative addition adducts.   One key difference between arenes and alkanes that has been suggested as the most likely explanation for the difference in their reactivities with transition metals is the presence of a π-electron system in the arenes. These π-electrons lead to enhanced reactivity with electrophilic reagents, and it was thought likely that the empty π-orbitals could lead to enhanced reactivity with nucleophilic reagents.

Chatt and Davidson in 1965 first suggested the interaction of the double bond of an arene with a metal center to explain the substitution behavior of Ru(dmpe)$_2$(naphthyl)H.   In explaining substitution reactions in which naphthalene was displaced from the complex by added ligands he proposed an equilibrium with the ruthenium(0) η^2-naphthalene complex, and that substitution occurred by way of ligand displacement in this Ru(0) species (Eq. 19).[39]

$$(19)$$

The η^2-arene structure soon found precedent in the X-ray structures of structurally characterizable η^2-arene complexes with a number of arenes.[40]   The more conventional complexes Ni[P(c-hexyl)$_3$]$_2$(η^2-anthracene)[41] and [CpRe(CO)(NO)(η^2-C$_6$H$_6$)]$^+$ [42] also are found to have dihapto arene ligands.

Klabunde and Parshall suggested this structural form as an intermediate in the C-H activation reactions of Cp$_2$TaH$_3$, Ir(PMe$_3$)$_2$H$_5$, and CpRh(C$_2$H$_4$)$_2$.   The equilibrium constants for η^2-arene complexation (K_1) and oxidative addition (K_2) were believed to compensate for one another, such that the product $K_1 \cdot K_2$ is approximately constant.   In this way the net effect of an electron-withdrawing substituent on an arene would increase K_1, but decrease K_2, resulting in an indifference of the overall activation reaction to substituent effects.[43]   Werner and Werner also postulated the presence of an η^2-arene species as an intermediate in the conversion of (η^6-C$_6$H$_6$)Os(C$_2$H$_4$)(PMe$_3$) to Os(PMe$_3$)$_4$(Ph)H.[44]

Studies of the compounds $(C_5Me_5)Rh(PMe_3)(aryl)H$ provided the first experimental evidence for this type of intermediate in C-H bond activation. These complexes can be synthesized by treatment of the chloro derivatives $(C_5Me_5)Rh(PMe_3)(aryl)Cl$ with $LiHBEt_3$ at room temperature. Surprisingly, both $(C_5Me_5)Rh(PMe_3)(p$-tolyl$)H$ and $(C_5Me_5)Rh(PMe_3)(m$-tolyl$)H$ were obtained in a 1:2 ratio when the *p*-tolyl chloride complex was reduced. When the hydride reagent was added to the more reactive THF complex $[(C_5Me_5)Rh(PMe_3)(p$-tolyl$)(THF)]^+$ at low temperature, the initially formed $(C_5Me_5)Rh(PMe_3)(p$-tolyl$)H$ was observed to slowly isomerize to the *m*-tolyl hydride complex at -10°C. Since temperatures of 60°C were required to exchange the toluene intermolecularly with arene solvent, *the rearrangement must have occurred intramolecularly, presumably by way of an* η^2*-toluene complex* (Scheme 4).[12]

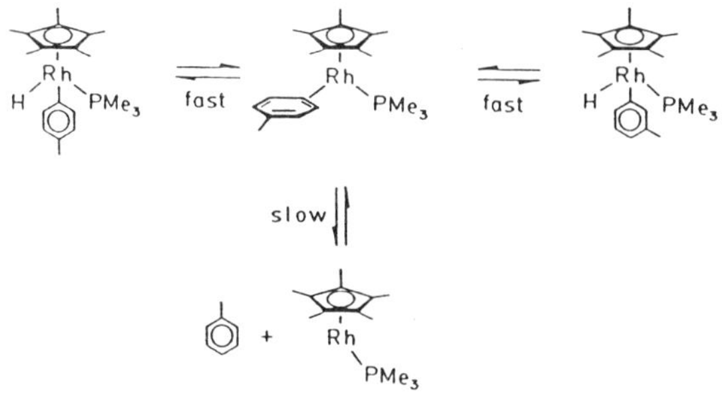

Scheme 4

Examination of $(C_5Me_5)Rh(PMe_3)(C_6D_5)H$ provided more details about the mechanism of the isomerization. The complex was prepared at low temperature with deuterium only in the aromatic positions of the molecule. At -10°C, the hydride was observed to migrate first to the *ortho*-phenyl position, then to the *meta*-phenyl position, and finally to the *para*-phenyl position. The mechanism proposed for this intramolecular isomerism involved a facile interconversion between the hydrido phenyl complex and the less stable η^2-benzene complex (Scheme 5), which involves an overall [1,2] shift of the carbon attached to the metal. Spin saturation transfer between hydrogen atoms in the aromatic hydrogen sites in this complex at 35°C provides additional evidence for the interchange of the bound aromatic carbon about once every 2-3 sec.[32]

Similarly, the complex $(C_5Me_5)Rh(PMe_3)(2$-$CH_3C_6H_3$-5-$CD_3)H$ was observed to isomerize to the isomer $(C_5Me_5)Rh(PMe_3)(2$-$CD_3C_6H_3$-5-$CH_3)H$. The methyl groups in the protio species $(C_5Me_5)Rh(PMe_3)(2,5$-$C_6H_3Me_2)H$ were found to interconvert rapidly enough to permit spin saturation studies between the two distinct methyl resonances at room temperature.[32]

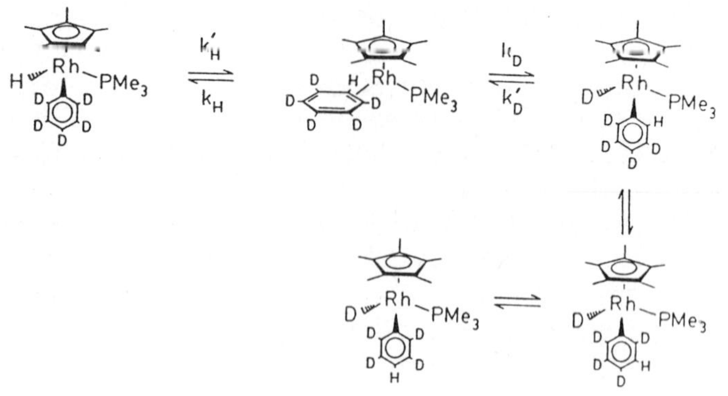

Scheme 5

The barrier to interconversion of isomers, for which the rate-determining step must be reductive elimination to directly form the η^2-arene complex, was found to be ~20 kcal/mol in these molecules. Activation parameters of $\Delta H^{\ddagger} = 16.3 \pm 0.2$ kcal/mol and $\Delta S^{\ddagger} = -6.3 \pm 0.8$ e.u. were obtained from the temperature dependence of the rate constant for interconversion of the *p*-xylyl isomers. The small negative value of ΔS was attributed to the restricted nature of the transition state for the intramolecular reductive elimination reaction.

These activation parameters stand in contrast to those for arene reductive elimination and dissociation determined by thermolysis of the hydrido phenyl complex in C_6D_6 solvent to give the perdeutero phenyl deuteride complex. By measuring the rate of this reaction as a function of temperature, the activation parameters $\Delta H^{\ddagger} = 30.5 \pm 0.8$ kcal/mol and $\Delta S^{\ddagger} = 14.9 \pm 2.5$ e.u. were obtained. The overall positive entropy of activation is expected for the formation of two particles from one in the rate determining transition state of the reaction.

With the arene 1,4-di-*tert*-butylbenzene oxidative addition to the aromatic C-H bond is sterically unfavorable, allowing the η^2-arene complex to be observed directly. The complex was generated by loss of alkane from $(C_5Me_5)Rh(PMe_3)(CD_2C_6D_{11})D$ at -30°C in the presence of the arene. The adduct was stable only below 0°C (Eq. 20). X-ray

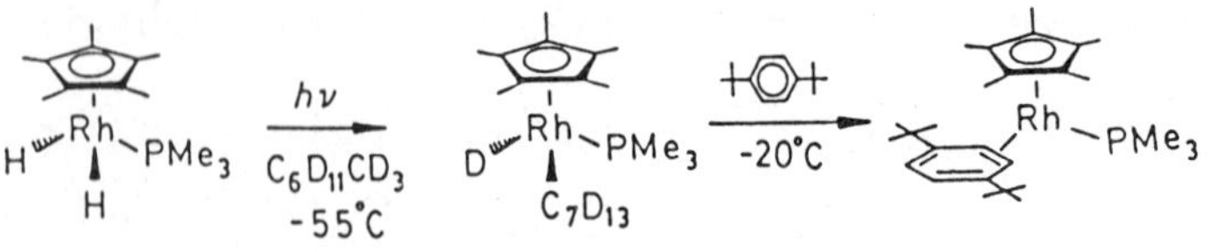

(20)

characterization of the aryl hydride complex (C_5Me_5) $Rh(PMe_3)(3,5-C_6H_3Me_2)H$ showed no evidence of an η^2 interaction.[45]

The importance of the η^2-arene intermediate becomes apparent on comparison with alkane activation. For the $(C_5Me_5)Rh(PMe_3)$ system, reductive elimination of propane from the hydrido propyl complex occurs at -17°C with a barrier of 18.6 kcal/mol. The barrier for reductive elimination and dissociation of benzene at this temperature is 26.7 kcal/mol. As mentioned earlier, the selectivity of the intermediate $(C_5Me_5)Rh(PMe_3)$ between propane and benzene was found to be quite small (0.6 kcal/mol). When these three barriers are combined on a free energy diagram, the reaction coordinate shown in Fig. 3 arises.

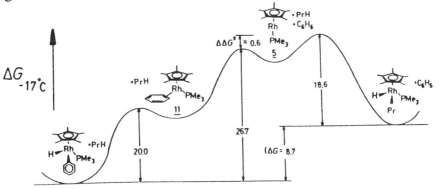

Figure 3.

Several conclusions about the activation of C-H bonds by rhodium can be made based upon this diagram. First, and most obvious, the phenyl hydride complex is thermodynamically preferred over the propyl hydride complex by about 9 kcal/mol. Second, the barrier (at -17°C) to reductive elimination of benzene (20 kcal/mol) is almost identical to that for reductive elimination of alkane (19 kcal/mol). The greater macroscopic stability of the aryl hydride complex can be attributed not to a slower rate of reductive elimination, but instead to the fact that the η^2-benzene complex that is formed undergoes oxidative addition to reform the hydrido phenyl complex much more rapidly than the η^2-benzene can dissociate. Third, in examining the rate-determining transition states for arene and alkane activation it is seen that the highest barrier along the arene activation reaction coordinate involves coordination of the metal to an isolated arene C=C bond, but does not cleave the C-H bond. Therefore, the ability of arenes to compete kinetically with alkanes has nothing to do with the strength of the arene C-H bond. In addition, the observed kinetic preference of the metal for electron-deficient arenes is accounted for in terms of the stronger π-complexes that would form with the low valent metal.

Finally, the difference in the C-H bond strengths of benzene and propane allow the estimation of the difference in the rhodium-phenyl and rhodium-propyl C-H bonds. The equilibrium between the hydrido phenyl complex and the hydrido propyl complex (Eq. 21) can be considered in terms of two separate activation reactions (Eqs. 22 and 23).

$$(C_5Me_5)Rh(PMe_3)(Ph)H + C_3H_8 \rightleftharpoons$$
$$(C_5Me_5)Rh(PMe_3)(propyl)H + PhH \qquad (21)$$

$$Ph-H + [(C_5Me_5)Rh(PMe_3)] \longrightarrow (C_5Me_5)Rh(PMe_3)(Ph)H \qquad (22)$$

$$C_3H_8 + [(C_5Me_5)Rh(PMe_3)] \longrightarrow (C_5Me_5)Rh(PMe_3)(propyl)H \qquad (23)$$

The ΔH for Reaction 21 (ΔH_{21}) can be approximated by the above calculated ΔG for the reaction ($\Delta G_{21} = 9$ kcal/mol), since the entropy change ΔS_{21} is probably small for this metathesis-type reaction. ΔH_{21} in turn is equal to $\Delta H_{23} - \Delta H_{22}$, which can be rewritten as the enthalpy difference of the C-H bonds and Rh-C bonds as indicated in Eqs. 24-26.

$$\Delta H_{21} = \Delta H_{23} - \Delta H_{22} = \Delta H(\text{Rh-propyl}) - \Delta H(\text{Rh-phenyl}) -$$
$$[D(\text{H-phenyl}) - D(\text{H-propyl})] \qquad (24)$$
or
$$\Delta H(\text{Rh-phenyl}) - \Delta H(\text{Rh-propyl}) = \Delta H_{21} + [D(\text{H-phenyl}) -$$
$$D(\text{H-propyl})] \qquad (25)$$
or
$$\Delta H(\text{Rh-phenyl}) - \Delta H(\text{Rh-propyl}) \cong 9 + 110 - 102 = 17 \text{ kcal/mol}$$
$$(26)$$

In other words, the rhodium-phenyl bond is approximately 17 kcal/mol stronger than the rhodium-propyl bond!

The large difference between C-H bond strengths and their M-C counterparts could account for the many examples of arene activation. In cases in which alkane activation is marginally favorable thermodynamically, arene activation will probably be quite favorable. This is certainly the case with the intermediate $(C_5Me_5)Rh(PMe_3)$. The alkane oxidative addition complexes were visible only at low temperature. Were the same experiments carried out at room temperature, there would have been no observations of alkane C-H oxidative addition.

Other evidence for η^2-arene intermediates can be found in the literature. Werner and Gotzig have observed a 1:1.75 kinetic distribution of *meta* : *para* tolyl isomers of $Ru(PMe_3)_2[P(OMe)_3]_2(tolyl)H$ upon reduction of the dichloride

$Ru(PMe_3)_2[P(OMe)_3]_2Cl_2$ in toluene. Furthermore, the kinetic distribution of isomers changed to a 1:1 thermodynamic distribution by an intramolecular process over several days in benzene solution. It is likely that the mechanism of this interconversion proceeds by way of an η^2-toluene complex as indicated in Eq. 27.[46]

$$(27)$$

The fact that the rhodium complexes activate arenes by way of an η^2-arene complex does not necessarily mean that all metals activate arenes by way of such a complex. One exception is the *cis*-Pt(PPh$_3$)$_2$(Ph)H complex, which loses benzene at -30°C. When the complex *cis*-Pt[P(C$_6$D$_5$)$_3$]$_2$(C$_6$D$_5$)H was prepared at -50°C and then allowed to warm to -30°C, only a single new resonance for C$_6$D$_5$H was observed to grow in at the expense of the hydride resonance in the ^{1}H-NMR spectrum. Were a reversible equilibrium between the d$_5$-phenyl hydride complex and the η^2-C$_6$D$_5$H complex occurring prior to dissociation of the arene, then three new resonances for the compounds with hydrogen in the *ortho, meta, and para* positions in the complex Pt[P(C$_6$D$_5$)$_3$]$_2$(C$_6$D$_4$H)D should have been observed as the sample was warmed.[47]

In comparison, Whitesides and co-workers have prepared Pt[PCy$_2$CH$_2$CH$_2$PCy$_2$](C$_6$H$_5$)D and observed that the phenyl ring hydrogen migrates into the bound hydride prior to elimination of benzene.[48] This observation is similar to that found with the (C$_5$Me$_5$)Rh complex and undoubtedly proceeds by way of an η^2-arene intermediate. In a related study, Periana and Bergman have reported the rapid rearrangement of a cyclopropyl hydride complex to a metallocyclobutane by an intramolecular pathway. In that cyclopropanes have substantial sp^2 character in their bonds, it is appropriate to mention these studies now. The complex (C$_5$Me$_5$)Rh(PMe$_3$)(c-propyl)H was prepared by either irradiation of (C$_5$Me$_5$)Rh(PMe$_3$)H$_2$ in liquid cyclopropane at -60°C or by warming a cyclopropane solution of (C$_5$Me$_5$)Rh(PMe$_3$)(propyl)H to -10°C.

Between 0 and -10°C the cyclopropyl hydride complex was observed to convert to a metallocyclobutane complex. In benzene solvent, a 1:1 mixture of the metallocyclobutane and phenyl hydride complex was formed (Eq. 28). This observation indicated that the rearrangement

$$ \qquad\qquad\qquad\qquad\qquad\qquad\qquad\qquad (28) $$

must be intramolecular, since benzene would have scavenged any free $(C_5Me_5)Rh(PMe_3)$ to form the phenyl hydride complex. An η^2-σ-complex of cyclopropane was postulated to account for this observation. X-ray structural characterization of the bromo cyclopropyl compound and metallocyclobutane showed no unusual features in the bonding.[49]

2.   Alkane Activation - Alkane Complexes

The mechanism of alkane activation is more difficult to study, since there are fewer molecules that activate alkanes and the alkyl hydride complexes are often unstable at room temperature. One exception to this general trend is the $(C_5Me_5)Ir(PMe_3)(R)H$ molecules studied by Bergman, which are stable up to above 100°C. Several mechanistic studies have been undertaken to elucidate features of the activation and elimination processes.

The thermolysis (130°C) of the cyclohexyl hydride complex $(C_5Me_5)Ir(PMe_3)(c\text{-}C_6H_{11})H$ in benzene was found to follow clean first-order kinetics producing cyclohexane and the phenyl hydride complex. When the reaction was carried out in neopentane solvent (a thermodynamically disfavored substrate) containing different concentrations of benzene, the rate of cyclohexane formation was found to be independent of the benzene concentration. A similar experiment in which the cyclohexyl complex was dissolved in a mixture of benzene-cyclohexane showed an inhibition of the reaction in direct proportion to the concentration of added cyclohexane. Thermolysis in C_6D_6 gave $(C_5Me_5)Ir(PMe_3)(C_6D_5)D$ at the same rate as the thermolysis in C_6H_6. Also, thermolysis of a mixture of $(C_5Me_5)Ir(PMe_3)(c\text{-}C_6H_{11})H$ and $(C_5Me_5)Ir(PMe_3)(c\text{-}C_6D_{11})D$ in benzene gave >90% d_0- and d_{12}-cyclohexane, indicating a highly intramolecular elimination pathway. The small amount of crossover is probably the result of a trace of a coordinatively unsaturated impurity, as addition of a small amount of PMe_3 or PPh_3 to the reaction medium prevents the scrambling.[35]

These observations are consistent with the mechanism in which reductive elimination of cyclohexane gives $(C_5Me_5)Ir(PMe_3)$. The unsaturated species can then either react with another C-H bond in solution or competitively backreact with the cyclohexane. The activation parameters for the reductive elimination were found to be $\Delta H^{\ddagger} = 35.6 \pm 0.5$ kcal/mol, $\Delta S^{\ddagger} = +10 \pm 2$ e.u. The small positive entropy of activation is consistent with a transition state in which two species are being formed from one, and is similar to that found earlier for arene elimination from $(C_5Me_5)Rh(PMe_3)(Ph)H$. The high enthalpy of activation attests to the strong Ir-C and Ir-H bonds that are present in this molecule.

In an effort to determine the kinetic isotope effect of alkane elimination, Bergman prepared the complex $(C_5Me_5)Ir(PMe_3)(c\text{-}C_6H_{11})D$ and examined its kinetic behavior upon heating. A surprising result was obtained, offering insight into the details of the alkane C-H bond activation mechanism by this compound. An isomerization to the α-deuterio hydride complex was observed competitive with the elimination of $C_6H_{11}D$!

Bergman suggested the formation of an alkane σ-complex to account for this [1,1] isomerization, as indicated in Scheme 6. If this

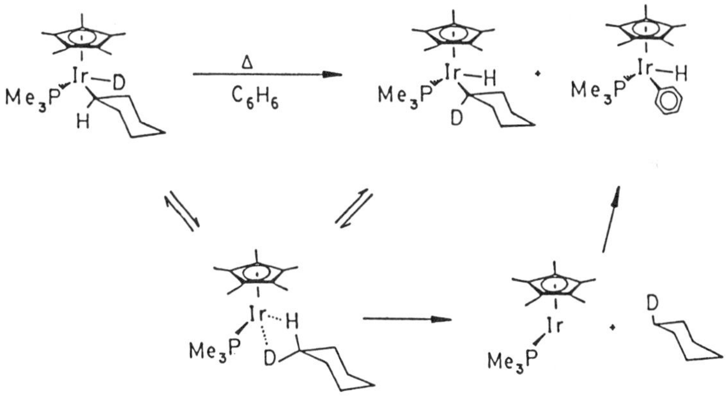

Scheme 6

species underwent readdition to one of the two C-H bonds competitively with dissociation of cyclohexane, then exchange of the hydrogen attached to the metal and the α-cyclohexyl hydrogen could occur intramolecularly.

Since the σ-complex of the alkane could be an intermediate formed in the process of activating an alkane, further studies of the nature of this species may help interpret the selectivities observed in the competitive alkane studies. It is possible that the σ-complexes involving secondary C-H bonds are weaker than those with primary C-H bonds.

If this were so, then even though the initial interactions of the metal were with a secondary C-H bond, the metal might migrate down a hydrocarbon chain to give the more stable primary complex, leading to the more stable primary alkyl hydrido complex.

In fact, evidence for this type of intramolecular rearrangement has been obtained by Bergman. The ^{13}C-labeled complex $(C_5Me_5)Rh(PMe_3)(^{13}CH_2CH_3)D$ is observed to rearrange to the complex $(C_5Me_5)Rh(PMe_3)(CH_2^{13}CH_2D)H$ at temperatures close to that at which ethane is eliminated. The integrity of the [13]CD unit is consistent with a [1,2] shift of the initially formed ethane σ-complex. A sequence involving reversible β-elimination would have been expected to scramble the deuterium over both of the carbons.[50]

Structural characterization of the iridium cyclohexyl complex displays no unusual features with regard to what would be expected for an alkyl hydride complex. The products of these reactions are true oxidative addition adducts and not alkane complexes, in contrast to the observations with H_2.

Saillard and Hoffmann have examined the interaction of methane with d^8 CpML and ML_4 fragments in an attempt to determine the nature of the interactions involved. The model trajectory employed involved an initial end-on approach of the C-H bond, followed by a linear transit to the idealized octahedral product geometry. They conclude that as the energy of the highest occupied molecular orbital (d_{yz}) of the metal increases, the activation energy for oxidative addition decreases. Other systems predicted to be active include $[Rh(benzene)]^-$ and $[Rh(CO)_2]^-$.[51]

Hofmann and Padmanabhan have examined the electronic structure of the d^8 CpML fragment and concluded that the bent molecule would have a singlet ground state, whereas the linear species would have a triplet ground state. Little was predicted about C-H bond activation or the interactions of hydrocarbons with the fragment, other than that the bent structure permitted a ground state oxidative addition as an allowed process.[52]

One last question about the activation of alkanes by $(C_5Me_5)ML$ complexes involves the lifetime of this coordinatively unsaturated intermediate. Rest et al. have examined the photolysis of $(C_5Me_5)Rh(CO)_2$, $CpIr(CO)_2$, and $(C_5Me_5)Ir(CO)_2$ in methane matrices at 12 K. Their findings indicate that the methane oxidative addition adducts $(C_5R_5)M(CO)(Me)H$ are formed readily at this temperature (R = Me, H). No evidence for the species $(C_5Me_5)M(CO)$ was observed.[53]

Furthermore, irradiation in methane matrices containing 5% [13]CO showed little incorporation of free CO into the starting dicarbonyl complexes, suggesting that CO dissociation was not occurring during the photolysis. Rest et al. suggested that perhaps ring slippage (η^5-$C_5Me_5 \rightarrow \eta^3$-$C_5Me_5$) was responsible for the formation of the coordinatively unsaturated species that activates C-H bonds.[53] As a result of the labeling and kinetic studies by Bergman, however, this ring-slippage mechanism can be ruled infeasible in solution.

Finally, Crabtree et al. have modeled the nature of the reaction coordinate for alkane activation by examining X-ray structural parameters of 17 organometallic structures that show agostic M···H-C interactions. The structures show a smooth relationship between the C-H-M bond angle and the radius of the bonding pair (distance from the point where the H and C covalent radii meet to the metal radius) as shown in Figure 4a. An idealized plot of the C-H bond trajectory using the curve in Figure 4a is shown in Figure 4b. Early in the reaction the C-H-M angle is close to 130°, but as the bond approaches the metal it swings down to 95° at a bonding pair radius of 0.4 Å. This trajectory minimizes both M-C and M-H steric interactions, and has been used to account for non linear M-H-M bridges between a 16-electron metal and an M-H bond (rather than a C-H bond).[54]

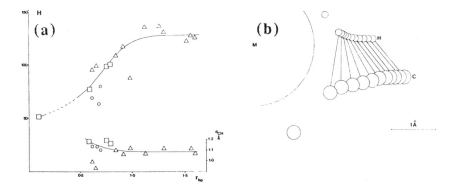

Figure 4. (a) is plot of both H (the angle C-H-M in deg) and d_{CH} (Å) against r_{bp}(Å), where $r_{bp} = d_{bp} - r_M$, d_{bp} = distance from the metal to the C-H bonding electron pair and r_M = covalent radius of the metal. Squares and hexagons refer to neutron data, triangles and circles to X-ray data. Hexagons and circles refer to α-CH bridges that are more constrained geometrically than β and higher types (squares and triangles); more weight is attached to the latter, particularly with regard to H. Only when r_{bp} falls appreciably below 1.0 does H fall below 130° and d_{CH} rise significantly above the value in the free C-H bond (ca. 1.08 Å). The calculated value of H for a cis alkyl hydride complex is also included; the corresponding value of d_{CH} (2.68 Å) is too large to include in this plot. (b) is the C-H + M   C-M-H trajectory as defined by the curve plotted in (a). The large curve corresponds to the surface of the metal atom as defined by the covalent radius of the metal. The smaller circles represent carbon and hydrogen atoms but are drawn with arbitrary radii smaller than the corresponding covalent radii, for clarity. The straight lines joining the C and the H atoms are the successive positions of the CH internuclear vector on varying r_{bp} by increments of 0.1 Å. The isolated circles are the final positions of the carbon and hydrogen atoms expected for a cis alkyl hydride complex, assuming normal M-C and M-H distances and an angle of 90° for C-M-H. The resulting trajectory shows the C-H bond approaching the metal with the C-H vector pointing toward the metal, and with an M-H-C angle of ca. 130°. The C-H bond then rotates and elongates before finally breaking. Reprinted with permission from *Inorg. Chem.* R. H. Crabtree (author) Copyright 1985, American Chemical Society.

3.   Alkane and Arene Activation - Isotope Effects

Isotope effects are used to probe transition states of reactions, and have been measured in both oxidative addition and reductive elimination reactions of alkanes and arenes. For reductive elimination of alkane from *cis*-Pt(PPh$_3$)$_2$RH, normal isotope effects for k_H/k_D of 3.3 (R = CH$_3$)[55] and 2.2 (R = CH$_2$CF$_3$)[56] were observed.

For oxidative addition, the intramolecular aromatic activation of the phenyl group in Ir(PPh$_3$)$_3$Cl showed a modest isotope effect with k_H/k_D = 1.4.[57] The intramolecular activation of the alkyl C-H bonds of a neopentyl group in Pt(PEt$_3$)$_2$(CH$_2$CMe$_3$)$_2$ shows an isotope effect of ~3.0.[58] Isotope effects of 1.9-2.9 were also seen in the benzene-D$_2$ H-D exchange reactions catalyzed by Cp$_2$NbH$_3$.[28] All of these normal isotope effects are consistent with processes in which the C-H bond is partially broken and the M-C and M-H bonds partially made either prior to or in the rate-determining step.

The more recent reactions of (C$_5$Me$_5$)Ir(PMe$_3$) and (C$_5$Me$_5$)Rh(PMe$_3$) with arenes and alkanes also show isotope effects that give information about the bonding in the transition states for these reactions. The studies with the rhodium system confirm the presence of an intermediate in the C-H activation reaction of benzene. Those of the iridium complex with alkanes are consistent with substantial C-H bonding character in the rate-determining transition state.

The photolysis of (C$_5$Me$_5$)Rh(PMe$_3$)H$_2$ in a 1:1 mixture of C$_6$H$_6$/C$_6$D$_6$ gave a k_H/k_D ratio of 1.06 ($\pm$ 0.05):1 for the kinetic isotope effect for selection between two arenes by (C$_5$Me$_5$)Rh(PMe$_3$). In a related experiment, irradiation of (C$_5$Me$_5$)Rh(PMe$_3$)H$_2$ at -40°C in a solution of 1,3,5-C$_6$H$_3$D$_3$ showed a 1.4:1 ratio of the complexes (C$_5$Me$_5$)Rh(PMe$_3$)(2,4,6-C$_6$D$_3$H$_2$)H and (C$_5$Me$_5$)Rh(PMe$_3$)(3,5-C$_6$D$_2$H$_3$)D. This corresponds to a k_H/k_D ratio of 1.4:1 for activation of a C-H bond when there is no intermolecular choice of which arene is to be activated (Scheme 7). The observed difference in kinetic isotope effects dependent upon the intramolecular vs intermolecular nature of the competition indicates that there must be two different rate-determining isotopic steps for the C-H bond activation of benzene, and offers strong support for the intermediacy of an η^2-benzene complex in C-H bond activation of arenes by (C$_5$Me$_5$)Rh(PMe$_3$).[59]

With regard to the pathway for reductive elimination of arene, the relative rates of exchange of the *meta*-xylene with benzene solvent upon thermolysis of (C$_5$Me$_5$)Rh(PMe$_3$)(3,5-C$_6$H$_3$Me$_2$)X (X = H or D) at 51°C showed a kinetic rate ratio k_H'/k_D' = 0.50 (primes denote reductive elimination). The *meta*-xylyl complexes were used to avoid scrambling of the carbon attached to the metal prior to loss of arene. Since the loss of arene included both reductive elimination to form an η^2-arene complex and dissociation of the coordinated arene, an independent determination of the kinetic isotope effect for only the reductive elimination portion of the reaction was made by monitoring the

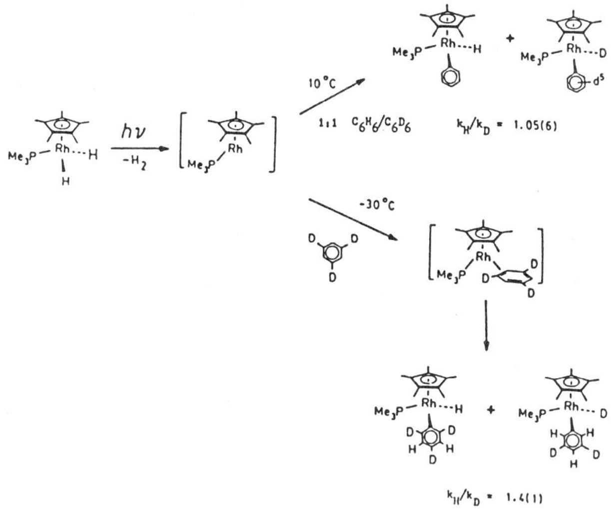

$10\,°C$

$1{:}1$ C_6H_6/C_6D_6 $k_H/k_D = 1.05(6)$

$h\nu$ $-H_2$

$-30\,°C$

$k_H/k_D = 1.4(1)$

Scheme 7

equilibrium between the positional isomers of $(C_5Me_5)Rh(PMe_3)$ $(C_6D_5)H$ (Scheme 5). From the equilibrium ratio of hydride and *ortho*-C_6D_4H species (1.35:1), the reductive elimination rate ratio k_H'/k_D' was independently determined to be 0.52 using the expression for the equilibrium constant shown in Eq. 29.[59]

$$K_{eq} = \frac{[\text{ortho-H}]}{[\text{hydrido-H}]} = 0.37 = \frac{k_H' \cdot k_D}{k_D' \cdot k_H} \tag{29}$$

The inverse isotope effect for the intramolecular reductive elimination reaction can be interpreted using the free energy diagram shown in Scheme 8. It shows the difference in H-D zero-point energies of the compounds, the magnitude of which is based upon M-H and C-H stretching frequencies. Also shown at the transition states is the saddle point nature of this region of the diagram, with the C-H or C-D stretch still partially intact. If the magnitude of the residual zero point energy left in the transition state (Δ_2) is less than that in the η^2-benzene complex (Δ_3) but greater than that in the phenyl hydride complex (Δ_1), then a normal isotope effect is expected for the intramolecular oxidative addition (the barrier for the H-containing material is less than that encountered by the D-containing material) and an inverse isotope effect is expected for the reductive elimination. The presence of a slightly weakened C-H bond in the transition state suggests that there is more η^2-arene character than aryl hydride character in this geometry, in accord with the Hammett postulate. Were there substantial reduction of the C-H bond order in the transition state, then normal isotope effects in both directions would be expected.[59]

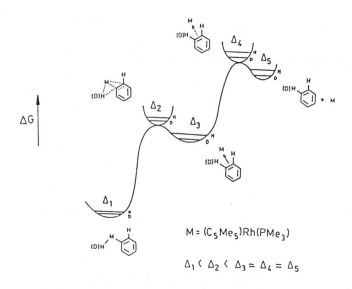

Scheme 8

It is interesting to note that Tolman et al. reported that thermolysis of Fe(dmpe)$_2$(naphthyl)H in a 1:1 mixture of C$_6$H$_6$/C$_6$D$_6$ did not show any isotope effect in product formation. This observation is similar to that found in the above reaction of (C$_5$Me$_5$)Rh(PMe$_3$), and may indicate the intermediacy of an η^2-arene complex in the arene activation reaction by iron.[33] Spin transfer experiments with a mixture of the positional isomers of Ru(dmpe)$_2$(C$_6$D$_5$)H [formed from Ru(dmpe)$_2$(naphthyl)H + C$_6$D$_5$H] did not show any magnetization transfer on the time scale of aromatic hydrogen relaxation ($T_1 \cong 35$ sec). This indicates that the reversible formation of an η^2-C$_6$D$_5$H complex must occur with a rate constant of < 0.02 sec^{-1}, if at all.[60]

The iridium complex studied by Bergman and co-workers does give isotope effects for both alkane activation and elimination. The kinetic selectivity for cyclohexane-neopentane (0.88) can be divided by the kinetic selectivity for cyclohexane-d$_{12}$-neopentane (0.64), giving a kinetic isotope effect of $k_H/k_D = 1.38$ for oxidative addition to cyclohexane. This is a relatively small value for a primary isotope effect in which a C-H bond is broken, but there are few other data with which the number can be compared. Halpern found a $k_H/k_D = 1.22$ for the oxidative addition of hydrogen to Vaska's complex, which was rationalized in terms of a reactant-like transition state. Perhaps the reaction coordinate for the activation of alkanes by iridium also shows a reactant-like transition state. If the formation of an alkane complex is the rate-determining (or close to rate-determining) step in alkane activation by $(C_5Me_5)Ir(PMe_3)$, then the transition state is indeed reactant-like and the small primary isotope can be accommodated.[35]

The kinetic isotope effect for reductive elimination of alkane shows an inverse isotope effect with the iridium complex. Thermolysis of $(C_5Me_5)Ir(PMe_3)(C_6D_{11})D$ in C_6D_6 at 130°C proceeds 1.43 times as fast as $(C_5Me_5)Ir(PMe_3)(C_6H_{11})H$, which corresponds to an inverse isotope effect of $k_H'/k_D' = 0.7 \pm 0.1$. The formation of a σ-alkane complex is one way to rationalize the inverse isotope effect. Bergman has suggested that there is a nearly fully formed C-H bond in this intermediate with an M-C-H interaction of the agostic variety. Although there are no structurally characterized alkane complexes, recent studies by Kubas et al. have revealed the presence of an analogous complex in which an intact hydrogen molecule coordinates to a metal in an η^2-fashion.[61] The isotope effect reactions along with the isomerizations mentioned earlier begin to form a consistent picture in which the interactions of intact alkanes with unsaturated metal complexes begin to look attractive.

References

1. Janowicz, A. H., Bergman, R. G., *J. Am. Chem. Soc.* **1981**, *103*, 2488.
2. Janowicz, A. H., Bergman, R. G., *J. Am. Chem. Soc.* **1982**, *104*, 352.
3. (a) Rathke, J. W., Muetterties, E. L., *J. Am. Chem. Soc.* **1975**, *97*, 3272. (b) Karsch, H. H., Klein, H. F., Schmidbaur, H., *Angew. Chem. Int. Ed. Engl.* **1975**, *14*, 637. (c) Werner, H., Werner, R., *J. Organomet Chem..* **1981**, *209*, C60. Werner, H., Gotzig, J., *Organometallics* **1983**, *2*, 547. (d) Chiu, K. W., Howard, C. G., Rzepa, H. S., Sheppard, R. N., Wilkinson, G., Galas, A. M. R., Hursthouse, M. B., *Polyhedron* **1982**, *1*, 441. (e) Brookhart, M., Cox, K., Cloke, F. G. N., Green, J. C., Green, M. L. H., Hare, P. M., Bashkin, J., Derome, A. E., Grebenik, P. D., *J. Chem. Soc. Dalton Trans.* **1985**, 423. (f) Gibson, V. C., Grebenik, P. D., Green, M. L. H., *J. Chem. Soc., Chem. Commun.* **1983**, 1101.
4. Berry, M., Cooper, N. J., Green, M. L. H., Simpson, S. J., *J. Chem. Soc. Dalton Trans.* **1980**, 29.
5. Cooper, N. J., Green, M. L. H., Mahtab, R., *J. Chem. Soc. Dalton Trans.* **1979**, 1557.
6. Berry, M., Elmitt, K., Green, M. L. H., *J. Chem. Soc. Dalton Trans.* **1979**, 1950.
7. Cloke, F. G. N., Green, J. C., Green, M. L. H., Morley, C. P., *J. Chem. Soc. Chem. Commun.* **1985**, 945.
8. Wax, M. J., Stryker, J. M., Buchanan, J. M., Kovac, C. A., Bergman, R. G., *J. Am. Chem. Soc.* **1984**, *106*, 1121.
9. Stoutland, P. O., Bergman, R. G., *J. Am. Chem. Soc.* **1985**, *107*, 4581.
10. Hoyano, J. K., Graham, W. A. G., *J. Am. Chem. Soc.* **1982**, *102* 3723.
11. Rausch, M. D., Gastinger, R. G., Gardner, S. A., Brown, R. K., Wood, J. S., *J. Am. Chem. Soc.* **1977**, *99*, 7870.
12. Jones, W. D., Feher, F. J., *J. Am. Chem. Soc.* **1982,** *104*, 4240.

13.   Jones, W. D., Feher, F. J.,*Organometallics* **1983**, *2*, 562.
14.   Periana, R. A., Bergman, R. G., *Organometallics* **1984**, *3*, 508.
15.   Hoyano, J. K., McMaster, A. D., Graham, W. A. G., *J. Am. Chem. Soc.* **1983**, *105*, 7190.
16.   Chetcuti, P. A., Hawthorne, M. F., *J. Am. Chem. Soc.* **1987**, *109*, 942.
17.   Bergman, R. G., Seidler, P. F., Wenzel, T. T., *J. Am. Chem. Soc.* **1985**, *107* 4358.
18.   Klahn-Oliva, A. H., Singer, R. D., Sutton, D., *J. Am. Chem. Soc.* **1986**, *108*, 3107.
19.   Jones, W. D., Maguire, J. A., *Organometallics* **1987**, *6*, 1301.
20.   Jones, W. D., Maguire, J. A., *Organometallics* **1986**, *5*, 590.
21.   Ghosh, C. K., Graham, W. A. G.,*J. Am. Chem. Soc.* **1987**, *109*, 4726.
22.   (a) Baker, M. V., Field, L. D., *J. Am. Chem. Soc.* **1987**, *109*, 2825; (b) Baker, M. V., Field, L. D., *J. Am. Chem. Soc.* **1986**, *108*, 7433; (c) Baker, M. V., Field, L. D., *J. Am. Chem. Soc.* **1986**, *108*, 7436.
23.   (a) Desrosiers, P. J., Shinomoto, R. S., Flood, T. C., *J. Am. Chem. Soc.* **1986**, *108*, 1346. (b) Desrosiers, P. J., Shinomoto, R. S., Flood, T. C., *J. Am. Chem. Soc.* **1986**, *108*, 7964.
24.   Hackett, M., Ibers, J. A., Jernakoff, P., Whitesides, G. M., *J. Am. Chem. Soc.* **1986**, *108*, 8094.
25.   Nemeh, S., Jensen, C., Binamira-Soriaga, E., Kaska, W. C. *Organometallics* **1983**, *2*, 1442.
26.   Werner, H., Hohn, A., Dziallas, M., *Angew. Chem. Int. Ed. Engl.* **1986**, *25*, 1090.
27.   Sakakura, T., Tanaka, M., *J. Chem. Soc., Chem. Commun.* **1987**, 758.
28.   Klabunde, U., Parshall, G. W., *J. Am. Chem. Soc.* **1972**, *94*, 9081.
29.   Green, M. L. H., Berry, M., Couldwell, C., Prout, K., *Nouv. J. Chim.* **1977**, *1*, 187.
30.   Berry, M., Elmitt, K., Green, M. L. H., *J. Chem. Soc. Dalton Trans.* **1979**, 1950.
31.   Cooper, N. J., Green, M. L. H., Mahtab, R., *J. Chem. Soc. Dalton Trans.* **1979**, 1557.
32.   Jones, W. D., Feher, F. J., *J. Am. Chem. Soc.* **1984**, *106*, 1650.
33.   Tolman, C. A., Ittel, S. D., Jesson, J. P., *J. Am. Chem. Soc.* **1979**, *101*, 1742.
34.   Bergman, R. G., *Science* **1984**, *223*, 902.
35.   Buchanan, J. M., Stryker, J. M., Bergman, R. G., *J. Am. Chem. Soc.* **1986**, *108*, 1537.
36.   Nolan, S. P., Hoff, C. D., Stoutland, P. O., Newman, L. J., Buchanan, J. M., Bergman, R. G., Yang, G. K., Peters, K. S., *J. Am. Chem. Soc.* **1987**, *109*, 3143.
37.   (a) Tsou, T. T., Loots, M., Halpern, J., *J. Am. Chem. Soc.* **1982**, *104*, 623. (b) Ng, F. T. T., Rempel, G. L., Halpern, J., *J. Am. Chem. Soc.* **1982**, *104*, 621.
38.   Wenzel, T., Bergman, R. G., *J. Am. Chem. Soc.* **1986**, *108*, 4856.
39.   Chatt, J., Davidson, J. M., *J. Chem. Soc.* **1965**, 843.
40.   Turner, R. W., Amma, E. L., *J. Am. Chem. Soc.* **1966**, *88*, 3243.
41.   Brauer, D. J., Kruger, C., *J. Am. Chem. Soc.* **1977**, *16*, 884.
42.   Sweet, J. R., Graham, W. A. G., *J. Am. Chem. Soc.* **1983**, *105*, 305.
43.   Klabunde, U., Parshall, G., *J. Am. Chem. Soc.* **1972**, *94* 9081.
44.   Werner, R.; Werner, H. *Angew. Chem. Int. Ed. Engl.* **1981**, *20*, 793-794.
45.   Jones, W. D., Feher, F. J., *Accts. Chem. Res.*, **1989**, in press.

46. Werner, H., Gotzig, J., *J. Organomet. Chem.* **1985**, *284*, 73.
47. Duttweiler, R., Ph.D. Thesis, University of Rochester, Rochester, NY, 1988.
48. Brainard, R. L., Nutt, W. R., Whitesides, G. M., *J. Am. Chem. Soc.* **1987**, *109*, in press.
49. Periana, R. A., Bergman, R. G., *J. Am. Chem. Soc.* **1984**, *106*, 7272.
50. Periana, R. A., Bergman, R. G., *J. Am. Chem. Soc.* **1986**, *108*, 7346.
51. Saillard, J. Y., Hoffmann, R., *J. Am. Chem. Soc.* **1984**, *106*, 2006.
52. Hofmann, P.; Padmanabhan, M. *Organometallics* **1983**, *2*, 1273.
53. Rest, A. J., Whitwell, I., Graham, W. A. G., Hoyano, J. K., McMaster, A. D., *J. Chem. Soc., Chem. Commun.* **1984**, 624.
54. Crabtree, R. H., Holt, E. M., Lavin, M., Morehouse, S. M., *Inorg. Chem.* **1985**, *24*, 1986.
55. Abis, L., Sen, A., Halpern, J., *J. Am. Chem. Soc.* **1978**, *100*, 2915.
56. Michelin, R. A., Faglia, S., Uguagliati, P., *Inorg. Chem.* **1983**, *22*, 1831.
57. Parshall, G. W., *Acc. Chem. Res.* **1970**, *3*, 139.
58. (a) Foley, P., Whitesides, G. M., *J. Am. Chem. Soc.* **1979**, *101*, 2732. (b) Moore, S. S., DiCosimo, R., Sowinski, A. F., Whitesides, G. M., *J. Am. Chem. Soc.* **1981**, *103*, 948.
59. Jones, W. D., Feher, F. J., *J. Am. Chem. Soc.* **1986**, *108*, 4814.
60. Woedy, L., Jones, W. D., unpublished observations.
61. Kubas, G. J., Ryan, R. R., Swanson, B. I., Vergamini, P. J., Wasserman, H. J., *J. Am. Chem. Soc.* **1984**, *106*, 451.

Chapter V

THE HOMOGENEOUS ACTIVATION OF CARBON-HYDROGEN BONDS BY HIGH VALENT EARLY d-BLOCK, LANTHANIDE, AND ACTINIDE METAL SYSTEMS

Ian P. Rothwell

Department of Chemistry
Purdue University
West Lafayette, Indiana 47907

I.   Introduction

The last few years have seen many dramatic and impressive developments in the study of homogeneous, transition metal-mediated activation of normally inert CH bonds.[1] The initial demonstration of the cyclometalation reaction in the 1960s[2,3] was rapidly followed by the development of systems able to oxidatively add arene CH bonds to electron-rich metal centers.[4-6] Over the next decade interest in the field expanded rapidly and a plethora of results were published concerning both the synthetic and mechanistic aspects of the intramolecular activation of sp^2 and sp^3 hybridized CH bonds as well as the intermolecular activation of aromatic CH bonds.[7-12] However, it was not until 1982 that the mild, intermolecular activation of saturated hydrocarbons, including methane, was conclusively demonstrated by the groups of Janowicz and Bergman[13] and Hoyano and Graham.[14] These "early" reports have been followed by the development of numerous systems (generated either thermally or photochemically) that have the ability to oxidatively add the CH bonds of saturated hydrocarbon substrates under very mild conditions.[15-20] Somewhat parallel with these developments over the last few years has been the recognition that an alternate method for CH bond activation by homogeneous transition metal systems was available. The rapidly developing field of high-valent early d-block, lanthanide, and actinide organometallic chemistry[21] was found to contain reactions involving the sometimes facile intramolecular activation of ligand CH bonds as well as the intermolecular activation of dihydrogen and aromatic CH bonds.[22] The high valency of the metal centers in these systems precluded an oxidative-addition pathway and so alternate mechanisms began to be proposed and evaluated. The dramatic demonstration by Watson in 1983[23] that even methane (the "Holy Grail" of CH bond activation research) could be functionalized under mild condition by homogeneous Lu(III) compounds gave added importance to this alternate method of hydrocarbon activation. In the last 4 years there have been many new developments in this field of CH bond activation. Particularly striking has been the emergence of a cohesive mechanistic picture of the way in which these reactions take place. This chapter is intended not only as a comprehensive review of the synthetic aspects of this emerging area of chemistry, but also as an evaluation of our present mechanistic knowledge and its implications for future work.

II.   Intramolecular Activation of CH Bonds of Metal-Alkyl Groups

1.   α-Hydrogen Abstraction

The inclusion of α-hydrogen abstraction processes in a body of work outlining CH bond activation reactions is well justified, at least in this author's opinion. The reaction has a number of characteristics common to more distal CH bond activations, except that the new metal-carbon bond that is formed is a π-bond. The formation of metal-alkylidene functional groups by an α-hydrogen abstraction process was initially identified by Schrock,[24] and some extensive reviews on the subject are available.[25,26] During the course of these thermal reactions the α-CH bond of the alkyl group is activated by the d⁰-metal center and the hydrogen atom is transferred to a suitable leaving group contained within the metal coordination sphere (Eq. 1). Theoretical work on this reactivity by Hoffmann and co-workers indicates the reaction proceeds through a concerted pathway in which the α-hydrogen is transferred to the leaving group with no formal change in metal oxidation state throughout.[27] A four-center, four-electron transition state is postulated with concomitant bond making and breaking (Eq. 1).[27] This pathway is

$$L_nM \begin{matrix} CH_2X \\ \diagup \\ \diagdown \\ R \end{matrix} \longrightarrow L_nM \begin{matrix} H \diagdown C \diagup X \\ \\ H \\ \diagdown R \diagup \end{matrix} \longrightarrow L_nM=CHX + RH \quad (1)$$

not only of direct relevance to more distal CH bond activations, but its reverse, the activation of aliphatic CH bonds by metal-carbon double bonds has been demonstrated and, hence, by microscopic reversibility, the two reactions are mechanistically coupled.

A number of photochemically induced α-hydrogen abstraction processes at d⁰-metal centers have also been found (Eqs. 2 and 3).[28,29]

$$(Ar'O)_2Ta(CH_3)_3 \xrightarrow{h\nu} (Ar'O)_2Ta(=CH_2)(CH_3) + CH_4 \quad (2)$$

$$(OAr' = 2, 6\text{-di-}tert\text{-butylphenoxide})$$

$$(Bu^tN)_2Re(CH_2SiMe_3)_3 \xrightarrow{h\nu} (Bu^tN)_2Re(=CHSiMe_3)(CH_2SiMe_3) + Me_4Si \quad (3)$$

A study of the photochemical reactivity of the tantalum compounds indicated that there is evidence for these photoreactions proceeding via initial homolytic cleavage of a metal-carbon bond followed by a hydrogen atom abstraction process.[30] It would, therefore, appear that there is a definite mechanistic difference between the thermal and photochemical formation of metal-alkylidene groups.

2. β-Hydrogen Abstraction

The intramolecular abstraction of a β-hydrogen from metal-alkyl groups is a common decomposition pathway and readily takes place at d^0-metal centers.[21] As far as CH bond activation is concerned, however, the formation of the hydrido-olefin product is a special case in that the alkyl group is also acting as its own leaving group (Eq. 4). Again a four-

$$\text{(4)}$$

center transition state can be envisioned (Eq. 4). Ground state interactions between d^0-metal centers and both α- or β-CH bonds of alkyl groups have been detected (Sec. VI. 2.A).

    In the case of organoactinide chemistry, a photochemically induced β-hydrogen abstraction process has been well characterized (Eq. 5).[31]

$$Cp_3Th\,(CH_2CH_2R) \xrightarrow{h\nu} Cp_3ThH + CH_2{=}CHR \qquad (5)$$

3. Metallo-Benzyne (*o*-Phenylene) Formation via *Ortho*-Hydrogen Abstraction from Metal-Aryl Groups

The ability of high-valent, early transition metal-aryl functional groups to undergo intramolecular activation of their *ortho*-CH bonds has been known for some time (Eq. 6).[32,33] The resulting metal-benzyne

$$L_nM(X)\,(C_6H_5) \longrightarrow L_nM(\eta^2\text{-}C_6H_4) + HX \qquad (6)$$

compounds have been shown to be extremely reactive species being able to activate a range of small molecules including even N_2. In a select few cases the η^2-C_6H_4 derivatives are sufficiently stable to be isolated and structural studies are completely supportive of the idea that the ligand is best considered as $C_6H_4{}^{2-}$ with the metal maintaining a d^o-configuration.[34,35] Furthermore, it has been shown that metal-benzyne intermediates have the ability to intermolecularly activate aromatic CH bonds (the reverse of their formation Eq. 6) as well as intramolecularly activate aliphatic CH bonds (see Sec. V.1). The isomerization of early transition metal-aryls via the reversible formation of a benzyne group has also been demonstrated and this contrasts with the situation on later transition metals in which such isomerization processes typically proceed via η^2-benzene intermediates.[36,37]

The few kinetic studies that have been carried out show the formation of benzyne ligands to have a significantly large, primary kinetic isotope effect (k_H/k_D) associated with deuterating the *ortho*-CH bonds that are activated (Eq. 7).[38]

$$Cp*_2Zr\,(C_6H_5)_2 \longrightarrow Cp*_2Zr\,(\eta^2\text{-}C_6H_4) + C_6H_6$$

$$\Delta H^{\ddagger} = 22.5\,(3)\,\text{kcal/mol} \tag{7}$$
$$\Delta S^{\ddagger} = -11.8\,(3)\,\text{e.u.}$$
$$k_H/k_D\,(70^{\circ}C) = 6.5\,(10)$$

4. γ-Hydrogen Abstraction in Alkyl Groups

It is generally assumed, although not conclusively proven, that the metal-center has minimal influence through the carbon side chain on γ-CH bonds. Hence, the activation of these CH bonds is separated slightly from the case of more proximal α- and β-CH bonds, and γ-CH bond activation is typically classified as distal.[39] Demonstrated cyclometalation reactions at these positions is restricted to alkyl groups of the type CH_2CMe_3, CH_2SiMe_3, and CH_2SnMe_3. Metalation of these groups at both mononuclear and dinuclear metal centers has been observed.[40,41] A detailed kinetic, thermodynamic, and structural study of the cyclometalation of these ligands at Th(IV) and U(IV) metal centers has been carried out by Marks and co-workers (Scheme 1).[42] The reactions involve the formation of a series of metallocyclobutane rings by the thermolysis of the corresponding $Cp*_2ThR_2$ or $Cp*_2UR_2$

substrates. Thermodynamic studies involving the careful measurement of heats of formation[43] show these reactions to be endothermic, but

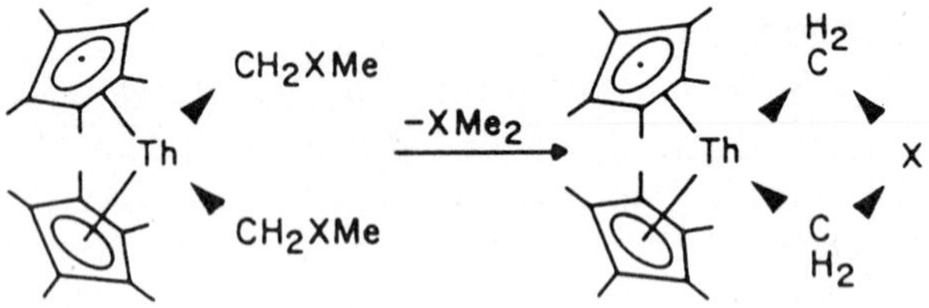

X	$\Delta H^{\ddagger} / Kcal\ mol^{-1}$	$\Delta S^{\ddagger}/e.u.$
CMe_2	21.2 (8)	−16 (2)
$SiMe_2$[a.]	25.1 (4)	−10 (1)
$CMeEt$	18.5 (7)	−24 (2)
$CMePh$	21.4 (8)	−20 (2)

a. See text, k_H/k_D (°C) = 10.0 ± 0.5 (85), 8.5 ± 0.5 (115)

Scheme 1

entropically driven to the metallocycle products.[42,43] Kinetic studies show the reactions to all be first order in solution with characteristically moderately negative entropies of activation (Scheme 1).[42] However, a detailed labeling study of the thermal reactivity of $Cp^*_2Th(CH_2SiMe_3)_2$ indicated that there are two competing first-order pathways leading to the observed products. The first, most dominant pathway at the temperatures used involves the direct combination of the γ-hydrogen atom with the alkyl leaving group to form $SiMe_4$. The alternate, minor pathway involves initial abstraction of a hydrogen atom from one of the Cp^* methyl group to form an intermediate "tuck-in" compound that then proceeds rapidly to the product by insertion of the γ-CH bond of the remaining $Th-CH_2SiMe_3$ group into the strained $Th-CH_2(Cp^*)$ bond (Scheme 1).[42] The fact that both of these pathways are operative means that simple interpretation of activation parameters obtained by kinetic methods is not straightforward. However, analysis of product mixtures form the thermolysis of $Cp^*_2Th[CH_2Si(CD_3)_3]_2$ over a range of temperatures allowed Marks and co-workers to conclude that the indirect, ring abstraction pathway would be a very minor component in the thermolysis of the undeuterated material.[42]

The metallocyclobutane rings formed in these reactions prove to be valuable substrates for the intermolecular activation of a range of CH bonds (see Sec. V.2). Structural studies involving both X-ray and neutron diffraction techniques on the substrate dialkyls prove to be interesting.[42,44] In the solid state there appears to be some distortions of the alkyl ligands in $Cp*_2Th(CH_2XMe_3)_2$ (X = C, Si). In particular there are close contacts between a methyl group of one alkyl and the α-carbon of the other. Although it is tempting to interpret the solid state data in terms of possible ground state electronic interactions between these groups, which would be of direct relevance to the observed CH bond activation chemistry, the authors point out the possible dominance of steric effect in these crowded molecules.[42] Furthermore, there is no spectroscopic evidence for such interactions carrying over into solution.

5. Activation of *Ortho*-(γ)-CH Bonds in Benzyl Ligands

The observation by Bulls et al. that the hafnium benzyl compound $Cp_2*Hf(CH_2Ph)_2$ undergoes the clean formation of toluene and a metallocyclobutene ring would appear at first to represent the direct, intramolecular metalation of a benzyl ligand with a benzyl leaving group. However, careful scrutiny of this reactivity by the authors has shown the reaction to proceed via a more elaborate (perhaps even devious) pathway involving two intermediates as shown (Scheme 2).[45]

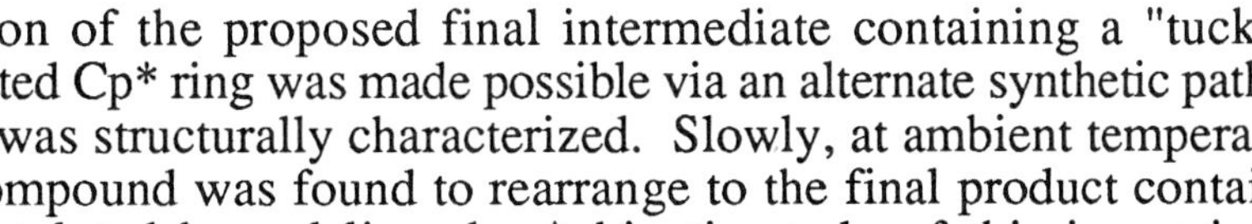

Scheme 2

Isolation of the proposed final intermediate containing a "tuck in," metalated Cp* ring was made possible via an alternate synthetic pathway and it was structurally characterized. Slowly, at ambient temperatures this compound was found to rearrange to the final product containing the metalated benzyl ligand. A kinetic study of this isomerization showed the reaction to be first order with a large primary kinetic isotope effect on deuterating the benzyl *ortho*-CH bonds. Furthermore, the use of a range of *para*-substituted benzyl ligands was found to have little effect on the rate of the ring closure reaction (Scheme 2).[45] This observation is particularly important from a mechanistic viewpoint (see Sec. VI).

6. Activation of δ-Alkyl or Aryl CH Bonds of Alkyl Groups.

The cyclometalation chemistry observed at later transition metal centers is dominated by the propensity for five-membered metallocycle formation.[7-12] Hence the cyclometalation of CH_2CMe_2Et ligands at Pt(II) was shown by Whitesides and co-workers to lead exclusively to a five-membered ring through activation of the δ-CH bond of the alkyl ligand.[40] In contrast, use of the same ligand at Th(IV) was shown by Marks and co-workers to lead only to the four-membered metallocycle by preferential activation of the γ-CH bond of a methyl group (Scheme 1).[42] This selectivity difference was ascribed to kinetic rather than thermodynamic control, with the more strained metallocycle being more rapidly formed. This argument was given support by the observed isomerization of the four-membered metallocycle formed by metalation of the CH_2SiMe_2Ph ligand into a more stable five-membered metallocycle in which the *ortho*-(δ)-CH bond of the aryl group has been activated (Eq. 8).[42]

$$Cp*_2Th\,(CH_2CMePhCH_2) \longrightarrow Cp*_2Th\,(CH_2CMe_2C_6H_4) \qquad (8)$$

7. Activation of Methyl CH Bonds in η^5-C_5Me_5 (Cp*) Ligands

The importance of the pentamethylcyclopentadienyl group as an ancillary ligand in organotransition metal chemistry cannot be understated.[46] However, in the field of CH bond activation by early d-block, lanthanide, and actinide metal centers, the η^5-C_5Me_5 group is often found to be anything but a purely supporting ligand. In numerous examples of both inter- and intramolecular CH bond activation (*vide supra* and *infra*), the reactions have been shown to proceed via an intermediate compound in which one of the η^5-C_5Me_5 methyl groups has been metalated.[38,48] In some cases the intermediate can be isolated. Structural studies of these compounds are entirely consistent with their being considered as containing "tucked-in," metalated $(\eta^5$-η^1-$C_5Me_4CH_2)^{2-}$ groups and not neutral, η^6-fulvene ligands in which the metal has formally undergone a two-electron reduction (see Sec. VI.1.A).

III. Intramolecular Activation of CH Bonds in Nitrogen Donor Ligands

1. β-CH Bonds in Dialkylamido Ligands

The cyclometalation chemistry of later transition metals is dominated by nitrogen donor ligands such as azobenzene, imines, substituted pyridines, and related molecules.[7-12] However, for the high-valent early transition metals it is the dialkylamido function that is found to exhibit an extensive and developing cyclometalation chemistry. Early work on the reactions of diethylamido groups with Group 5 metal halides was found to lead not to the expected homoleptic amido derivatives but to compounds containing the metal-imido function M-NEt formed by overall dealkylation (Eq. 9).[47,48] Careful study

(a) $TaCl_5 + 5 \, LiNEt_2 \longrightarrow (Et_2N)_3Ta(\eta^2\text{-EtNCHMe}) + Et_2NH + 5 \, LiCl$ (9)

(b) $(Et_2N)_3Ta(\eta^2\text{-EtNCHMe}) \longrightarrow (Et_2N)_3Ta =NEt + CH_2=CH_2$

showed that the first step in this transformation involves the activation of the β-CH bond of the amido group leading to an η^2-imine intermediate and diethylamine. Thermolysis of this azametallocyclopropane ring then results in formation of the imido ligand and ethylene (Eq. 9).[48]

Mayer et al.[49] have shown that this type of reaction can also take place with the smaller dimethylamido group. Hence solutions of $Cp^*Ta(NMe_2)Me_3$ formed from $Cp^*TaClMe_3$ and $LiNMe_2$ slowly undergo loss of methane in solution with formation of a cyclometalated, η^2-imine complex (Eq. 10). A kinetic study showed a first-order

$$Cp^*Ta (NMe_2) Me_3 \longrightarrow Cp^*Ta (\eta^2\text{-MeNCH}_2) Me_2 + CH_4$$ (10)

$$k_H/k_D \, (34°C) = 9.7$$

dependence for the reaction with $k_H/k_D = 9.7$ at 34°C when using $N(CD_3)_2$.[49]

Work by Nugent et al. has indicated that in fact this type of reaction is common for a number of d^0-metal dimethylamido complexes.[50] However, as a result of the high reactivity of the generated, intermediate η^2-imines, the reaction was inferred by labeling studies. Hence, in the presence of various catalysts $M(NMe_2)_x$ (M = Zr, x = 4; Nb, Ta, x = 5 and W, x = 6) the amine $DN(CH_3)_2$ has the deuterium label scrambled into the methyl group.[50] It was also found possible to insert olefins into the strained, intermediate azametallocyclopropane, allowing aminomethylation of olefins to be catalyzed (Eq. 11).[50]

$$L_nM[N(CH_3)_2]_2 \; \longrightarrow \; L_nM[\eta^2\text{-MeNCH}_2] \; \xrightarrow{DN(CH_3)_2}$$

$$L_nM[NCH_2D)\,(CH_3)]\,[N(CH_3)_2] + HN(CH_3)_2 \qquad (11)$$

2. Activation of *Ortho*-(β)-CH Bonds in Pyridine Ligands

The coordinatively very unsaturated compounds Cp^*_2MX (M = Sc,[51] Lu;[52] X = CH_3, H) will react with pyridine to generate compounds containing cyclometalated pyridine rings with the elimination of HX (Eq. 12). In the case of the scandium derivative a single crystal X-ray

$$Cp^*_2ScX + NC_5H_5 \; \longrightarrow \; Cp^*_2Sc\,(\eta^2\text{-CN-C}_5H_4N) + HX \qquad (12)$$

$$(X = H, CH_3)$$

diffraction study has confirmed the bonding through both C and N.[51] For X = H the reaction can be readily reversed as evidenced by the selected deuterium incorporation on treating the metalated pyridine compound $Cp^*_2Sc(C,N\text{-}\eta^2\text{-C}_5H_4N)$ with D_2.[51] Furthermore, pyridine exchange is facile as evidenced by this compound's reaction with C_5D_5N to produce the complex $Cp^*_2Sc(C,N\text{-}\eta^2\text{-C}_5D_4N)$ and pyridine-d_1.[51] This reaction presumably involves initial formation of a pyridine adduct, accommodated into the coordination sphere by the metalated pyridine becoming purely carbon bound. Hydrogen transfer can then take place via a symmetrical transition state. The interconversion of the metalated pyridine from C,N bound to purely C-bound can be related to the $\eta^2\text{-} \longrightarrow \eta^1\text{-}$ bonding mode change of isoelectronic iminoacyl groups.[53] In the case of addition of pyridine to Cp^*_2ScMe a pyridine adduct is detected spectroscopically prior to methane loss and metalation.[51] This metalation chemistry of pyridine contrasts markedly with the intramolecular reduction of pyridine found with the less sterically crowded hydride compound $(Cp_2LuH)_x$.[54]

The α-metalation of 2-substituted pyridines and quinolines by Cp_2TiR complexes has been reported by Klei and Teuben.[55] Although stoichiometrically related, the fact that the cyclometalation occurs at a Ti(III), d^1-metal center clouds the mechanistic picture slightly. However, this in itself does not rule out that related mechanistic pathways are operative.

3.   Activation of γ-CH Bonds in Dialkylamido Ligands

The use of the sterically demanding ligand bis(trimethylsilyl)amide, $N(SiMe_3)_2$, by Bradley and others has allowed the stabilization of an extensive series of transition metal complexes containing the metal in a low coordination environment.[56,57] However, in 1974 Bennett and Bradley demonstrated that it was also possible to activate the γ-CH bonds in this ligand under mild conditions giving rise to cyclometalated products containing a four-membered metallocycle (Eq. 13).[58] A reinvestigation of this reaction by Simpson and Andersen shows that the metalation step probably occurs in an intermediate, monosubstituted complex by loss of HCl.[59] Furthermore the Zr(IV) analog was also obtained, only this time by treating Schwartz's reagent, $Cp_2ZrH(Cl)$

$$Cp_2TiCl_2 + 2 LiN(SiMe_3)_2 \longrightarrow Cp_2Ti[N(SiMe_3)SiMe_2CH_2] + 2 LiCl$$
$$+ NH(SiMe_3)_2 \qquad (13)$$

with $LiN(SiMe_3)_2$ to yield the metallocycle and H_2 (Eq. 14).[59]

$$Cp_2Zr(H)Cl + NaN(SiMe_3)_2 \longrightarrow Cp_2Zr[N(SiMe_3)SiMe_2CH_2] + NaCl + H_2 \qquad (14)$$

In later studies Andersen and co-workers studied the thermal reactivity of the bis alkyls of formulas $R_2M[N(SiMe_3)_2]_2$ (M = Ti, Zr, Hf).[60,61] For M = Zr and Hf loss of both alkyl groups as alkane on thermolysis was observed, leading to a dimeric product linked by alkylidene bridges (Scheme 3).[60] The dimeric nature of the products was confirmed by X-ray crystallography. The reaction presumably proceeds via initial activation of a γ-CH bond of a $N(SiMe_3)_2$ ligand. Loss of the second equivalent of alkane then involves an α-hydrogen abstraction process. However, the dimeric products make it uncertain whether either or both of these steps are uni- or bimolecular in nature. When equimolar mixtures of $Me_2Zr[N(SiMe_3)_2]$ and $Et_2Hf[N(SiMe_3)_2]$ were heated a 1:2:1 mixture of Zr_2, ZrHf and Hf_2 containing products were produced.[60] On treatment with dmpe the dimers undergo both

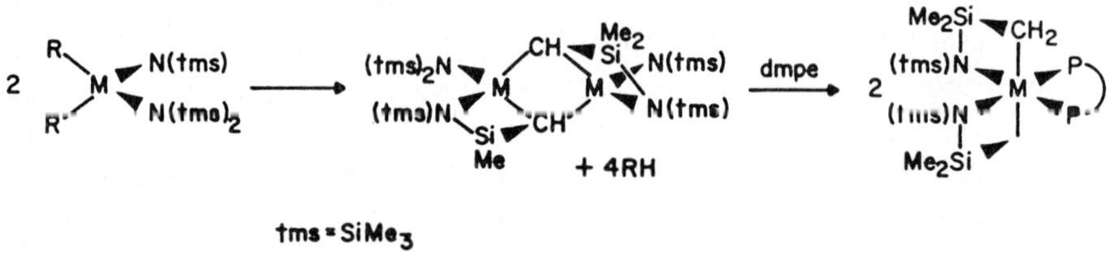

$$tms = SiMe_3$$

Scheme 3

cleavage and rearrangement to produce bismetalated, mononuclear products as shown (Scheme 3).[61] The lack of any thermal reactivity of the titanium derivatives is mechanistically significant as it implies rather strongly that a radical pathway is not operative (see Sec. VI.1.B). The fact that the lack of metalation in $R_2Ti[N(SiMe_3)_2]$ is a kinetic problem is confirmed by the ready formation of the metalated titanium dimer on reduction of $Cl_2Ti[N(SiMe_3)_2]$ by 2 equiv of Na/Hg.[60]

Studies of the actinide metal derivatives of the $N(SiMe_3)_2$ ligand by Andersen and co-workers has also allowed facile cyclometalation chemistry to be identified.[62] Furthermore, the monohydride derivative $HU[N(SiMe_3)_2]_3$ can undergo reversible metallocycle formation in the presence of added D_2, allowing the silylamido groups to be catalytically deuterated (Scheme 4).[62]

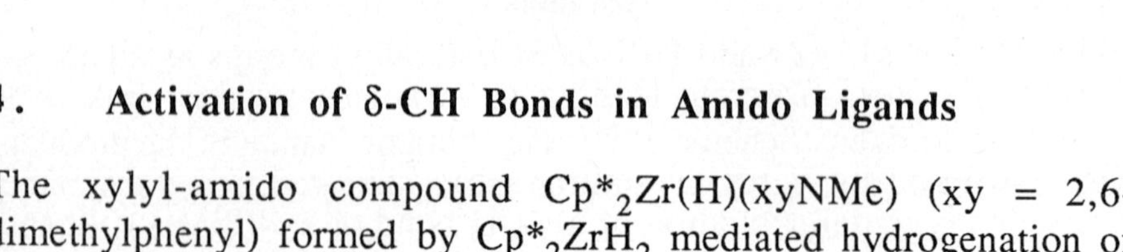

$$tms = SiMe_3$$

Scheme 4

4.   Activation of δ-CH Bonds in Amido Ligands

The xylyl-amido compound $Cp^*_2Zr(H)(xyNMe)$ (xy = 2,6-dimethylphenyl) formed by $Cp^*_2ZrH_2$ mediated hydrogenation of xylylisocyanide undergoes facile loss of H_2 and formation of a five-membered metallocycle (Eq. 15).[63] The reaction involves activation of the benzylic δ-CH bonds with elimination of H_2.

$$Cp^*_2Zr(H)\,(MeNC_6H_3Me_2\text{-}2,6) \longrightarrow Cp^*_2Zr(MeNC_6H_3Me\text{-}CH_2) + H_2 \quad (15)$$

IV.   Intramolecular Activation of CH Bonds in Oxygen and Sulfur Donor Ligands

1.   Activation of β- and γ-CH Bonds in Alkoxide and Thiophenoxide Ligands[64]

An exciting observation by Nugent et al. that various early transition metal alkoxides will catalyze the scrambling of the ^{2}H label of CH_3CH_2OD into the methyl group initially led to speculation that a pathway involving reversible activation of the γ-CH bonds of alkoxide ligands was operative.[50]   However, a careful reinvestigation of these systems concluded that processes involving initial β-CH bond activation and formation of organic carbonyl compounds led to a better interpretation of the observed results.[65a]   In particular, the multiple deuterium incorporation into the γ-positions was explained (via keto-enol tautomerism).

In a recent study, Buchwald et al. have reported the intramolecular activation of the β-CH bonds in thioalkoxide ligands. Hence, thermolysis of the thiobenzyl compounds $Cp_2Zr(CH_3)(SCH_2Ar)$ in the presence of PMe_3 leads to the elimination of methane and formation of the η^2-thioaldehyde, $Cp_2Zr(\eta^2\text{-}SCHAr)(PMe_3)$.[65b]

2.   Activation of Aliphatic ε-CH Bonds in Aryloxide Ligands

In the last few years there has been increasing interest in the use of sterically very demanding alkoxide,[66] aryloxide,[67] and related oxygen donor functionalities[68] as ancillary ligands in early transition metal chemistry.   However, as is the case in many areas of chemistry, the chosen "spectator" ligands are not averse to getting involved in the action.   Hence, use of the supposedly innocent, sterically demanding ligand 2,6-di-*tert*-butylphenoxide in order to support early transition metal organometallic chemistry is complicated by its sometimes facile cyclometalation.[22,69]   In the case of the Group 4 and 5 metals, mixed alkyl, aryloxides such as $M(OAr')_2(CH_2Ph)_2$ (M=Ti, Zr)[70,71] and $Ta(OAr')_2(CH_3)_3$[72] exhibit thermal instability characterized by the loss of alkane and the formation of new organometallic products containing six-membered metallocycle rings (Scheme 5).   The six-membered ring is formed by activation and cleavage of one of the CH bonds of one of

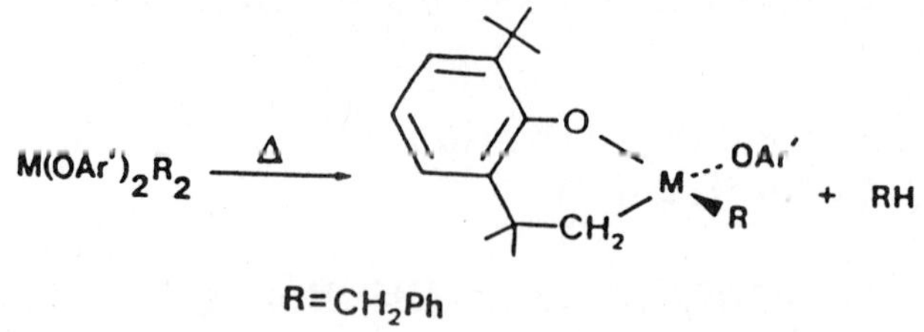

$$M(OAr')_2R_2 \xrightarrow{\Delta}$$

$$R = CH_2Ph$$

M	$\Delta H^{\ddagger}$/ Kcal.mol^{-1}	$\Delta S^{\ddagger}$/e.u.
Ti	23·0±0·7	−13± 3
Zr	21·6±1·0	−19± 3

Scheme 5

the aryloxide *tert*-butyl groups. In the case of tantalum, the trimethyl derivative Ta(OAr')$_2$(CH$_3$)$_3$ has been shown to undergo two sequential ring closure reactions with the elimination of two equivalents of methane (Scheme 6).[72] Kinetic studies of the thermolysis of the benzyl

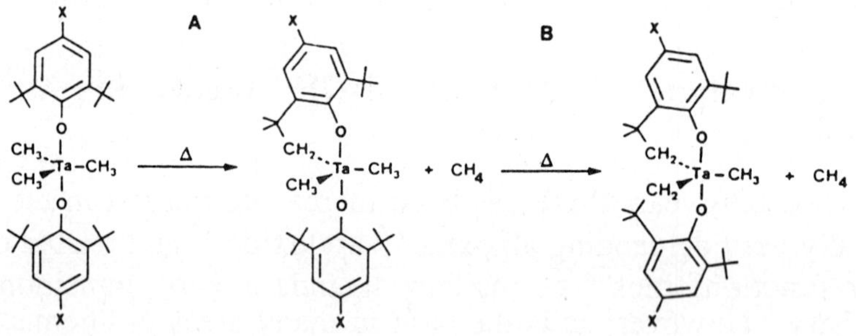

	$\Delta H^{\ddagger}$/ Kcal.mol^{-1}	$\Delta S^{\ddagger}$/e.u.
A	26·4±1·0	−7± 3
B	29·5±1·0	−6± 3

Scheme 6

compounds $M(OAr')_2(CH_2Ph)_2$ (M=Ti, Zr) show the ring closure reactions to be first order and that there is only a small dependence of the rate on the nature of the metal, Ti>Zr.[71] Furthermore, the use of a somewhat limited series of substituted benzyl leaving groups on zirconium indicated a negligible substituent effect on the rate of the reaction (Eq. 16). The use of the specifically deuterated phenoxide

$$Zr(OAr')_2(CH_2C_6H_4\text{-}X)_2 \longrightarrow Zr(OC_6H_3Bu^tCMe_2CH_2)(OAr')(CH_2C_6H_4\text{-}X)$$
$$+ CH_3C_6H_4X \qquad (16)$$

10^5k at 114°C: X=H, 32.3; 4Me, 39.2; 4F, 45.8; 3F, 18.4

ligand $OC_6D_2(4\text{-}CH_3)\text{-}2,6\text{-}t\text{-}C_4D_9$ (OAr*) has also allowed primary kinetic isotope effects (k_H/k_D) to be measured (Schemes 5 and 6).[69]

In the case of the tantalum trimethyl compound labeling studies using $Ta(OAr')_2(CD_3)_3$ and $Ta(OAr^*)_2(CH_3)_3$ also show the absence of pathways via methylidene intermediates.[69,72] However, in direct contrast to this thermal reactivity the photolysis of the trimethyl leads to the essentially quantitative formation of the methylidene compound $Ta(OAr')_2(=CH_2)(CH_3)$ and one equivalent of methane (Eq. 2 and Scheme 7). Furthermore, at room temperature this complex undergoes

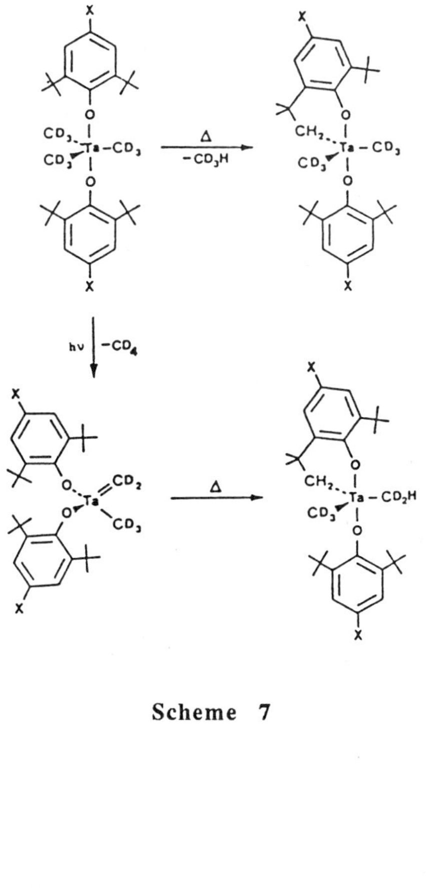

Scheme 7

isomerization involving intramolecular addition of an aliphatic CH bond of an aryloxide *tert*-butyl group across the tantalum-carbon double bond of the methylidene function to produce the monocyclometalated product also formed by thermolysis of $Ta(OAr')_2Me_3$.[28,71] Labeling studies clearly differentiate these two pathways (Scheme 7). Kinetic studies show the ring closure reaction of the methylidene compounds to be enthalpically facile but to be severely inhibited entropically (see Sec. VI. 2.C). Stable examples of related mixed alkyl, alkylidene compounds of tantalum are obtained on treating $Ta(OAr')_2Cl_3$ with $LiCH_2SiMe_3$ or $Mg(CH_2Ph)_2$.[73] In the case of the trimethylsilylmethylidene analog the product has been structurally characterized.

Treatment of the trichloride substrate $Ta(OAr')_2Cl_3$ with phenyllithium (3 equiv) leads directly to a monocyclometalated derivative and 1 equiv of benzene.[74,75] This reaction does not proceed via a triphenyl intermediate. Instead, the act of CH bond activation takes place after only one phenyl group has been introduced into the coordination sphere (Scheme 8).[75] Further thermolysis of this monometalated diphenyl compound leads to a second elimination of benzene and formation of a bismetalated product. Although kinetically a first-order reaction, labeling studies show that this second metalation step consists of two directly competing pathways (Scheme 9).[75] The first involves a direct loss of the phenyl leaving group with the

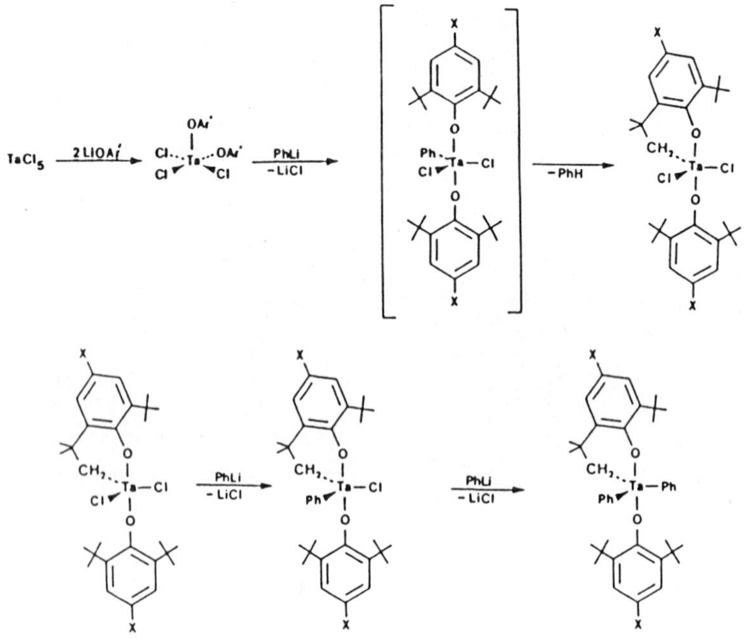

Scheme 8

atom of the aryloxide *tert*-butyl side chain. The second pathway involves the formation of an intermediate benzyne (*o*-phenylene) complex that then isomerizes to the bismetalated product. Both steps of this latter pathway involve the activation of CH bonds. The second step is significant as it represents the activation of a distal aliphatic CH bond by one side of the strained three-membered metallocycle ring of the *o*-phenylene functionality. A combination of kinetic measurements and product analysis from deuterated substrates has allowed the rates of the two pathways to be quantitated as well as the values of k_H/k_D to be obtained (Scheme 9). A further complication of this reactivity arises through the fact that the biscyclometalated product can reversibly form the *o*-phenylene intermediate at elevated temperatures (Scheme 9). This is most readily demonstrated in the observed isomerization of aryl ligands in the bis-metalated compound $Ta(OC_6H_3Bu^tCMe_2CH_2)_2$ (aryl)

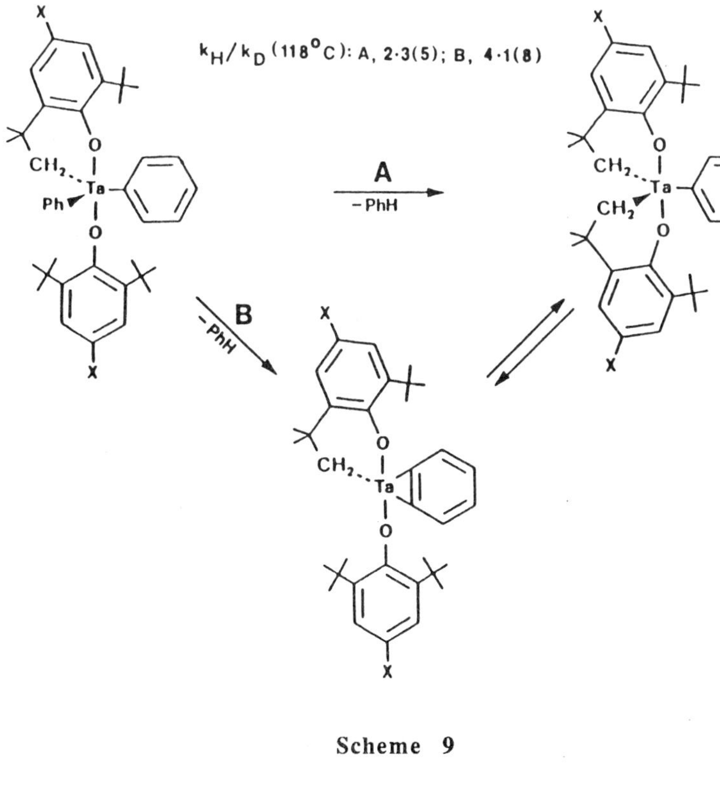

Scheme 9

as well as the scrambling of ^2H labels throughout all of the aryloxide aliphatic and Ta-Ph positions on extended thermolysis of $Ta(OC_6H_3Bu^tCMe_2CH_2)_2(C_6D_5)$.[75] As with the zirconium-benzyl compounds above, the introduction of substituents into the aryl leaving group ring was found to have a minimal effect on the rate of these reactions (Eq. 17).[75]

$$\text{Ta}(OC_6H_2Bu^t\text{-}4Me\text{-}CMe_2CH_2)(OAr'\text{-}4Me)(C_6H_4\text{-}X)_2 \longrightarrow$$
$$\text{Ta}(OC_6H_2Bu^t\text{-}4Me\text{-}CMe_2CH_2)_2(C_6H_4\text{-}X) + C_6H_5\text{-}X \qquad (17)$$

10^5k at 118°C: X = H, 7.3; 3Me, 10.4; 4Me, 8.9; 3F, 14.3; 4F, 9.6

3.   Activation of Aromatic ε-CH Bonds in Aryloxide Ligands

The facile intramolecular activation of the aliphatic side chain observed for 2,6-di-*tert*-butylphenoxide ligands has led to an investigation of the possibility for related reactivity involving side chain aromatic CH bonds. Specifically the ligand 2,6-diphenylphenoxide (OAr") has been shown to also undergo facile six-membered metallocycle formation at Group 5 and 6 metal centers.[69,76,77] Although this reactivity is stoichiometrically related to that seen for 2,6-di-*tert*-butylphenoxide, the aromatic substrate offers the potential for different reaction pathways to

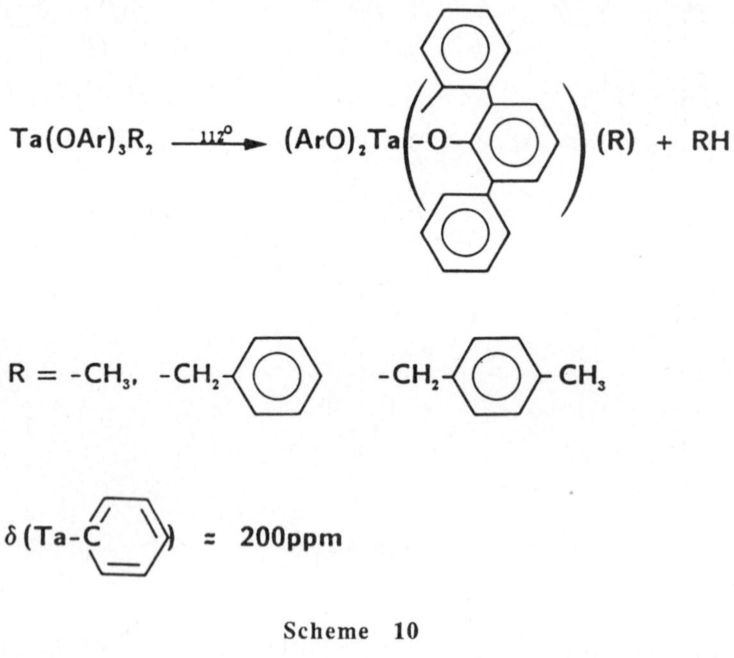

$$\text{Ta}(OAr)_3R_2 \xrightarrow{\text{uz}^o} (ArO)_2Ta\left(-O-\bigcirc\right)(R) + RH$$

$$R = -CH_3, -CH_2-\bigcirc \qquad -CH_2-\bigcirc-CH_3$$

$$\delta(\text{Ta-C}\,\bigcirc) = 200\text{ppm}$$

Scheme 10

be operative.[69] In the case of tantalum the bis- and tris-alkyls of stoichiometry Ta(OAr")$_3$R$_2$ and Ta(OAr")$_2$R$_3$ (R = CH$_3$, CH$_2$Ph) have been isolated and shown to undergo clean ring closure reactions on thermolysis to form the expected metallocycle ring and 1 equiv of alkane with first-order kinetics (Scheme 10).[76]   A labeling study using $Ta(OAr'')_3(CD_3)_2$ shows that the monocyclometalated product is formed directly with no methylidene intermediate.   The metalated product has been structurally characterized and again an almost planar six-membered ring is readily accommodated at the metal as a result of the large Ta-O-C angles.[69,76]   A kinetic study of this ring closure reaction again shows moderately negative entropies of activation (Scheme 10).[76]

V.   Intermolecular Activation of Hydrocarbon CH Bonds

1.   Activation of Aromatic Hydrocarbons by Benzyne Intermediates

In a number of cases it has been shown that the *ortho*-hydrogen abstraction reaction of diphenyl complexes leading to metal benzyne (*o*-phenylene) derivatives plus eliminated benzene can be reversed (Eq. 6).[78,79]   This was first demonstrated by Erker, for compounds of the type Cp_2ZrAr_2.[78]   Hence the thermolysis of bistolyl derivatives in benzene was found to result in the formation of mixed phenyl, tolyl derivatives and toluene.   Product analysis was carried out by photochemically induced coupling of the aryl group to produce biphenyl molecules that could be readily characterized.[78]

   Marks and co-workers have shown that the stoichiometrically related diphenyl $Cp*_2U(C_5H_5)_2$ will also undergo exchange with aromatic solvents.[79]   The benzyne pathway is confirmed by carrying out the thermolysis in C_6D_6 whereupon $Cp*_2U(C_6D_5)(C_6H_4D)$ and C_6H_6 are generated (Eq. 18).   The monodeuterated phenyl group contains the ^2H label exclusively in the *ortho* position.[79]

$$Cp*_2U (C_6H_5)_2 \xrightarrow[-C_6H_6]{} Cp*_2U (\eta^2\text{-}C_6H_4) \xrightarrow[+C_6D_6]{} Cp_2U (C_6H_4D) (C_6H_5) \quad (18)$$

2.   Hydrocarbon Activation by $Cp*_2Th(CH_2CMe_2CH_2)$

As mentioned previously, Marks and co-workers have shown that a series of metallocyclobutane rings can be readily formed by thermolysis of dialkyl substrates such as $Cp*_2Th(CH_2XMe_3)_2$ (X = C, Si) (Scheme

1).[42]   In the case of the parent, thoracyclobutane formed by cyclometalation of neopentyl groups, careful thermodynamic measurements have indicated the presence of considerable amounts of strain in the four-membered ring.[43] This results in a particularly low bond disruption energy for the opening up of one side of the metallocycle. Fendrick and Marks have capitalized on the high energy of this material by demonstrating its sometimes facile stoichiometric functionalization of hydrocarbon molecules.[80] Some of these reactions are shown in Scheme 11. The lack of reaction with cyclohexane, presumably   on   steric   grounds,   makes   available   a   convenient

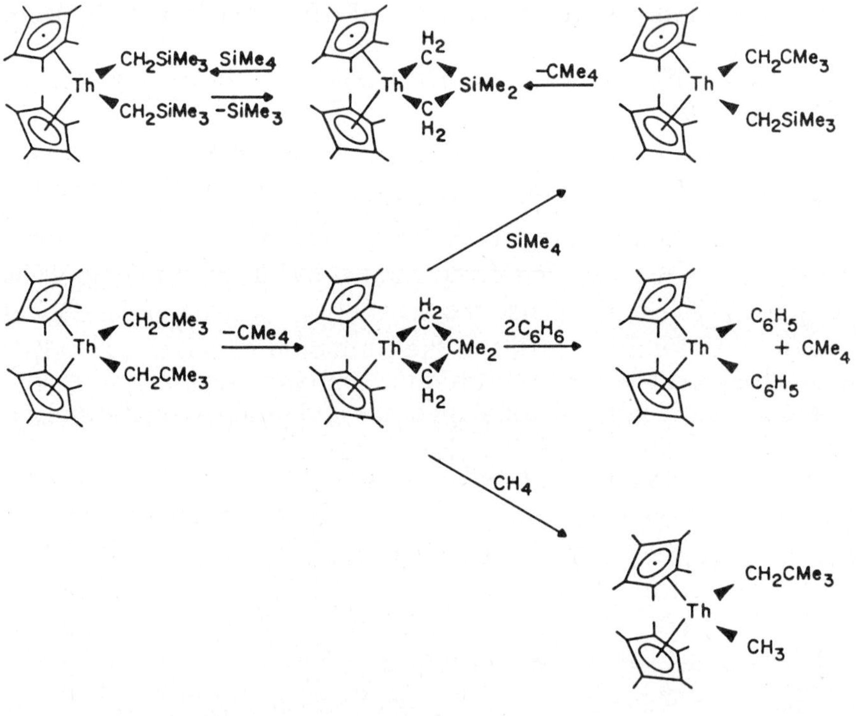

Scheme 11

hydrocarbon solvent in which the reactions can be carried out and monitored by NMR techniques. The most dramatic result involves the direct reaction with methane at 60°C to yield the ring opened, monomethyl compound $Cp^*_2Th(CH_2CMe_3)(CH_3)$ in ~ 50% yield. A primary kinetic isotope effect, k_H/k_D of 6±2 at 60°C was measured by reacting the thoracyclobutane substrate with CD_4. The ring opened neopentyl group was shown to contain one ^2H label, confirming the lack of involvement of the C_5Me_5 rings.[80] Reaction with ethane was studied briefly and shown to take place more slowly than the methane reaction. However, no ethyl products were detected and it is unsure whether the final products obtained were simply the result of the simple thermolysis of the thoracyclobutane substrate or the presence of β-CH bonds in the expected ethyl derivatives that would allow facile decomposition.[80] With the β-elimination stabilized EMe_4 (E = Si, Sn), substrates clean ring opening takes place rapidly at ambient temperatures. A kinetic study of the Me_4Si reaction showed a first-order dependence on both the metallocyclobutane and TMS substrate.[80] The ring opening was faster for the tin compound.

The reactions with cyclopropane and benzene are characterized by a rapid initial ring opening to generate intermediate, mixed alkyl complexes. A subsequent reaction then involves displacement of the neopentyl group generated in the first step by a second equivalent of hydrocarbon to yield biscyclopropyl and bisphenyl compounds, respectively (Scheme 11).[80] The less strained silicon containing metallocycle, $Cp^*_2Th(CH_2SiMe_2CH_2)$ is found to be less reactive than its carbon analog. However, the compound will undergo ring cleavage with Me_4Si to regenerate the $Cp^*_2Th(CH_2SiMe_3)_2$ substrate from which it can be formed (Scheme 11).[80,81] The thermodynamic aspects of this equilibrium situation are of mechanistic significance (see Sec. VII).[81]

3. Hydrocarbon Activation by Cp^*_2MX (M = Lu, Y; X = H, CH_3)

In a series of communications[23,52,82] and a review,[83] Watson and co-workers have outlined the remarkable reactivity of the hydride and methyl compounds Cp^*_2LuH, Cp^*_2LuMe and their yttrium analogs. Teuben et al. have also outlined some reaction chemistry of Cp^*_2YR compounds in general.[84] The methyl compounds exist in a dimer-monomer equilibrium involving a single methyl bridge (see Sec. VI.2. A). In cyclohexane-d_{12} solvent, the hydride and methyl compounds were found to rapidly insert into a wide range of sp^2 and sp^3-hybridized CH bonds (Scheme 12).[23,52,82,83] The degenerate, self-exchange with methane was demonstrated by the use of labeled $^{13}CH_4$.[23] Over a period of hours at 70°C the formation of $Cp^*_2Lu^{13}CH_3$ was observed by ^1H-NMR. Kinetic studies showed both a unimolecular as well as bimolecular pathway. The unimolecular path was thought to involve initial metalation of the Cp* ring to form a "tuck-in" intermediate complex that then rapidly reinserted a molecule of methane.[83] The predominant, bimolecular step, first order in both metal and CH_4, involves direct exchange of methyl groups. The yttrium derivative $Cp*_2YCH_3$ was found to react even more rapidly.[83] With ethane and propane, reactions were found to be slower than with methane. However, the expected alkyl products were not detected because of facile β-hydrogen transfer. The β-hydrogen lacking substrate Me_4Si reacted cleanly on warming to yield the stable trimethylsilylmethyl derivative and methane (Scheme 12).[83]

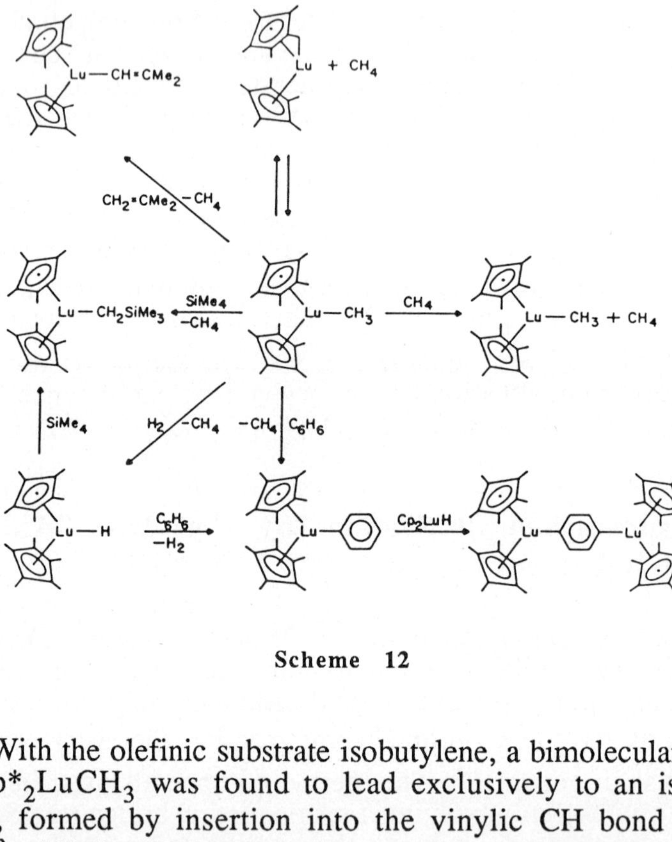

Scheme   12

With the olefinic substrate isobutylene, a bimolecular reaction with $Cp*_2LuCH_3$ was found to lead exclusively to an isobutenyl product, formed by insertion into the vinylic CH bond (Scheme 12).[82,83] Allylic CH bond activation was also observed for other olefins such as propylene. Rapid reactions of both the hydride, $Cp*_2LuH$ and the methyl, $Cp*_2LuCH_3$ with benzene was also observed.[52,83] In the case of the hydride, formation of a phenyl intermediate is reversible allowing H-D scrambling among benzene molecules to be catalyzed. On standing, solutions of the phenyl compound undergo a slower reaction leading to a *para*-dimetalated phenyl complex (Scheme 12). With the methyl compound, benzene was found to react both in a unimolecular (via a "tuck-in" intermediate) as well as bimolecular fashion (Scheme 12).[52,83]

Teuben et al. have reported that the hydride $(Cp*_2YH)_n$ will smoothly catalyze H-D exchange into various sp^2 and sp^3 CH bonds at room temperature.[84] In the case of the substrate $(Me_3Si)_2CH_2$ no exchange into the methylene proton positions is seen even at 100°C, presumably for steric reasons. Because H-D exchange into the methyl groups of C_5Me_5 ligands is relatively slow, the involvement of a "tuck-in" intermediate in these exchanges is disfavored.[84]

4. Hydrocarbon Activation by $Cp*_2ScR$ (R = H, Alkyl, Aryl)

The ability of the organoscandium compounds $Cp*_2ScR$ (R = H, R, Ar) to activate a wide range of CH bonds in hydrocarbon substrates has been studied in great detail by Thompson et al.[51] and Bercaw.[85] Structural and spectroscopic studies show the simple hydride to be oligomeric, whereas the thf adduct $Cp*_2ScH(thf)$ as well as the alkyl and aryl derivatives are monomeric. Although single crystal structure determinations of $Cp*_2ScCH_3$ and $Cp*_2ScCH_2CH_3$ confirmed their monomeric nature, severe disorder precluded obtaining any evidence for "agostic" interactions in these very electron-deficient (14-electron) compounds.[51]

The base-free monohydride, $(Cp*_2ScH)_n$, was shown to be a potent catalyst for H-D exchange between D_2 and aromatic hydrocarbons. With toluene-d_8 substrate treated with $Cp*_2ScH/H_2$, exchange into the aromatic ring was ~ 70 times faster than into the methyl side chain. Incorporation of ^1H into the *meta/para* positions also was faster than into the *ortho-* position although a statistical distribution of label within the aromatic ring resulted on extended thermolysis (80°C, 48 hr). A range of other substrates was found to be deuterated, albeit much slower than benzene, by $Cp*_2ScD/D_2/C_6D_6$ mixtures. Although 1° CH bonds of hydrocarbons, e.g., CH_4, $SiMe_4$, methyl groups of $CH_3CH_2CH_3$ and CH_3CH_2Ph were found to exchange, the incorporation of ^2H label into 2° aliphatic CH bonds was much slower. Hence cyclohexane proved to be a valuable, inert solvent for studying this reactivity. The equilibrium between $Cp*_2ScH$ and $Cp*_2ScC_6H_5$ in C_6H_6 solvent (Eq. 19) was used to measure the relative strengths of the Sc-H and Sc-Ph bonds, the former being found to have a bond dissociation energy 1.5(4) kcal/mol larger than the latter.[51]

$$Cp*_2ScH + C_6H_6 \rightleftharpoons Cp*_2ScC_6H_5 + H_2 \qquad (19)$$

The methyl compound $Cp*_2ScCH_3$ was found to undergo direct reaction with $^{13}CH_4$ to produce $Cp*_2Sc^{13}CH_4$ at 70°C in cyclohexane- d_{12} (Scheme 13). The reaction is slower than with the Y, Lu analogs previously discussed[83] and is found to follow second-order

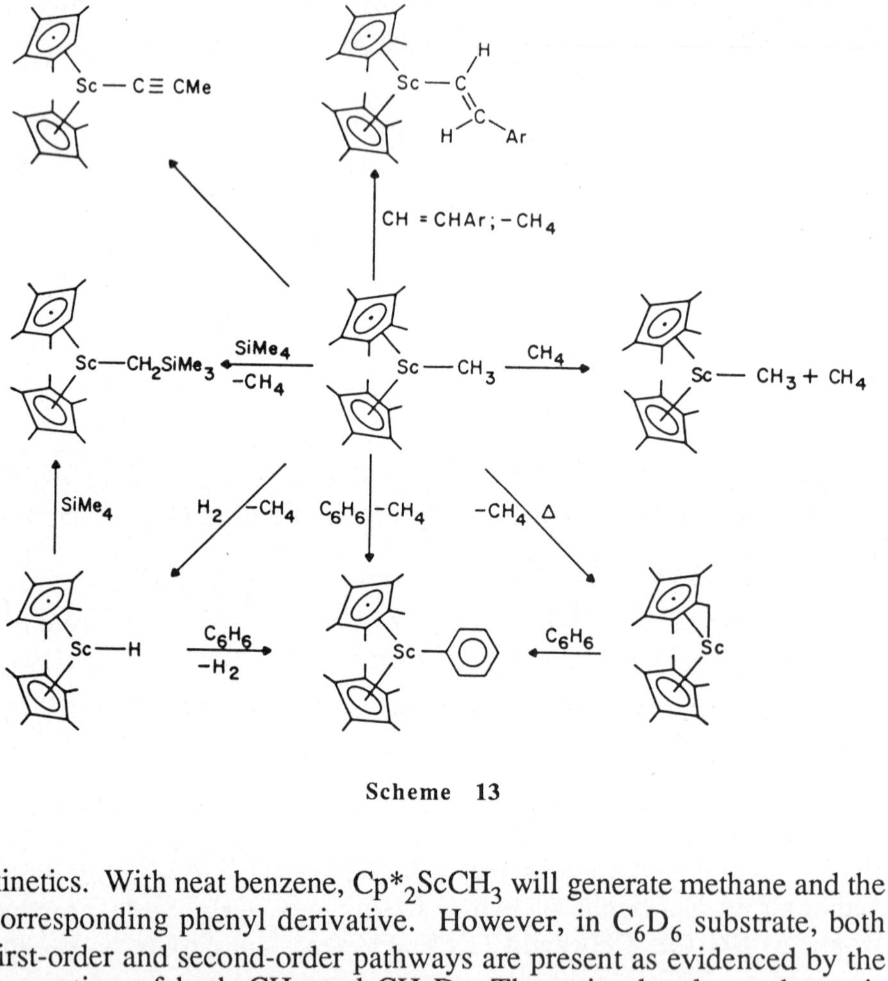

Scheme 13

kinetics. With neat benzene, $Cp*_2ScCH_3$ will generate methane and the corresponding phenyl derivative. However, in C_6D_6 substrate, both first-order and second-order pathways are present as evidenced by the generation of both CH_4 and CH_3D. The unimolecular pathway is believed to proceed via a "tuck-in" intermediate that can also be obtained by thermolysis of $Cp*_2ScCH_3$ in cyclohexane-d_{12} (Scheme 13). The products of the reaction of olefins with $Cp*_2ScCH_3$ are highly dependent on the olefin structure.[51] Ethylene is rapidly polymerized whereas propene undergoes insertion to produce an isobutyl group that then undergoes further reaction with propene to generate a trans vinyl complex $Cp*_2Sc(CH=CHMe)$ and isobutane. With isobutylene and styrene as substrates, methane is eliminated directly with the formation of metal-vinyl ligands (Scheme 13). The rate of reaction of $Cp*_2ScCH_3$

with styrenes containing electron-donating (OMe) or withdrawing (CF_3) *para* substituents were found to be almost identical (Eq. 20). Similarly

$$Cp*_2ScCH_3 + H_2C=CHAr \longrightarrow trans\text{-}Cp*_2Sc(CH=CHAr) + CH_4$$

(20)

$10^4 k/sec^{-1} M^{-1}$ at 60°C; Ar = $4\text{-}CF_3C_6H_4$, 8.5(5); Ar = $4\text{-}OMeC_6H_4$, 12.5(7)

the rate of reaction with substituted benzenes was only slightly modified by the introduction of substituents (NMe_2, CH_3, CF_3) into the aromatic ring (Eq. 21).[51] In this latter study the use of $(Cp*\text{-}d_{15})_2ScCH_3$ as

$$(Cp*\text{-}d_{15})_2ScCH_3 + 4\text{-}XC_6H_5 \longrightarrow (Cp*\text{-}d_{15})_2ScC_6H_4\text{-}4X + CH_4$$

(21)

$10^5 k/sec^{-1} M^{-1}$ at 80°C: X = NMe_2, 3.2(1); CH_3, 3.4(1); H, 3.3(1); CF_3, 1.4(1)

metal reagent allowed the importance of the indirect, unimolecular pathway to be evaluated by the ratio of CH_3D/CH_4 generated. The lack of any dramatic substituent effect as well as the almost statistical ratio of attack at the *meta* and *para* positions of toluene prove to be of mechanistic significance (see Sec. VI.2. D).

A much more rapid reactivity was found when using propyne as hydrocarbon substrate. The acetylide $Cp*_2Sc(C\equiv CMe)$ is formed both with $Cp*_2ScCH_3$ and $Cp*_2ScH$ (no insertion observed) (Scheme 13).[51]

VI. Mechanistic Aspects of CH Bond Activation at d⁰-Metal Centers

One of the most impressive aspects of the recent rapid development of early transition metal organometallic chemistry associated with CH bond activation has been the careful attention to mechanistic details by the majority of groups in the area. This has resulted in a significant body of important kinetic and other data being accumulated in the literature. These data present a consistent and cohesive picture of these reactions taking place via heterolytic pathways involving a multicenter transition state. However, although this picture is now generally accepted, it is important to review the arguments against other pathways that may indeed be viable mechanisms in systems yet to be developed.

1. Redox Pathways

A. Oxidative-Addition, Reductive-Elimination Sequences

The d^0-electronic configurations of metal centers such as M(III) (M = Sc, Lu, Y), M(IV) (M – Ti, Zr, IIf), M(V) (M – Nb, Ta), and Th(IV) preclude pathways involving direct oxidative-addition of the CH bonds as the initiation step. However, polyhydride compounds such as Cp_2TaH_3 that formally contain a d^0-metal center will activate and scramble aromatic CH bonds via pathways involving initial reductive-elimination of H_2 to produce an unsaturated Ta(III) center that then can oxidatively add the aromatic CH bond.[4-6,86] Further strong evidence for the potency of d^2-metal centers such as Ti(II) and Ta(III) for at least intramolecularly, oxidatively adding CH bonds comes from carrying out "classical" reduction of chloride substrates such as $Cl_2Ti[N(SiMe_3)_2]_2$,[60] $Cl_3Ta(OAr')_2$, $Cl_2Ta(OAr'')_3$,[69] and $Cl_2Ta(OR)_3$ (OR = silox)[66] (Eqs. 22 and 23). In the case of the silox compound the

(a) $Ta(OAr')_2Cl_3$ + 2 Na/Hg $\longrightarrow$ $Ta(OC_6H_4Bu^tCMe_2CH_2)_2Cl$ + H_2 + 2 NaCl

(b) $Ta(OAr'')_3Cl_2$ + 2 Na/Hg $\longrightarrow$ $Ta(OC_6H_3PhC_6H_4)_2(OAr'')$ + H_2 + 2 NaCl   (22)

(OAr' = 2,6-di-*tert*-butylphenoxide; OAr'' = 2,6-diphenylphenoxide)

$Ta(OR)_3Cl_2$ + 2 Na/Hg $\longrightarrow$ $Ta(OR)_3$ + 2 NaCl $\longrightarrow$ $Ta(OSiBu^t_2CMe_2CH_2)(H)(OR)_2$ (23)
(OR = $OSiBu_3^t$, Silox)

$Ta(OR)_3$ intermediate is insoluble prior to intramolecular oxidative-addition of a CH bond.[66] However, in the case of the d^0-alkyl substrates discussed that typically undergo cyclometalation reactions or intermolecular hydrocarbon activation, there is no evidence (product analysis) to suggest that there is any significant pathway occurring via initial reductive-elimination from the metal center taking place to generate such reactive d^2-metal centers.

In the case of the very important metal reagents $Cp*_2MX$ (M = Sc, Y, Lu; X = H, R, Ar) that can readily functionalize hydrocarbons, an intramolecular reductive-elimination involving transfer of X to a C_5Me_5 ring is certainly conceivable. This scenario would generate a reactive $(\eta^4-C_5Me_5X)M(\eta^5-C_5Me_5)$ intermediate that could oxidatively add RH prior to product formation and elimination of HX. However, the cases in which first-order kinetics is present for reactions involving C_5Me_5 ancillary ligands, labeling studies combined with product analysis rule out this pathway in favor of intermediates of the "tuck-in" kind.

Also in this context special mention has to be made of those reactions (e.g., Schemes 2^{48} and 7^{72}) that proceed via an initial α-hydrogen abstraction process to produce an alkylidene intermediate. The general consensus of opinion based both on reactivity and theoretical studies views such molecules as indeed containing alkylidene (CR_2^{2-}) ligands.[24-27] Hence the process of α-hydrogen abstraction does not change the metal from its initial d^0-electronic configuration. An alternate analysis, however, might view the new functionality as a neutral carbene ligand (CR_2), formally giving the metal a d^2-electron count. In this scenario it might further be proposed that CH bond activation is now viable via an initial oxidative-addition to the metal center. The observed product would then be formed by a rapid hydrogen atom migration converting the $M=CR_2$ unit into an alkyl group. It is difficult to completely negate this possible pathway. However, based on all of the known chemistry and spectroscopy of early transition metal alkylidenes of this type there is no evidence to suggest that there is any significant electron density at the metal center available for such reactions.[24-27] A similar scenario holds for *o*-phenylene (benzyne) as well as "tuck-in" (Sec. II.7) intermediates. Structural as well as reactivity studies indicate that the η^2-C_6H_4 group is best considered as $C_6H_4^{2-}$ strongly σ-bound to the metal in a strained metallocyclopropene ring, as against a π-bound benzyne, C_6H_4.[34,35] Even for acetylene bound to metal centers such as Zr(II) there is good evidence for a picture involving Zr(IV) and $C_2R_2^{2-}$.[87,88]

B. Homolytic Pathways

Given the fact that initial homolysis of metal alkyl bonds is used to explain the sometimes explosive thermal decomposition of a number of early transition metal per-alkyl compounds,[89,90] it is not surprising that a homolytic pathway is given some consideration for intra- and intermolecular CH bond activation reactions involving alkyl leaving groups (Scheme 14). However, a large number of observations on a diverse range of different systems is inconsistent with a pathway involving initial, thermal M-R bond homolysis followed by intra- or intermolecular hydrogen atom abstraction (Scheme 14). Although not all of these data can be reviewed here, the types of arguments used can be briefly outlined.

$$L_nM-R \longrightarrow L_nM{\cdot} + R{\cdot} \xrightarrow{R'H} L_nM-R'$$

Scheme 14

The most straightforward argument against a radical pathway for intermolecular hydrocarbon activation comes from the typical bimolecular form of the rate equation, rate = $k_{obs}[L_nMR][R'H]$. A preequilibrium, solvent caged, radical pair might confuse this simple analysis. However, analysis of the activation parameters associated with those systems studied by kinetics shows $\Delta H^{\ddagger}$ values that are consistently lower than one would expect for complete homolysis of M-R bonds.[91] In the case of the Th(IV) systems studied by Marks and co-workers,[42] hard thermodynamic data on bond disruption energies for cyclometalation substrates (Scheme 1) as well as the metallacyclobutane (Scheme 11) are available,[43] while it is possible to estimate values for other early transition metal systems based on literature values.[91] In all cases, even with pessimistic estimates of M-R bond strengths, the mild conditions typically needed to effect these reactions would not be expected to cause any disruption of these bonds. Furthermore, the thermal stability of related molecules that do not contain cyclometalatable ancillary ligands argues against any unforeseen weaknesses in the metal-alkyl bond. Hence the thermal stability of $Ta(OAr-2,6Me_2)_2Me_3$ at temperatures of 150-200°C for weeks contrasts with the first ring closure of $Ta(OAr')_2Me_3$ that occurs readily at 70°C.[69] A dramatic example of this type of argument is shown in the stability of the titanium compound $Me_2Ti[N(SiMe_3)_2]_2$ at 190°C for extended periods whereas the complexes $R_2M[NSiMe_3)_2]_2$ (M = Zr, Hf) readily lose alkane at 160°C (Scheme 3).[60] If a rate-determining homolytic cleavage of the M-R bond was important, one would expect a much more rapid reaction for M = Ti given the demonstrated weakness of bonds to this metal compared to zirconium and hafnium. The lack of CH bond activation in the titanium complex can probably be associated with a problem with the CH bond approaching the much smaller metal center for activation.

The clean nature of nearly all of these reactions coupled with the lack of any solvent participation, even when good hydrogen atom donors are used as reaction media, also tends to argue against radical intermediates. Again a "caged" radical intermediate could be invoked to account for the lack of solvent participation. However, the photochemical reactivity of some of these alkyl substrates has been shown to be dominated by the sometimes highly efficient photolytic cleavage of M-R bonds. In cases in which studies have been made, no related CH bond activation products are generated photochemically.[69] In the case of the dibenzyls $M(OAr')_2(CH_2Ph)_2$ (M=Ti, Zr) (Scheme 5) photolysis efficiently generates benzyl radicals that can either be quenched with suitable hydrogen atom donors, or else allowed to dimerize.[69] Only reduced metal compounds are generated with no evidence for radical abstraction of hydrogen atoms from the *tert*-butyl side chains. In the case of $Ta(OAr')_2Me_3$, irradiation into the alkyl to metal charge transfer bands causes bond homolysis with a quantum efficiency of ~ 100%.[72] However, no metalated products are generated directly. Instead α-hydrogen abstraction to produce a methylidene compound takes place (Scheme 7). This then does thermally convert to a monometalated compound. However, labeling studies show the thermal and photochemical routes to be different (Scheme 7).[28,72]

2. Multicenter, Heterolytic Pathway

All of the data obtained on these and related systems can best be accommodated into a heterolytic pathway proceeding via a four-center, four electron transition state (Scheme 15). A much related pathway

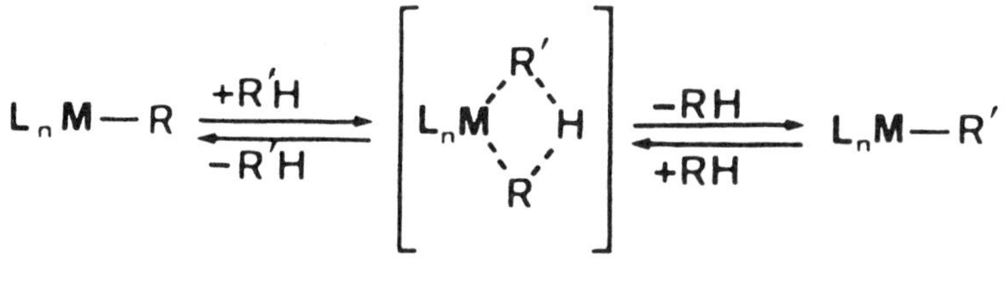

$$L_nM-R \; \underset{-R'H}{\overset{+R'H}{\rightleftharpoons}} \; \left[L_nM \overset{R'}{\underset{R}{\diamondsuit}} H \right] \; \underset{+RH}{\overset{-RH}{\rightleftharpoons}} \; L_nM-R'$$

Scheme 15

has been proposed for hydrogenolysis reactions involving early metal-alkyl bonds[92-95] and alkyllithium reagents[96] and it seems reasonable that other types of reactions such as cleavage of M-R bonds by more protic reagents may also be related. This pathway allows the reaction to proceed in a concerted fashion with no change in the formal oxidation state of the metal (Scheme 5). Theoretical analysis of these reaction pathways has been reported.[97,98]

The most important evidence in support of the multicenter pathway for these reactions comes from kinetic data. In the case of intermolecular reactions that are not complicated by the formation of intermediates, second-order kinetics is obeyed. Activation parameters show negative values of $\Delta S^{\ddagger}$ consistent with the bimolecular nature of the reaction. A consistent characteristic of the intramolecular cyclometalation reactions that have been examined kinetically is the presence of a moderate to large negative value of the entropy of activation (Eq. 7, Schemes 1, 5, and 6).[22] Given that the kinetic data are typically obtained in nonpolar, hydrocarbon solvents these values cannot be ascribed to changes in solvation on proceeding to the transition state. In order to explain these data it is typically argued that the attainment of the multicenter transition state in which the ligand geometry is highly constrained represents a severe loss of vibrational and particularly rotational entropy. Entropic factors can similarly be used to explain the importance of steric crowding in the metal coordination sphere in order to ensure facile cyclometalation rates. This necessity for bulky ligands was first recognized by Shaw[99] and used to design, with some rationality, phosphine ligands that would more readily metallate. Restricting the movement of the ligand to be metalated in the ground state, it is argued, will reduce the entropic loss on going to the even more structured transition state.[99] This idea was recognized very early in organic chemistry in which ring closure reactions were found to be enhanced by the presence of substituents on the carbon skeleton. This phenomenon is generally referred to as the "gem-dimethyl" or "Thorpe-Ingold" effect.[100]

Further strong support for the proposed heterolytic pathway comes from a growing body of measurements of k_H/k_D for these reactions. In all cases primary kinetic isotope effects are seen on deuterating either the CH bonds of the ligand or hydrocarbon substrate consistent with the CH bond breaking taking place in the transition state. Furthermore, the values of k_H/k_D are sometimes significantly large (Eq. 7, Schemes 1 and 9) compared to classical values calculated for linear CH bond breaking.[101] Interpretation of the magnitudes of these primary KIE's is, however, made difficult by the nonclassical geometry of the proposed transition state.[102]

If at this stage the proposed multicenter pathway is accepted on the basis of this evidence, a plethora of important points are raised concerning more specific characteristics of the transition state and the intimate mechanism that leads to it. In the next few sections some of these points are elaborated on, hopefully with an open mind concerning the possible presence of other as yet unforeseen mechanisms which may equally explain the data obtained for these CH bond activation reactions.

A.   "Agostic" Precursors to CH Bond Activation

A crucial point is whether or not the transition metal center initially interacts with and activates the CH bond prior to the concerted bond cleavage (Scheme 16). There is now extensive proof in the literature (both structural and spectroscopic) of the ability of high-valent, electron-deficient, early transition metal centers to undergo ground electronic interactions (bonding) with normally inert CH bonds.[103-108] The term agostic, first proposed by Brookhardt and Green, is typically used to describe this situation. In such cases the electron-deficient, Lewis acidic metal is hindered from seeking more classical donor atom lone pairs of electrons, and hence CH bonds of suitable alkyl fragments of coordinated ligands are forced to act as electron donors to the metal center. In the case of Ti(IV), Green has shown that such interactions are possible in the ground state between α- and β-CH bonds of alkyl groups.[109] The close approach of more "distal" CH bonds to related

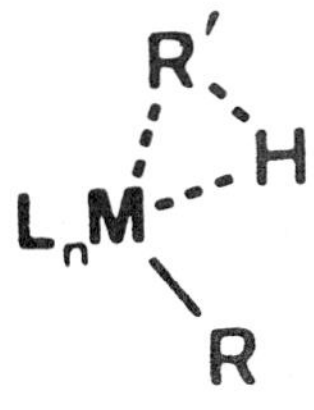

$$R' \cdots H$$
$$L_nM$$
$$R$$

Scheme 16

metal centers has also been observed. In certain derivatives of 2,6-*tert*-butylphenoxide (OAr'), *tert*-butyl methyl groups have been shown to be located closer to the metal center than would be expected on the basis of van der Waals radii.[69,110] In the case of the bridging alkylidyne complex $(Ar'O)_2Ta(\mu\text{-}CSiMe_3)_2Ta(CH_2SiMe_3)_2$ the distances to hydrogen atoms were refined as 2.64(6) and 2.84(9) Å.[110] However, in the absence of any confirmatory spectroscopic data it is impossible to conclude whether an agostic situation is present in these compounds or not.

A series of structural studies on very electron-deficient ytterbium(II) complexes of $N(SiMe_3)_2$ ligands by Andersen and co-workers has identified a different type of metal-CH interaction.[111-113] In the compounds $[N(SiMe_3)_2]Yb(dmpe)$,[111] $NaYb[N(SiMe_3)_2]_3$,[112] and $Yb_2[N(SiMe_3)_2]_4$[113] there are definite indications of bonding between the $SiMe_3$ methyl groups and the metal center. However, the primary interaction in all cases appears to be between Yb and the carbon center so that the CH_3 groups appear to be bridging between Yb and Si. Solution spectra and solid state infrared data failed to indicate the maintenance of these interactions in solution and did not indicate the presence of anomalous CH bond vibrations.[111-113]

These three-center, two-electron MHC bonding situations are very attractive as precursors leading to the proposed four-center transition state. The removal of electron density from the σ-orbital of a CH bond by the electron-poor metal would be expected to lead to an increase in the acidity of the bond. In fact, in the absence of a leaving group participation one might speculate on the possibility of a direct metalation of the carbon atom by the metal center (S_E2) with loss of a proton to a suitable external base. This proton loss exo to the complex could take place either with or without retention of configuration at carbon (Scheme 17).[114] For the systems being considered in this work, there is strong kinetic evidence against a stepwise reaction aided by an external base. However, the four-center transition state can be easily viewed as the intramolecular analog of this process with the transfer of this protic hydrogen to the alkyl leaving group "base" (see Sec. VI.2.D).

The intermolecular counterpart of the "agostic" interaction of

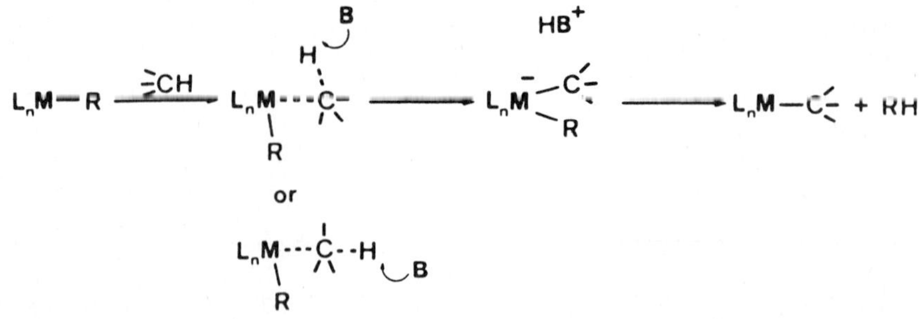

Scheme   17

ligand CH bonds with metal centers (hydrocarbon complexation) is understandably not known under ambient conditions.[115,116] The weakness of such bonding will preclude, in the absence of chelate effects, the routine observation of such fragile species. However, there are a number of interesting organolanthanide compounds that have been studied that, it has been argued, give some insight into how a molecule such as methane would interact (prior to CH bond breaking) with electron-deficient metal centers. As mentioned earlier, the highly unsaturated monomers, $Cp^*_2MCH_3$ (M = Y, Lu) readily condense to form dimers in both solution and the solid state.[83] Structural studies show the presence of one unique methyl group bridging between the two metal centers with an almost linear $M-CH_3-M$ angle. This situation can be considered as representing the donation of electron density from one of the $M-CH_3$ groups to the other metal center. The linearity of the interaction, as compared to more classical bridging methyl groups (e.g., in Al_2Me_6), means that the three CH bonds are actively involved in the electron donation. This concept has been elaborated by Burns and Andersen in the synthesis and study of the complex $Cp^*_2Yb(\mu-Me)BeCp^*$.[117] Here there is again a linear $Yb-H_3C-Be$ bond. The structure has been interpreted as representing the coordination of a "functionalized" methane, H_3CX (X = $BeCp^*$) donating electron density to the highly electron deficient Cp^*_2Yb metal center.[117]

B.   Importance of the Metal Center and Ancillary Ligation

A high electrophilicity of the metal center is clearly an essential requisite for this type of reactivity to be feasible. The electrophilicity will be a consequence not only of the actual chemical nature of M and its oxidation state but also be "tuned" somewhat by the ancillary ligands bound to it. It is also crucial that vacant frontier molecular orbitals of the correct symmetry be available for accepting the electron density of the CH bond to be activated. Again ancillary ligation will also be critical here, sterically maintaining mononuclearity without tying up too many orbitals or overcrowding the coordination sphere. In the area of CH bond activation the only clear cut reactivity comparison of metals involves methane exchange involving the $Cp^*_2MCH_3$ (M = Sc, Y, Lu) substrates.[51,83] The reactivity parallels the predicted electrophilicity of the metal center, with rates of reaction being Y > Lu > Sc. The Y complex reacts ~ 250 times faster than for Sc and ~ 5 times faster than the Lu analog. In other systems that are available, steric effects cloud the underlying electronic effects. However, the mechanistically related heterolytic activation of H_2 by d^o-metal and related metal alkyl bonds, the so called hydrogenolysis reaction, has been shown to be insensitive to steric effects. The rate of hydrogenolysis of a series of lanthanum alkyls, Cp^*_2LnR was found to increase with decreasing ionic ratios of the Ln^{3+} ion, also paralleling the expected electrophilicity of the metal ion.[118] Similarly it has been shown that the presence of π-donating ancillary ligation can seriously retard the rates of hydrogenolysis of Th(IV) alkyl bonds.[95] This can be argued to be a consequence of reducing the electrophilicity of the metal center, the π-donor ligand pushing up in energy the frontier molecular orbitals needed in the reaction. However, the π-donor ligand also strengthens the Th-R bond.[95]

C. Importance of the Leaving Group

The vast majority of these CH bond activation reactions, either intra- or intermolecular, involve alkyl leaving groups leading to the elimination of alkane. An energetic assessment of the four-center, four-electron transition state (Scheme 15) shows that the new metal carbon σ-bond is being formed at the direct expense of the M-R (leaving group) σ-bond. Simultaneously a CH bond is being both broken and formed as the hydrogen atom is transferred to generate alkane. Besides the necessary presence of a significant KIE (k_H/k_D) for this process, a logical conclusion to be drawn is that if everything else is held constant, the reactions will be facilitated by the presence of a weaker M-R bond. Although thermodynamic data on bond dissociation energies are not as plentiful as one would like, the available data on high valent early d-block and related metal systems[43,91] indicate the following trend for values of $D(L_nM-R)$; R = Ph > CH_3 ~ CH_2SiMe_3 > CH_2Ph ~ CH_2CMe_3. Hence in the absence of steric effects, neopentyl and benzyl ligands should be better leaving groups to effect CH bond activation. Although data is sketchy, this trend would appear to be followed. Hence in the cyclometalation of 2,6-di-*tert*-butylphenoxide (OAr') at Ti(IV) and Zr(IV) metal centers, benzyl leaving groups result in cyclometalation rates much faster than for identical reactions involving methyl ligands.[69] Similarly at Ta(V) the compounds $Ta(OAr'')_3R_2$ (OAr'' = 2,6-diphenylphenoxide) undergo metalation with R = CH_2Ph faster than with R = CH_3.[76]

    This thermodynamic argument can also be used to explain the potency of strained metallocycle rings such as in thoracyclobutanes and metal-*o*-phenylenes for intermolecular CH bond activation. Marks and co-workers have shown that the first bond disruption enthalpy for the thoracyclobutane ring of $Cp^*_2Th(CH_2CMe_2CH_2)$ is only 59 kcal/mol compared to values of ~ 70-80 kcal/mol for terminal alkyls.[43,119] Furthermore, the lower reactivity of the less strained $Cp^*_2Th(CH_2SiMe_2CH_2)$ ring is paralleled in a higher (70 kcal/mol) value for the bond disruption energy needed to open the ring.[43,119] However, although the first bond disruption energy of the thoracyclobutane is lower than for unstrained alkyls, it is still nowhere near values that would allow simple homolytic ring opening under the conditions in which CH bond activation by this compound takes place.[80]

    Although thermodynamic data are lacking, it seems entirely that that the first bond disruption enthalpy of the metallacyclopropane ring of metal-*o*-phenylene compounds (Secs. I.3 and V.1) is also lower than for terminal alkyl or aryl groups.

    The higher reactivity of alkylidene functional groups over their saturated alkyl counterparts[48,72] can be readily rationalized in turns of the 4-center, 4-electron transition state theory. It can be seen that for the activation of a CH bond by an alkylidene compound, the reaction proceeds through a transition state in which the new metal-carbon σ-bond is being formed only at the expense of the original metal-carbon π-bond (Scheme 18). Although hard thermodynamic data are unavailable it seems entirely reasonable that the π-component of the metal-carbon double bond is significantly weaker than a metal-carbon σ-bond. Hence the reaction of a CH bond with an alkylidene group should be

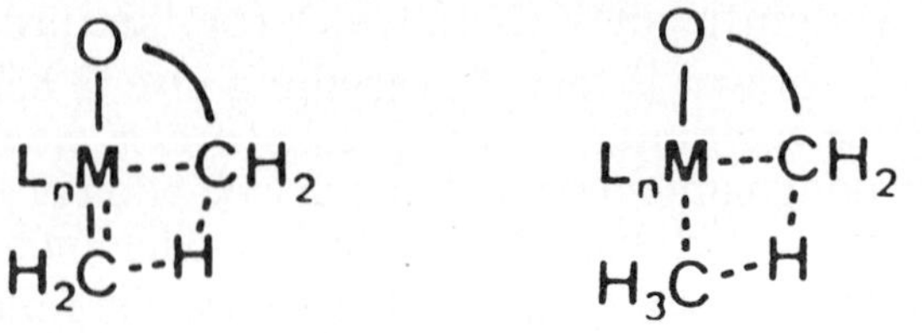

Scheme 18

enthalpically favored over σ-bond exchange.[72] Kinetic data on the facile, intramolecular addition of the aliphatic CH bonds of

2,6-di-*tert*-butylphenoxide ligands to tantalum-methylidene groups (Scheme 19) indeed show the reactions to be enthalpically favored over

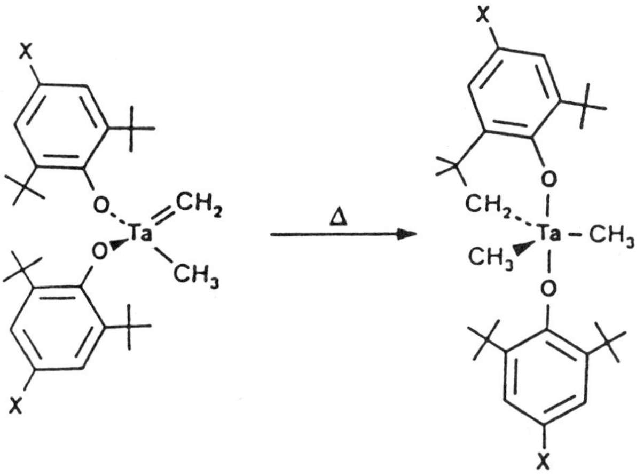

X	$\Delta H^{\ddagger}$/ Kcal.mol^{-1}	$\Delta S^{\ddagger}$/e.u.
H	14·3±1·0	−31±6
OMe	15·4± 1·0	−28±6

Scheme   19

related reactions involving tantalum-methyl leaving groups.[72] However, the kinetic data also show the reaction to be severely inhibited entropically (Scheme 19).   Whether this entropic inhibition is a characteristic of CH bond activation by alkylidenes in general, or purely an artifact of this particular system, is as yet unknown.[72]

   In the case of the mechanistically related hydrogenolysis of metal-alkyl bonds (Scheme 15) again a dependence of the rate on the strength of the M-R bond is predicted. A very recent study by Lin and Marks on the hydrogenolysis of a series of organoactinide alkyls shows a good correlation of the rate of the reaction with known values of bond disruption enthalpies.[95] The presence of a KIE (k_H/k_D) value of 2.5(4) at 30°C for the reaction of Cp*$_2$Th(CH$_2$Bu^t)(OBu^t) with H$_2$-D$_2$ is consistent with H-H bond breaking in the transition state.[95] This value is larger than values of ~ 1-1.3 reported for the oxidative-addition of H_2-D_2 to later transition metal centers.[120-122]

A number of studies have attempted to gain more insight into the electronic nature of these transition states by introducing electronically noninnocent substituents into the alkyl and aryl leaving group and watching the effect on the reaction rate. In the case of the cyclometalation of 2,6-di-*tert*-butylphenoxide ligands at Zr(IV) with substituted benzyl leaving groups (Eq. 16), only a negligible substituent effect was seen.[71] This was interpreted in terms of there being little charge separation in the 4-center, 4-electron transition state.[71] Similar studies with substituted phenyl leaving groups at Ta(V) also showed little dependence of the rate on a series of *meta-* and *para*-substituents (Eq. 17).[75] In this latter case the lack of a dramatic substituent effect argues strongly against a stepwise pathway for the reaction (Scheme 20) in which electrophilic displacement of the metal by a proton from the activated CH bond takes place via a "Metallo-Wheland" intermediate. A study of the rate of hydrogenolysis of the two substrates $Cp^*_2Th(C_6H_5)(OBu^t)$ and $Cp^*_2Th(C_6H_4\text{-}4NMe_2)(OBu^t)$ by Lin and Marks (Eq. 24) does show an acceleration in this rate of the reaction on

$$Cp^*_2Th(OBu^t)(Ar) + H_2 \longrightarrow Cp^*_2Th(OBu^t)(H) + ArH$$

$$(24)$$

$10^4 k/sec^{-1} M^{-1}$ at 60°C: Ar = C_6H_5, 8.8(10); Ar = C_6H_4-4NMe$_2$, 44(3)

introducing the electron-releasing NMe$_2$ group.[95] However, the use of σ^+ values generates a Hammet dependence with a ρ-value of only -0.4,[95] compared to values of -4 or greater for classical electrophilic substitution.[123] These results are complemented by studies involving the intramolecular and intermolecular activation of aromatic CH bonds, to which they must be mechanistically related by microscopic reversibility.

D.   Importance of the Substrate CH Bond

A consistent picture of the dependence of the rate of CH bond activation on the substrate CH bond has emerged for the Cp^*_2MX (M = Sc, Y, Lu) systems.[23,51,83] The most extensively reported system is the scandium one by Bercaw and co-workers[51] and the results can be generalized as follows: sp-CH > sp^2-CH > sp^3-CH (1° > 2°). Hence acetylenic CH bonds, e.g., HC≡CMe, react much more rapidly than do olefinic or alkane CH bonds. The much lower reactivity of 2°-CH bonds over their 1° counterparts is a characteristic of all of the systems capable of intermolecularly activating hydrocarbon substrates (Sec. V). This selectivity has been rationalized on steric grounds. An important consequence of this is the use of cyclohexane or cyclohexane-d_{12} as a relatively inert solvent for carrying out reactivity/kinetic studies. For the thoracyclobutane system developed by Fendrick and Marks,[80] some parallels in reactivity are evident. However, olefinic substrates do not lead to ring opening via CH bond activation. Instead a ring expansion takes place to produce a metallocyclohexane ring via insertion of the olefinic bond into one side of the initially strained thoracyclobutane ring.[80] The strained, sp^2-hybridized CH bonds of cyclopropane were found to be as reactive as aromatic CH bonds. Hence treatment of $Cp*_2Th(CH_2CMe_2CH_2)$ with c-C_3H_6 leads to a biscyclopropyl compound and CMe_4 (Scheme 11).[80]

In considering the higher reactivity of unsaturated hydrocarbons with these highly electron-deficient metal centers it is necessary to consider pathways involving initial electrophilic attack at the π-system. For aromatic substrates this would lead to a stepwise pathway proceeding via an arenium ion or "Metallo-Wheland" intermediate. This pathway would be identical to these shown to be operative for the substitution of aromatic CH bonds by Hg^{2+}, Pb^{4+},[124] and possibly Pd^{2+}.[22] This pathway is the exact reverse of the one previously considered for the inter- or intramolecular activation of CH bonds by M-Ar leaving groups (Scheme 20). Again as yet there is little support in the literature for these pathways being operative with d^o-metal compounds. Both the intramolecular activation of the aromatic CH bonds of benzyl groups (Scheme 2) by Hf(IV)[48] or the intermolecular activation of arenes by Sc(III)[51] have been shown to possess rates of reactions negligibly sensitive to substituents in the aromatic rings (Eq. 21). This contrasts with the high sensitivity of aromatic mercuration to substituents, with a ρ-value of -4 (vs σ$^+$)[125] Similarly Bercaw and co-workers have shown that the reaction of $Cp*_2ScCH_3$ with the

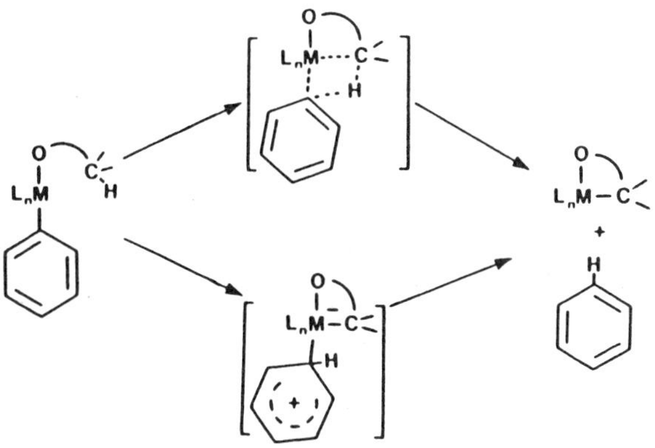

Scheme 20

substituted styrenes $H_2C=CHAr$ (Ar-C_6H_4-4CF_3; C_6H_4-4OMe) occurs at essentially the same rate (Eq. 20). It was hence concluded from these studies that a concerted pathway was operative for the cleavage of the sp^2-CH bonds with no rupture by the metal of the hydrocarbon substrate π-system. This characteristic reactivity of these d^o-metal systems (say compared to Hg^{2+}) may very well have its origin in steric effects and not electronic factors. It can be argued that formation of arenium ions in the coordination sphere of a Cp*$_2$MX unit is difficult to accommodate.[48,51] Meanwhile the concerted pathway in which the olefinic-hydrocarbon substrate can slip in sideways appears to be sterically much more feasible. The development of future d^o-metal systems able to activate CH bonds that do not contain such steric constraints may very well reveal reactivity more consistent with a "classical" electrophilic substitution pathway.

Given this consistent picture of a concerted pathway for these reactions via a 4-center, 4-electron transition state, how can the observed selectivity of the metal for various CH bonds be rationalized. Based on his results with the Cp*$_2$ScX system, Bercaw and co-workers prefer a nonpolar transition state, in which the deciding factor is the ability of the hydrocarbon substrate to overlap most efficiently with the bridging metal and hydrocarbon atom centers, and thereby lower the transition state energy.[51] Increasing amounts of s-character of the CH bond should increase this overlap and hence the explanation for the rate increase sp > sp^2 > sp^3. A more polar transition state is favored by Fendrick and Marks[80] (and recognized by Bercaw and co-workers[51]) to explain results obtained with organothorium systems. Here the important characteristic will be the acidity of the hydrocarbon CH bond. A good correlation can be seen between hydrocarbon reactivity/selectivity and gas-phase acidities,[80,126] and again the trend sp > sp^2 > sp^3 is rationalized.

VII. Intramolecular versus Intermolecular CH Bond Activation

The use of sterically demanding ligands in order to control electronic and coordinative unsaturation at metal centers has led to the development of exciting and diverse chemistry over the last decade. However, it can be seen that the "totally innocent" bulky ligand that does not offer alternate reaction pathways via intramolecular activation of CH bonds is as yet unknown. Hence, all d^o-metal systems developed that have the ability to intermolecularly activate hydrocarbon CH bonds also exhibit related intramolecular ligand reactivity under certain conditions (Sec. V). Given this interplay of inter- vs intramolecular CH bond activation typically demonstrated, it is important to discuss the factors that are important in controlling this reactivity. Early thermodynamic assessments focused, rightly so, on the bimolecular nature of intermolecular CH bond activation compared to its intramolecular (cyclometalations) counterpart.[127] Estimates of the entropic inhibition of the intermolecular reaction by ~ 30 e.u. allow estimates of the rate differences of the two reactions assuming substrate CH bond enthalpies are constant.[127] However, the picture is certainly much more complex, particularly given the ring strains sometimes encountered for product metallocycles formed in cyclometalation reactions. To date only a few systems have been reported in which direct thermodynamic assessments of intra- vs intermolecular CH bond activation can be made.[128] A recent report by Marks and co-workers[81] is particularly relevant to this chapter. It was found that the cyclometalation of the dialkyl substrate $Cp*_2Th(CH_2SiMe_3)_2$ to generate a metallocyclobutane ring could be reversed in the presence of excess Me_4Si (Scheme 11). Thermochemical measurements of the heats of formation of the two metal compounds combined with activation parameters obtained kinetically allowed a detailed profile of the reaction to be made.[81] Although the intermolecular ring opening reaction was, as expected, inhibited entropically, it was, however, found to possess an enthalpic advantage. This enthalpic advantage of the intermolecular reaction was accountable to the amount of ring strain contained in the metallocycle formed by cyclometalation.[43,81]

References

1. See other chapters in this book and Shilov, A. E., *Activation of Saturated Hydrocarbons by Transition Metal Complexes*. D. Reidel, Dordrecht MA, 1984.
2. Kleiman, J. P., Dubeck, M., *J. Am. Chem. Soc.* **1963**, *85*, 1544.
3. Cope, A. C., Friedrich, E. C., *J. Am. Chem. Soc.* **1968**, *90*, 909.
4. Chatt, J., Davidson, J. M., *J. Chem. Soc.* **1965**, 843.
5. Parshall, G. W., *Catalysis* **1977**, *1*, 334.
6. Parshall, G. W., *Acc. Chem. Res.* **1975**, *8*, 113; **1970**, *3*, 139.
7. Bruce, M. I., *Angew. Chem., Int. Ed. Engl.* **1977**, *16*, 73.
8. Dehand, J., Pfeffer, M., *Coord. Chem. Rev.* **1976**, *18*, 327.
9. (a) Crabtree, R. H., *Chem. Rev.* **1985**, *85*, 245; (b) *Chemtech*, **1982**, *12*, 506.
10. Constable, E. C., *Polyhedron* **1984**, *3*, 1037.
11. Halpern, J., *Inorg. Chim. Acta* **1985**, *100*, 41.
12. Webster, D. E., *Adv. Organomet. Chem.* **1977**, *15*, 147.
13. Janowicz, A. H., Bergman, R. G., *J. Am. Chem. Soc.* **1982**, *104*, 352.
14. Hoyano, J. K., Graham, W. A. G., *J. Am. Chem. Soc.* **1982**, *104*, 3723.
15. (a) Janowicz, A. H., Periana, R. A., Buchanan, J. M., Kovac, C. A., Stryker, J. M., Wax, M. J., Bergman, R.G., *Pure Appl. Chem.* **1984**, *56*, 13. (b) Bergman, R. G., Science, **1984**, *233*, 902. (c) Buchanan, J. M., Stryker, J. M., Bergman, R. G., *J. Am. Chem. Soc.*, **1986**, *108*, 1537.

16.   (a) Rest, A. J., Whitwell, I., Graham, W. A. G., Hoyano, J. K., McMaster, A. D., *J. Chem. Soc., Chem. Commun.* **1984**, 624; (b) Hoyano, J. K., McMaster, A. D., Graham, W. A. G., *J. Am. Chem. Soc.* **1983**, *105*, 7190.

17.   Jones, W. D., Feher, F. J., *J. Am. Chem. Soc.* **1984**, *106*, 1650.

18.   (a) Berry, M., Elmitt, K., Green, M. L. H., *J. Chem. Soc. Dalton Trans.* **1979**, 1950; (b) Green, M. L. H., *Pure. Appl. Chem.* **1978**, *50*, 27.

19.   Crabtree, R. H., Mellea, M. F., Mihelcic, J. M., Quirk, J. M., *J. Am. Chem. Soc.* **1982**, *104*, 107.

20.   (a) Faller, J. W., Felkin, H., *Organometallics* **1985**, *4*, 1488 and references therein; (b) Zeiher, E. H. K., DeWit, D. G., Caulton, K. G., *J. Am. Chem. Soc.* **1984**, *106*, 706.

21.   Abel, E., Stone, F. G. A., Wilkinson, G., Eds., *Comprehensive Organometallic Chemistry*. Pergamon Press, New York, 1982.

22.   (a) Rothwell, I. P., *Polyhedron* **1985**, *4*, 177. (b) The well explored cyclopalladation reaction involving aromatic CH bonds is thought to involve an electrophilic substitution pathway; see Parshall, G. W., *Acc. Chem. Res.* **1970**, *3*, 139; (c) Electrophilic cyclopalladation may also take place without formation of an arenium ion, see Deeming, A. J., Rothwell, I. P., *Pure Appl. Chem.* **1980**, *52*, 649.

23.   Watson, P. L., *J. Am. Chem. Soc.* **1983**, *105*, 6491.

24.   Schrock, R. R., *J. Organomet. Chem.* **1986**, *300*, 96.

25.   Schrock, R. R., *Acc. Chem. Res.* **1979**, *12*, 98.

26.   Grubbs, R. H., in *Comprehensive Organometallic Chemistry* (G., Wilkinson, F. G. A., Stone, E. W., Abel eds. Chapter 54. Pergamon Press, Oxford, 1982.

27.   Rabaa, H., Saillard, J. -Y., Hoffmann, R., *J. Am. Chem. Soc.* **1986**, *108*, 4327 and references therein.

28.   Chamberlain, L. R., Rothwell, A. P., Rothwell, I. P., *J. Am. Chem. Soc.* **1984**, *106*, 1847.

29.   Edwards, D. S., Bland, L. V., Ziller, J. W., Churchill, M. R., Schrock, R. R., *Organometallics* **1983**, *2*, 1505.

30.   Chamberlain, L. R., Rothwell, I. P., *J. Chem. Soc. Dalton Trans.* **1987**, 163.

31.   Bruno, J. W., Kalina, D., Mintz, E. A., Marks, T. J., *J. Am. Chem. Soc.* **1982**, *104*, 1860.

32.   (a) Erker, G., Kropp, K., *J. Am. Chem. Soc.* **1979**, *101*, 3659; (b) Kropp, K., Erker, G., *Organometallics* **1982**, *1* 1246; (c) Erker, G., *J. Organomet. Chem.* **1977**, *134*, 189; (d) Berkovich, E. G., Shur, V. G., Vol'pin, M. E., Lorenz, B. E., Rummel, S., Wahren, M., *Chem. Ber.* **1980**, *113*, 70; (e) Kolominikov, I. S., Lobceva, T. S., Gorbachevska, V. V., Aleksandrov, G. G., Vol'pin, M. E., *J. Chem. Soc., Chem. Commun.* **1971**, 972.

33.   (a) Bockel, C. P., Teuben, H. J., de Liefde Meijer, H. J., *J. Organomet. Chem.* **1974**, *81*, 371; (b) Rausch, M. D., Mintz, E. A., *J. Organomet. Chem.* **1980**, *190*, 65.

34.   (a) McLain, S. J., Schrock, R. R., Sharp, P. R., Churchill, M. R., Youngs, W. J., *J. Am. Chem. Soc.* **1979**, *101*, 263; (b) Churchill, M. R., Youngs, W. J., *Inorg. Chem.* **1979**, 18, 1697.

35.   Buchwald, S. L., Watson, B. T., Huffman, J. C., *J. Am. Chem. Soc.* **1986**, *108*, 7411.

36. Isomerization of aryl groups via the formulation of η^2-arene complexes has been well documented, see Jones, W. D., Feher, F. J., *J. Am. Chem. Soc.* **1986**, *108*, 4814 and discussions therein.

37. Sweet, J. R., Graham, W. A. G., *Organometallics* **1983**, 2, 135.

38. Schock, L. E., Brock, C. P., Marks, T. J., *Organometallics* **1987**, 6, 232.

39. γ-CH bonds are referred to as "distal"; nevertheless they too have to come into close proximity of the metal for activation.

40. (a) DiCosimo, R., Moore, S. S., Sowinski, A. F., Whitesides, E. M., *J. Am. Chem. Soc.* **1982**, *104*, 124; (b) Tulip, T. H., Thorn, D. L., *J. Am. Chem. Soc.* **1981**, *103*, 2448.

41. Andersen, R. A., Jones, R. A., Wilkinson, G., *J. Chem. Soc. Dalton Trans.* **1978**, 446.

42. Bruno, J. W., Smith, G. M., Marks, T. J., Fair, C. K., Schultz, A. J., Williams, J. M., *J. Am. Chem. Soc.* **1986**, *108*, 40 and references therein.

43. Bruno, J. W., Stecher, H. A., Morss, L. R., Sonnenberger, D. C., Marks, T. J., *J. Am. Chem. Soc.* **1986**, *108*, 7275.

44. Bruno, J. W., Marks, T. J., Day, V. W., *J. Organomet. Chem.* **1983**, *250*, 237.

45. (a) Bulls, A. R., Schaefer, W. P., Serfas, M., Bercaw, J. E., *Organometallics* **1987**, *6*, 1219; (b) The first two steps of this sequence have also been observed on thermolysis of $Cp^*_2Ti(CH_3)_2$: see McDade, C., Green, J. C., Bercaw, J. E., *Organometallics* **1982**, *1*, 1629.

46. For example see any recent copy of the Division of Inorganic Chemistry abstracts for the ACS national meeting.

47. Airoldi, C., Bradley, D. C., Vuru, G., *Trans. Met. Chem.* **1979**, *4*, 64.

48. Takahashi, Y., Onoyama, N., Ishikaway, Y., Motojra, S., Sugiayma, K., *Chem. Lett.* **1978**, 525.

49. Mayer, J. M., Curtis, C. J., Bercaw, J. E., *J. Am. Chem. Soc.* **1983**, *105*, 2651.

50. Nugent, W. A., Overall, D. W., Holmes, S. J., *Organometallics* **1983**, *2*, 161.

51. Thompson, M. E., Baxter, S. M., Bulls, A. R., Burger, B. J., Nolan, M. C., Santarsiero, B. D., Schaefer, W. R., Bercaw, J. E., *J. Am. Chem. Soc.* **1987**, *109*, 203.

52. Watson, P. L., *J. Chem. Soc. Chem. Commun.* **1983**, 276.

53. Chamberlain, L. R., Durfee, L. D., Fanwick, P. E., Kobriger, L., Latesky, S. L., McMullen, A. K., Rothwell, I. P., Folting, K., Huffman, J. C., Streib, W. E., Wang, R., *J. Am. Chem. Soc.* **1987**, *109*, 390.

54. Evans, W. J., Meadows, J. H., Hunter, W. E., Atwood, J. L., *J. Am. Chem. Soc.* **1984**, *106*, 1291.

55. Klei, E., Teuben, J. H., *J. Organomet. Chem.* **1981**, *214*, 53.

56. Chamberlain, L. R., Rothwell, I. P., Huffman, J. C., *J. Am. Chem. Soc.* **1982**, *104*, 7338.

57. Lappert, M. F., Power, P. P., Sanger, A. R., Srivastava, R. C., *Metal and Metalloid Amides.* Wiley, New York, 1980.

58. Bennett, C. R., Bradley, D. C., *J. Chem. Soc., Chem. Commun.* **1974**, 29.

59. Simpson, S. J., Andersen, R. A., *Inorg. Chem.* **1981**, *20*, 3627.

60. Planalp, R. P., Andersen, R. A., Zalkin, A., *Organometallics* **1983**, *2*, 16.

61. Planalp, R. P., Andersen, R. A., *Organometallics* **1983**, *2*, 1675.

62. Simpson, S. J., Turner, H. W., Andersen, R. A., *Inorg. Chem.* **1981**, *20*, 2991.

63. Wolczanski, P. T., Bercaw, J. E., *J. Am. Chem. Soc.* **1979**, *101*, 6450.

64. For the purposes of this chapter, the position of CH bonds being activated is given along the backbone relative to the metal and not the heteroatom donor. Hence activation of β^-, γ^-, δ^- or ε-CH bonds results in the formation of 3^-, 4^-, 5^-, or 6-membered metallocycles.

65. (a) Nugent, W. A., Zubyk, R. M., *Inorg. Chem.* **1986**, *25*, 4604; (b) Buchwald, S. L., Nielson, R. B., Dewan, J. C., *J. Am. Chem. Soc.* **1987**, *109*, 1590.

66. Lubben, T. V., Wolczanski, P. T., *J. Am. Chem. Soc.* **1987**, *109*, 424 and references therein.

67. (a) Fanwick, P. E., Kobriger, L., McMullen, A. K., Rothwell, I. P., *J. Am. Chem. Soc.* **1986**, *108*, 8095; (b) Durfee, L. D., Fanwick, P. E., Rothwell, I. P., Folting, K., Huffman, J. C., *J. Am. Chem. Soc.* **1987**, *109*, 4720; (c) Chamberlain, L. R., Rothwell, I. P., Huffman, J. C., *Inorg. Chem.* **1984**, *23*, 2575; (d) Duff, A. W., Kamarudin, R. A., Lappert, M. F.; Norton, R. J., *J. Chem. Soc. Dalton Trans.* **1986**, 489 and references therein.

68. Murray, B. D., Hope, H., Power, P. P., *J. Am. Chem. Soc.* **1985**, *107*, 169 and references therein.

69. Rothwell, I. P., *Acc. Chem. Res.* **1988**, *21*, 153.

70. Latesky, S. L., McMullen, A. K., Rothwell, I. P., Huffman, J. C., *Organometallics* **1985**, *4*, 902.

71. Latesky, S. L., McMullen, A. K., Rothwell, I. P., Huffman, J. C., *J. Am. Chem. Soc.* **1985**, *107*, 5981.

72. Chamberlain, L. R., Rothwell, I. P., Huffman, J. C., *J. Am. Chem. Soc.*, **1986**, *108*, 1502.

73. Chamberlain, L. R., Rothwell, I. P., Folting, K., Huffman, J. C., *J. Chem. Soc. Dalton Trans.* **1987**, 155.

74. Chamberlain, L. R., Rothwell, I. P., *J. Am. Chem. Soc.* **1983**, *105*, 1665.

75. Chamberlain, L. R., Kerschner, J. L., Rothwell, A. P., Rothwell, I. P., Huffman, J. C., *J. Am. Chem. Soc.* **1987**, *109*, 6471.

76. (a) Chesnut, R. W., Durfee, L. D., Fanwick, P. E., Rothwell, I. P., Folting, K., Huffman, J. C., *Polyhedron* **1987**, *6*, 2019; (b) Chesnut, R. W., Fanwick, P. E., Steffey, B. L., Rothwell, I. P., *Polyhedron*, **1988**, *9*, 753.

77. Kerschner, J. L., Fanwick, P. E., Rothwell, I. P., *J. Am. Chem. Soc.* **1987**, *109*, 5840.

78. Erker, G., *J. Organomet. Chem.* **1977**, *134*, 189.

79. Fagan, P. J., Manriquez, J. M., Maalta, E. A., Seyam, A. M., Marks, T. J., *J. Am. Chem. Soc.* **1981**, *103*, 6650.

80. Fendrick, C. M., Marks, T. J., *J. Am. Chem. Soc.* **1986**, *108*, 425.

81. Smith, G. M., Carpenter, J. D., Marks, T. J., *J. Am. Chem. Soc.* **1968**, *108*, 6805.

82. (a) Watson, P. L., *J. Am. Chem. Soc.* **1982**, *104*, 337; (b) Watson, P. L., Roe, D. C., *J. Am. Chem. Soc.* **1982**, *104*, 5471.

83. Watson, P. L., Parshall, G. W., *Acc. Chem. Res.* **1985**, *18*, 51 and references therein.

84. Teuben, J., Schaverien, C. J., Dewan, J. C., Schrock, R. R., *J. Am. Chem. Soc.* **1986**, *108*, 2771
85. Thompson, M. E., Bercaw, J. E., *Pure Appl. Chem.* **1984**, *56*, 1.
86. Baudry, D., Ephritikine, M., Felkin, H., Holmes-Smith, R., *J. Chem. Soc., Chem. Commun.* **1983**, 788.
87. Buchwald, S. L., Lum, R. T., Dewan, J. C., *J. Am. Chem. Soc.* **1986**, *108*, 7441.
88. Negishi, E., Holmes, S. J., Tour, J. M., Miller, J. A., *J. Am. Chem. Soc.* **1985**, *107*, 2568.
89. Davidson, P. J., Lappert, M. F., Pearce, R., *Chem. Rev.* **1976**, *76*, 219.
90. Schrock, R. R., Parshall, G. W., *Chem. Rev.* **1976**, *76*, 243.
91. Halpern, J., *Acc. Chem. Res.* **1982**, *15*, 238 and references therein.
92. Brothers, P. J., *Prog. Inorg. Chem.* **1981**, *28*, 1.
93. (a) McAlister, D. R., Erwin, D. K., Bercaw, J. E., *J. Am. Chem. Soc.* **1978**, *100*, 5966; (b) Mayer, J. M., Bercaw, J. E., *J. Am. Chem. Soc.* **1982**, *104*, 2157; (c) Chiu, K. W., Jones, R. A., Wilkinson, G., Galas, A. M. R., Hursthouse, M. B., Abdul-Malik, K. M., *J. Chem. Soc. DaltonTrans*, **1981**, 1204.
94. (a) Gell, K. I., Schwartz, J., *J. Am. Chem. Soc.* **1978**, *100*, 3246; (b) Gell, K. I., Posin, B., Schwartz, J., *J. Am. Chem. Soc.* **1982**, *104*, 1846; (c) Evans, W. J., *Adv. Organomet. Chem.* **1985**, *24*, 131.
95. Lin, Z., Marks, T. J., *J. Am. Chem. Soc.* in press.
96. Vitale, A. A., San Filippo, J., *J. Am. Chem. Soc.* **1982**, *104*, 7341 and references therein.
97. Saillard, J.-Y., Hoffmann, R., *J. Am. Chem. Soc.* **1984**, *106*, 2006.
98. Steigerwald, M. L., Goddard, W. A., III, *J. Am. Chem. Soc.* **1984**, *106*, 308.
99. Shaw, B. L., *J. Am. Chem. Soc.* **1975**, *97*, 3856.
100. (a) Beesley, R. M., Ingold, C. K., Thorpe, J. F., *J. Chem. Soc.* **1915**, *107*, 1080; (b) Ingold, C. K., *J. Chem. Soc.* **1921**, *119*, 305.
101. Wiberg, K. B., *Physical Organic Chemistry*, p. 376. Wiley, New York, 1964.
102. Chiao, W.-B., Saunders, W. H., Jr., *J. Am. Chem. Soc.* **1978**, *100*, 2802.
103. Brookhart, M., Lamanna, W., Humphrey, M. B., *J. Am. Chem. Soc.* **1982**, *104*, 2117.
104. Cotton, F. A., Stanislowski, A. G., *J. Am. Chem. Soc.* **1974**, *96*, 5074.
105. Colvert, R. B., Shapley, J. R., *J. Am. Chem. Soc.* **1978**, *100*, 7726.
106. William, J. M., Brown, R. K., Schultz, A. J., Stucky, G. D., Ittel, S. D., *J. Am. Chem. Soc.* **1978**, *100*, 7407.
107. Crabtree, R. H., Holt, K. M., Larin, M., Morehouse, S. M., *Inorg. Chem.* **1985**, *24*, 1986.
108. Brookhardt, M., Green, M. L. H., *J. Organomet. Chem.* **1983**, *250*, 395.
109. Dawoodi, Z., Green, M. L. H., Metetwa, V. S. B., Prout, K., Schultz, A. J., Williams, J. M., Koetze, T. F., *J. Chem. Soc. DaltonTrans.* **1986**, 1629.
110. Fanwick, P. E., Ogilvy, A. E., Rothwell, I. P., *Organometallics* **1987**, *6*, 73.
111. Tilley, T. D., Andersen, R. A., Zalkin, A., *J. Am. Chem. Soc.* **1982**, *104*, 3725.
112. Tilley, T. D., Andersen, R. A., Zalkin, A., *Inorg. Chem.* **1984**, *23*, 2271.
113. Boncella, J. M., Tilley, T. D., Andersen, R. A., *Inorg. Chem.* in press.

114. Jensen, F. R., Rickborn, B., *Electrophilic Substitution of Organomercurials*, McGraw-Hill, New York, 1968.
115. Billups, W. E., Kowarski, M. M., Hauge, R. H., Margrave, J. L., *J. Am. Chem. Soc.* **1980**, *102*, 7393.
116. Turner, J. J., Burdett, J. K., Perutz, R. N., Poliakoff, M., *Pure Appl. Chem.* **1977**, *49*, 271.
117. Burns, C. J., Andersen, R. A., *J. Am. Chem. Soc.* in press.
118. Jeske, G., Lauke, H., Mauvermam, H., Schumann, H., Marks, T. J., *J. Am. Chem. Soc.* **1985**, *107*, 8111.
119. Marks, T. J., Fragala, I. L., eds., *Fundamental and Technological Aspects of Organo-f-Element Chemistry.* NATO ASI Series, D. Reidel, Boston, 1985.
120. Zhov, P., Vitale, A. A., San Fillippo, J., Saunders, W. H., *J. Am. Chem. Soc.* **1985**, *107*, 8049.
121. Chock, P. B., Halpern, J., *J. Am. Chem. Soc.* **1966**, *88*, 3511.
122. Kitaura, K., Obara, S., Morokuma, K., *J. Am. Chem. Soc.* **1981**, *103*, 2891.
123. Sykes, P., *A Guidebook to Mechanism in Organic Chemistry*, 5th Ed. Longman, London, 1981.
124. Norman, R. O. C., Taylor, R., *Electrophilic Substitution in Benzoid Compounds*, p. 200. Elseiver, London, 1965.
125. Brown, H. C., Goldman, G., *J. Am. Chem. Soc.* **1962**, *84*, 1650.
126. Bartmess, J. E., McIver, R. T., in *Gas Phase Ion Chemistry* M. F. Bowers, ed.), Vol. 2, p. 87. Academic Press, New York, 1979.
127. See for example p. 107 of Shilov, A. E., Shteinman, A. A., *Coord. Chem. Rev.* **1977**, *24*, 97.
128. Jones, W. D., Feher, F. J., *J. Am. Chem. Soc.* **1985**, *107*, 620.

Chapter VI

ALKANE FUNCTIONALIZATION BY CYTOCHROMES P-450 AND BY MODEL SYSTEMS USING O_2 OR H_2O_2

D. Mansuy and P. Battioni

Laboratoire de Chimie et Biochimie
Pharmacologiques et Toxicologiques
Université René Descartes
Paris, France

I.   Introduction

Monooxygenases catalyze the insertion of one oxygen atom derived from a reductive activation of dioxygen, into a substrate, according to Eq. 1.

$$RH + O_2 + 2e^- + 2H^+ \longrightarrow ROH + H_2O \qquad (1)$$

A few monooxygenases able to hydroxylate C-H bonds of alkanes are dependent on nonheme iron-proteins. For instance, this is the case for the methane monooxygenase from *Methylococcus capsulatus* [1] as well as for the monooxygenase from *Pseudomonas oleovorans,* which hydroxylates octane into 1-octanol.[2] However, in most cases, the hydroxylation of unactivated C-H bonds of alkanes in almost all living organisms such as mammals, microorganisms, or plants is catalyzed by cytochrome P-450-dependent monooxygenases.[3] The cytochromes P-450 are *b*-type cytochromes containing iron-protoporphyrin IX at the active site. They are involved in many steps of the biosynthesis and biodegradation of endogenous compounds such as steroids, fatty acids, or prostaglandins. They also play a central role in the oxidative metabolism of exogenous compounds such as drugs and environmental products allowing their elimination from living organisms. The cytochromes P-450 are able to catalyze not only the insertion of an oxygen atom into C-H bonds but also the transfer of this oxygen atom to double or triple bonds, to aromatic rings, and to heteroatoms of various substrates (for recent reviews on cytochrome P-450 reactions, see for instance refs. 4 and 5). Recently, chemical model systems using iron- or manganese-porphyrins as catalysts have been shown to be able to reproduce most cytochrome P-450 reactions (for recent reviews, see refs. 6-9). This chapter will be restricted to C-H bond hydroxylation reactions. The characteristics of these reactions catalyzed by cytochromes P-450, which are to be reproduced in model systems, will be reviewed in Sec. II. Sections III and IV will be devoted to a description of the metalloporphyrin model systems described so far for selective hydrocarbon hydroxylation under mild conditions using either alkylhydroperoxides, H_2O_2, or O_2 as oxygen atom sources.

II.   Characteristics   and   Mechanisms   of   Alkane Hydroxylation by Cytochrome P-450

1.   The Catalytic Cycle of Dioxygen Activation and Substrate Hydroxylation by Cytochrome P-450

From many spectroscopic studies performed on various cytochrome P-450-dependent monooxygenases, it is now clear that there is a common mechanism by which these enzymes from very diverse origins activate dioxygen and are able to transfer one oxygen atom into substrates. The corresponding catalytic cycle is given in Figure 1. In

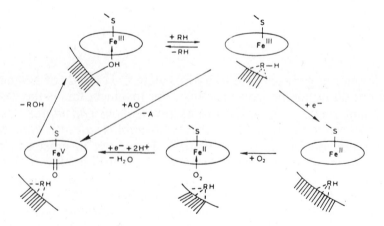

Figure 1.   Catalytic cycle of cytochrome P-450.

the resting state of cytochrome P-450, two ferric complexes are in equilibrium: a hexacoordinate low-spin complex bearing two axial ligands, a cysteinate and an OH-containing residue, and a pentacoordinate high-spin complex with the cysteinate as the only axial ligand. The binding of a substrate that generally occurs on a protein hydrophobic site near the heme leads to a shift of this equilibrium toward the pentacoordinate high-spin state. The existence and structure of these two ferric complexes of the catalytic cycle have been confirmed recently by an X-ray structure analysis of cytochrome P-450 from *Pseudomonas putida* either in its substrate-free ferric low-spin state or in its substrate (camphor)-bound high-spin state.[10,11] The high-spin cytochrome P-450-substrate complex is then reduced by one electron coming from NADPH or NADH via an electron-transfer chain. The resulting high-spin pentacoordinate ferrous complex is able to bind various ligands such as CO, nitrogenous bases, and dioxygen. Binding of O_2 leads to an hexacoordinate low-spin ferrous complex. Model iron-porphyrin complexes for the four first intermediates of the cytochrome P-450 catalytic cycle have been prepared and completely characterized by various spectroscopic techniques including X-ray structure determinations (for reviews on that subject, see refs. 12-14).

The last intermediate complex of the catalytic cycle is the real oxygenating species. It is derived from the Fe(II)-O_2 complex by a one-electron reduction. However, its lifetime is so short that thus far it has not been observed and studied by any spectroscopic technique. Our knowledge about its possible nature and, in a more general manner, about the final steps of the catalytic cycle is based on indirect evidence coming from studies on the characteristics of the oxidations catalyzed by cytochrome P-450 and on comparisons with better known active oxygen complexes derived from other hemeproteins or from iron-porphyrins. Based on these data, the most likely structure for the active oxygen complex is a high-valent iron-oxo structure that could derive from an heterolytic cleavage of the O-O bond of a possible Fe(III)-O-OH intermediate formed upon a one-electron reduction and protonation of the Fe(II)-O_2 complex. This high-valent iron-oxo complex is able to transfer its oxygen atom into a wide range of substrates leading to various kinds of reactions such as the hydroxylation of C-H bonds, the epoxidation of alkenes, the epoxidation and hydroxylation of aromatic rings, and the transfer of an oxygen atom to compounds containing nitrogen, sulfur, phosphorus, and iodine atoms.[4] In agreement with the iron-oxo structure proposed for the P-450 active complex, single oxygen atom donors such as H_2O_2, C_6H_5IO, or $NaIO_4$ may replace O_2 and NADPH for the cytochrome P-450-catalyzed oxidations of many substrates[4,5] leading to a considerably shortened catalytic cycle (Fig. 1). A high-valent porphyrin-iron-oxo complex corresponding *formally* to a Fe(V)=O structure has been prepared by reaction of *meta*-chloroperbenzoic acid with Fe (tetramesityl porphyrin = TMP)(Cl)[15] and studied by ^{1}H-NMR, EPR, Mössbauer, and EXAFS spectroscopy.[15-17] All its characteristics are compatible with a (porphyrin radical cation)Fe(IV)=O structure and similar to those of horseradish peroxidase compound I. Interestingly, this complex transfers its oxygen atom to alkenes almost quantitatively, as the equivalent complex of cytochrome P-450 does.

2. Mechanisms of Alkane Hydroxylation by the Active Oxygen Complex of Cytochrome P-450

The formation of alcohols by insertion of an oxygen atom into C-H bonds catalyzed by cytochromes P-450 was first thought to occur by a concerted mechanism, mainly because of the stereospecificity of such reactions in the biosynthesis of endogenous compounds such as steroid hormones. However, evidence accumulated in the last decade favors a nonconcerted mechanism involving

1. a hydrogen abstraction by the active oxygen complex of cytochrome P-450 that seems to have a free radical like reactivity, and
2. a rapid collapse of the free radical derived from the alkane with the Fe(IV)-OH species leading to the hydroxylated product (Fig. 2).

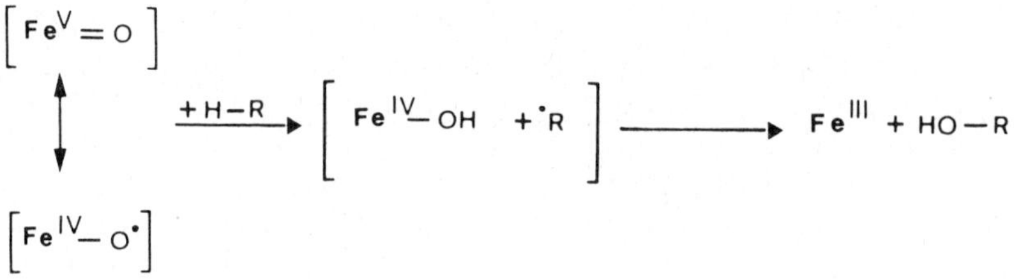

$$\left[Fe^V = O\right] \quad \xrightarrow{+H-R} \quad \left[Fe^{IV}-OH \quad + \,^\bullet R\right] \quad \longrightarrow \quad Fe^{III} + HO-R$$
$$\updownarrow$$
$$\left[Fe^{IV}-O^\bullet\right]$$

Figure 2. Possible mechanisms for alkane hydroxylation by cytochrome P-450.

A first major argument in favor of this mechanism is the *partial loss of stereochemistry* observed during the hydroxylation of certain alkanes. This was first observed in the hydroxylation of *exo, exo, exo, exo*-2,3,5,6-tetradeuteronorbornane by a purified rabbit liver microsomal cytochrome P-450.[18] An analysis of the *exo*- and *endo*-2-norborneols formed in this reaction showed a significant amount of epimerization during this process and suggested a partial epimerization of an intermediate carbon free radical in the enzyme-substrate cage.[18] Similar results were described for the *Pseudomonas putida* cytochrome P-450-dependent hydroxylation of camphor analogs containing deuterium at either the 5-*exo* or 5-*endo* position.[19]   These results suggested that hydrogen abstraction could occur from either the *exo* or *endo* position at carbon 5 of camphor but that the oxygen was stereospecifically added to only the *Re* face to give 5-*exo*-hydroxycamphor as the only product.   More recently, a study of the hydroxylation of four isotopically substituted phenylethane substrates by a single isozyme of rabbit liver microsomal cytochrome P-450, called P-450 LM$_2$, showed that 25–40% of the hydroxylation events resulted in benzylic alcohols with a configuration opposite to that of the original hydrocarbon substrate.[20]   These partial losses of stereochemistry are occurring since epimerization of the intermediate carbon radicals competes with their capture by Fe(IV)-OH inside the enzyme cage.

A second major argument in favor of the nonconcerted mechanism of Figure 2 concerns the *rearrangements that were observed during the oxidation of several alkanes.* Allylic alcohol isomers were observed during the hydroxylation of alkenes containing allylic C-H bonds.[21,22] Presumably, these regioisomers are derived from the delivery of the OH group of Fe(IV)-OH to the two ends of an allylically delocalized radical (Fig. 3). A very recent study of cytochrome P-450-

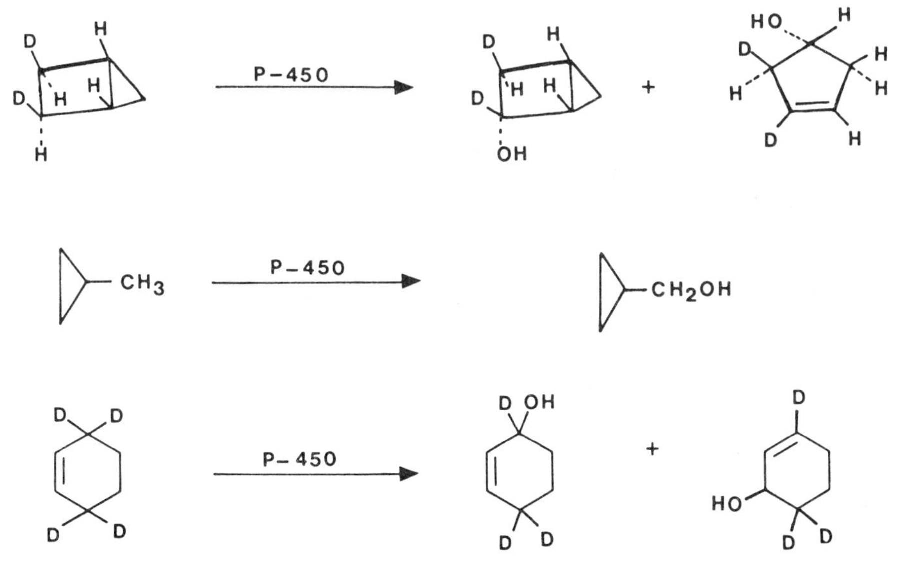

Figure 3.   Hydroxylation of C-H bonds by cytochrome P-450 that may lead to rearrangements.

dependent oxidations of alkanes containing the cyclopropyl group allowed the estimation of the rate of collapse of intermediate carbon radicals with the Fe(IV)-OH complex inside the enzyme cage.[23] The enzymatic hydroxylation of methylcyclopropane and nortricyclane leads, respectively, to cyclopropylmethanol and nortricyclanol without formation of the possible rearranged products 3-buten-1-ol and norborn-5-en-2-ol.   On the contrary, microsomal oxidation of bicyclo[2.1.0]pentane yields the rearranged product 3-cyclopenten-1-ol besides *endo*-2-hydroxybicyclo[2.1.0]pentane (Fig. 3).[23] From what is known of the rates of rearrangement of cyclopropylmethyl radicals, it was deduced that the [R$^{\bullet}$+ Fe(IV)-OH] pair formed in P-450-catalyzed hydroxylations collapses at a rate in excess of 1×10^9 sec^{-1}.

A third characteristic of alkane hydroxylations by cytochromes P-450 is in good agreement with the nonconcerted mechanism of Figure 2. Many of these hydroxylations were found subject to *large internal isotope effects*. Such effects (k_H/k_D around 10) have been observed for the hydroxylation of 1,1-d$_2$-1,3-diphenylpropane,[24] norbornane,[18] phenylethane,[20] 1,1,1-d$_3$-octane,[25,26] cumene,[27] and 7-ethoxycoumarin[28] by liver microsomal cytochromes P-450. Since such large intramolecular isotope effects have been observed in the hydroxylation of various alkanes by different cytochromes P-450, it is consistent with a hydrogen abstraction-recombination mechanism that is the *consequence of the chemical reactivity of the high-valent iron-oxo complex but that is independent of specific apoprotein structure*. Accordingly, hydroxylation of several alkanes by model systems using PhIO and an iron-porphyrin catalyst was also found to occur with such large internal isotopic effects (see below).

These data strongly support the nonconcerted mechanism of Figure 2 for C-H bond hydroxylation by the different cytochromes P-450. However, the detailed mechanism of the recombination step and of alcohol formation is not well understood. A possible mechanism would be the formation of a carbon-oxygen bond during this recombination by an oxidative OH ligand transfer from the Fe(IV) to the carbon radical.[18] This transfer would involve a one-electron reduction of the iron regenerating the ferric resting state (Fig. 4). Another

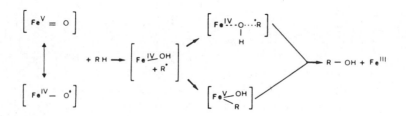

Figure 4. Possible mechanisms for the second step of alkane hydroxylation by cytochromes P-450.

possibility would be, as previously proposed,[29,30a] a combination of the carbon radical with the iron itself leading to a high-valent σ-alkyl intermediate complex which would involve the σ-alkyl and OH ligands in *cis* position on the iron (Fig. 4). Reductive elimination of these two *cis* ligands would lead to the hydroxylated product and regenerate cytochrome P-450-Fe(III). There is now ample evidence for the existence of stable complexes of cytochromes P-450 or of iron-porphyrins containing iron-σ-alkyl or iron-carbene bonds.[8,30] However, the best evidence for the involvement of such complexes during C-H bond hydroxylation came from the oxidative metabolism of 1,3-benzodioxole derivatives. These compounds act as suicide-substrates for cytochromes P-450 by formation of stable P-450-iron-metabolite complexes.

   Indirect evidence suggests that these stable complexes involve an iron-carbene bond (Fig. 5).[13,30,31] Moreover, model iron-porphyrin

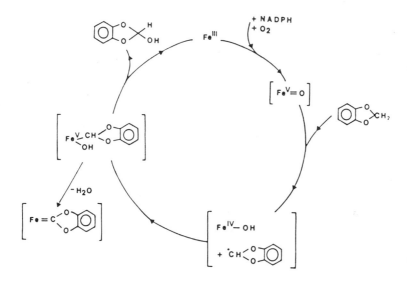

Figure 5. Catalytic cycle of 1,3-benzodioxole oxidation by cytochrome P-450.

complexes containing the proposed iron-benzodioxole-derived carbene bond have been prepared and characterized.[32] A possible catalytic cycle for the oxidation of benzodioxole derivatives by cytochromes P-450 is given in Figure 5. It explains not only the formation of product derived from C-H bond hydroxylation but also that of the observed iron-metabolite complexes. The partition between direct OH ligand transfer and intermediate formation of a σ-alkyl complex shown in Figure 4 would depend greatly upon the hindrance of the carbon radical, the direct OH ligand transfer being favored in the case of hindered carbon radicals. Concerning the possible involvement of high-valent iron complexes containing σ-alkyl or carbene ligands, it is noteworthy that the formation of such complexes has been invoked recently to explain the deuterium-hydrogen exchange observed during the epoxidation of *trans*-1-deuteropropene by cytochrome P-450[33] and the formation of *N*-vinylprotoporphyrin IX during the oxidative metabolism of a sydnone derivative.[34,8]

3. Some Properties of Alkane Hydroxylation by Cytochrome P-450

In order to build accurate chemical model systems for alkane hydroxylation by cytochrome P-450, it is important to know some properties of the enzymatic reactions in addition to their detailed mechanisms. An important property is the chemio- and regioselectivity of these enzymatic systems. In fact, some cytochromes P-450 are highly regioselective, such as those involved in steroid hormone biosynthesis or in the ω-hydroxylation of lauric acid and of linear alkanes.[35] However, several cytochromes P-450 involved in the detoxification of xenobiotics, such as many of those present in liver microsomes, are not so regioselective. For instance, the rat liver cytochrome P-450 isozymes induced by phenobarbital are able to accommodate a wide range of substrates and to oxidize them in several positions. With such relatively unselective enzymes allowing substrate mobility inside the active site, the regioselectivity of alkane hydroxylation may be in great part governed by the intrinsic chemioselectivity of the iron-oxo active species. In that case, tertiary C-H bonds are preferentially hydroxylated relative to secondary CH_2 bonds, and methyl groups are by far less reactive.[36] In a more general manner, whereas the C-H bond hydroxylation mechanism by cytochromes P-450 is largely dependent upon the intrinsic reactivity of the iron-oxo active species but independent of specific apoprotein structure, the product regio- and stereoselectivities are mainly dependent on apoprotein structure. This has been recently illustrated by a study of the ω-hydroxylation of racemic, (*R*)- and (*S*)-2-phenylpropane-1,1,1-d_3 by microsomal cytochromes P-450 from rats either untreated or pretreated with phenobarbital or β-naphthoflavone.[27] It was shown that the magnitude of the intramolecular isotope effect associated with this reaction was consistent with a hydrogen atom abstraction-recombination mechanism involving a reasonably symmetrical transition state irrespective of the specific cytochrome P-450 isozyme catalyzing the reaction. Product stereoselectivity was found to be primarily a function of the isozyme structure and the chirality of the active site.

    When trying to build model systems, it is important to keep in mind two other properties of cytochrome P-450-dependent hydroxylation of alkanes. First, the rates of such reactions, which are dependent upon the nature of the isozyme and of the alkane, vary from about 1 to 100 turnovers per minute[35b,36] Second, the yields of hydroxylated products based on NADPH (or NADH) consumed for the monooxygenation reaction (see Eq. 1) are again dependent on the nature of the isozyme and of the alkane and vary from 50 to 100%.[37]

From all the presently available data, alkane hydroxylation by cytochrome P-450 should be viewed as follows (Fig. 6), as recently illustrated with phenylethane as substrate:[20]

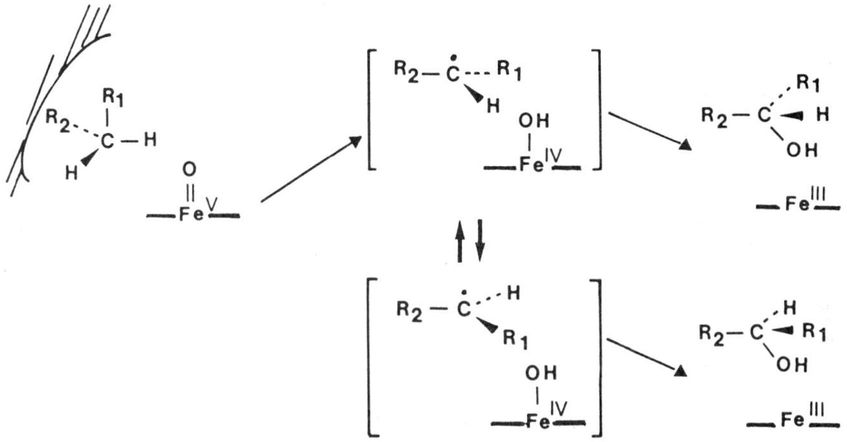

Figure 6. Tentative stereochemical view of the different steps of cytochrome P-450-dependent hydroxylation of alkanes.

• Alkane positioning determined by the protein active site: in the case of phenylethane and P-450 LM$_2$, the local protein structure is such that pro-*S* benzylic hydrogen abstraction is favored 4 to 1. With isozyme leaving enough substrate freedom in the active site, the electrophilic Fe=O species will react preferentially with tertiary C-H bonds and then with secondary CH$_2$ bonds.

• Approach of an alkane C-H bond parallel to the heme plane as deduced from model studies[38] and as indicated by the position of the 5-*exo* C-H bond of camphor relative to the heme plane in the camphor-cytochrome P-450$_{cam}$ complex.[10]

• Abstraction of a hydrogen atom by the Fe=O species explaining a large internal isotopic effect.

• Possible rearrangement of the intermediate radical as in the case of some cyclopropyl-containing alkanes[23] or possible movement of this radical in the active site to present the opposite face to the Fe(IV)-OH species (Fig. 6). This movement inside the [radical-Fe(IV)-OH] pair explains the partial inversion of configuration occurring in phenylethane hydroxylation.[20]

• Recombination of the original carbon radical or a rearranged isomer with the oxygen or iron atom of the Fe(IV)-OH species, leading to the alcohol. The stereospecificity of the hydroxylation reaction will naturally depend upon the rate of recombination of the radical with the Fe(IV) complex compared to that of the rearrangement or movement of the radical inside the active site. This will depend on the intrinsic reactivity of the radical and on the restrictions imposed by the protein active site on the radical movement. In the case of very reactive primary radical intermediates formed during ω-hydroxylation of octane, recombination should be very fast leading to a complete retention of configuration.[25]

In the case of the benzylic hydroxylation of phenylethane by P-450 LM$_2$, because of the more stable benzylic radical and the relatively large active site of the isozyme, the reaction occurs with 25-40% inversion of configuration.[20]

Thus, in the design of metalloporphyrin-based model systems for alkane hydroxylation, the following key characteristics must be obtained:

• The involvement of *high valent iron-oxo* intermediates, equivalent to Fe(V)=O formally, leading to *large internal isotope effects* k_H/k_D (around 10) and to *preferential hydroxylation of tertiary C-H bonds* in the absence of steric constraints

• *Rates* around *1-100 turnovers* per minute and *yields* based on the reducing agent (if O$_2$ itself is used as an oxygen atom donor) *between 50 and 100%*.

• Good regio- and stereoselectivities if some groups maintained in the vicinity of the porphyrin lead to some preferential positioning of the alkane and to some constraints to the intermediate radical movements.

III.   Model Systems Using H$_2$O$_2$

1.   Introductory Remarks: Model Systems Using Oxidants Containing a Single Oxygen Atom or Using Alkylhydroperoxides

The long catalytic cycle of cytochrome P-450 using O$_2$ and NADPH appears difficult to mimic particularly because of a possible fast reaction of the strongly oxidizing agent Fe(V)=O with the reducing agent in excess. These two agents are separated in the enzymatic system since the reducing equivalents of NADPH are transferred to the heme only via an electron transfer chain.[3] Their separation in a chemical model system appears more difficult to perform. This is the reason why the modeling of the shortened catalytic cycle using single oxygen atom donors (Fig.

1) first attracted the attention of chemists. Groves et al.[39] first demonstrated that iron-*meso*-tetraarylporphyrins are able to catalyze the transfer of the oxygen atom of PhIO into many substrates in a manner very similar to cytochrome P-450 (for recent reviews see refs. 6-8). For instance, alkane hydroxylation occurs with high isotopic effects (k_H:k_D around 13 for cyclohexane),[38] retention of configuration of the C-H bond of *cis*-decalin,[38,40] and preferential reaction on tertiary CH bonds.[38] Unfortunately simple iron-tetraarylporphyrins such as iron-tetraphenylporphyrin chloride, Fe(TPP)Cl, are oxidatively destroyed during alkane hydroxylation. Major improvements of the catalytic activities of these model systems were obtained by using tetraarylporphyrins containing halogen-substituted aryl groups such as the pentafluorophenyl[41] and the 2,6-dichlorophenyl[42] groups. The iron-porphyrin, Fe(TDCPP = tetra-2,6-dichlorophenylporphyrin)Cl, (Fig. 7), catalyzes alkene epoxidation by C$_6$F$_5$IO with an initial rate as

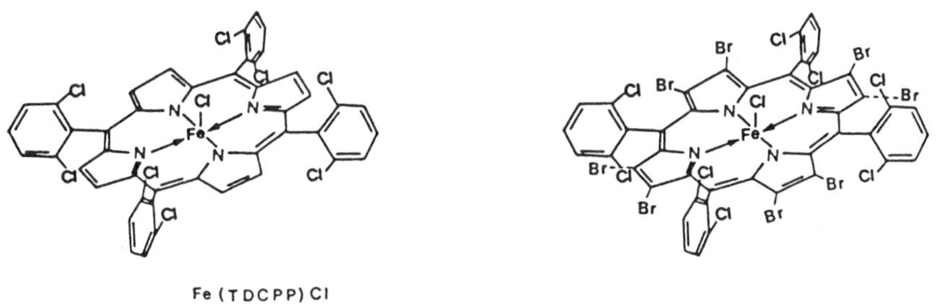

Fe (TDCPP) Cl

Figure 7. Formula of iron-polyhalogenophenyl porphyrins used as stable biomimetic catalysts for hydrocarbon oxidation.

high as 300 turnovers per second,[43] and more than 100,000 mols of epoxide are obtained per mol of catalyst without appreciable destruction of this catalyst. More recently, the even more robust Fe[tetrakis(2,6-dichlorophenyl)octabromoporphyrin]Cl complex was found able to hydroxylate norbornane with a 75% yield based on starting C$_6$F$_5$IO and without loss of catalyst[44] (Fig. 7). From the various metalloporphyrins tested as catalysts, Fe(III)- and Mn(III)-porphyrins appeared to be best.[7,8,38,40,45]

In the design of modified metalloporphyrin catalysts for selective oxidations, efforts have been made toward regioselective and asymmetric hydroxylations of alkanes. With the very hindered Mn(III)[tetrakis(2,4,6-triphenyl)phenylporphyrin] catalyst, the hydroxylation of linear alkanes occurs mainly in ω-1 position but also to a great extent in the ω-position[46] (Fig. 8). By comparison, the hydroxylation of *n*-heptane catalyzed by the nonhindered catalyst Mn(TPP)Cl occurs mainly in the ω-2 and ω-1 positions but almost not in the ω-position. The hydroxylation of phenylethane by PhIO

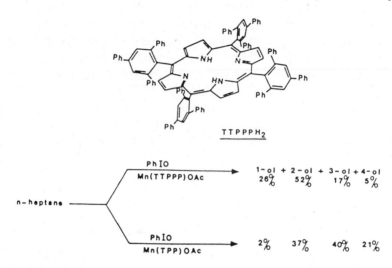

Figure 8. Control of the regioselectivity of alkane hydroxylation by the use of hindered Mn-porphyrins.

catalyzed by an iron-porphyrin bearing chiral "basket-handles" on both sides of the porphyrin ring (Fig. 9)[47] leads to a mixture of

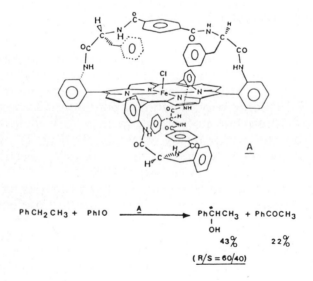

Figure 9. Hydroxylation of phenylethane by PhIO catalyzed by a chiral iron-porphyrin

phenyl-1-ethanol and acetophenone (J. P. Renaud, P. Battioni, and D. Mansuy, unpublished results). The obtained phenyl-1-ethanol is not racemic and appears as a 40:60 mixture of the (*S*) and (*R*) enantiomers. However, it was found that the oxidation of racemic phenyl-1-ethanol under identical conditions led to a faster oxidation of the (*S*)-enantiomer. Thus, it seems that the enantiomeric excess observed during phenylethane hydroxylation is mainly the result of a further enantioselective oxidation of phenyl-1-ethanol and not of an asymmetric hydroxylation of the benzylic C-H bonds of phenylethane.

Other oxidants containing a single oxygen atom such as hypochlorite, $KHSO_5$, or tertiary amine-oxides associated with Mn- or Fe-porphyrin catalysts have led to results similar to those obtained with PhIO (for a review, see refs. 7-9 and Chapter VII of this book). On the contrary, simple Mn- or Fe(porphyrin)Cl complexes fail to catalyze alkane hydroxylation by alkylhydroperoxides by a cytochrome P-450-like mechanism. Actually, Mn(porphyrin)Cl complexes lead to a slow decomposition of alkylhydroperoxides,[48,49] and Fe(porphyrin)Cl complexes lead to a fast decomposition of alkylhydroperoxides but without good epoxidation of alkenes.[50,51] It is noteworthy that alkanes are oxidized by cumylhydroperoxide in the presence of Fe-porphyrin catalysts, to the corresponding alcohols and ketones, with good yields (Eq. 2).[50,51] However, in these systems, the active species is not linked to the metal and appears to be a cumyl$O^\bullet$ or cumyl$OO^\bullet$ radical derived from a homolytic cleavage of the O-O or OH bond of the starting hydroperoxide.[46,50] Two explanations have been given for this

$$\text{Fe(TPP)Cl} \atop \text{PhC(CH}_3\text{)}_2\text{OOH}$$

$$\bigcirc \xrightarrow[\text{PhC(CH}_3\text{)}_2\text{OOH}]{\text{Fe(TPP)Cl}} \bigcirc\!\!-\text{OH} \; + \; \bigcirc\!\!=\!\text{O} \quad (2)$$

40%   20%

poor ability of Fe(porphyrin)Cl complexes to catalyze hydrocarbon monooxygenation by alkylhydroperoxides via Fe(V)=O intermediates as active species. First, it was proposed that Fe(porphyrin)Cl complexes led to a homolytic "Fenton-like" cleavage of the O-O bond of alkylhydroperoxides with formation of an alkoxy radical as active species instead of the expected Fe(V)=O intermediate (Eq. 3).[50,52] Very recently, it was shown that Fe(porphyrin)Cl complexes led to a heterolytic cleavage of the O-O bond of alkylhydroperoxides and the lack of efficient transfer of its oxygen atom to hydrocarbons was explained by a faster reaction of the iron-oxo species with ROOH (Eq. 4) than with hydrocarbons.[53] This fast reaction would explain the formation of ROO$^\bullet$ and RO$^\bullet$ in such systems[50,53] as well as of ROOR and O_2.[49,53,54] It would be at the origin of many problems encountered with the use of alkylhydroperoxides as oxygen atom donors (by comparison with ArIO) in biomimetic systems based on Fe- or Mn-porphyrins.

$$Fe(III) + ROOH \longrightarrow Fe(IV)-OH + RO\bullet \qquad (3)$$

$$Fe(III) \xrightarrow[-\ ROH]{+\ ROOH} Fe(V){=}O \xrightarrow{+\ ROOH} Fe(IV)-OH + ROO\bullet \qquad (4)$$

It was also shown that the ability of Fe- or Mn-porphyrins to catalyze alkene epoxidation by alkylhydroperoxides is considerably improved by the use of imidazole as cocatalyst. For instance, under conditions for which CumOOH is unable to epoxidize 2-methyl-hept-2-ene in the presence of Fe(TPP)Cl, it leads to a 17% epoxidation yield upon addition of imidazole.[51]

Addition of catalytic amounts of imidazole to the Mn(TPP)Cl-cumylhydroperoxide system makes it able to epoxidize cyclooctene, *cis*-stilbene, and 2-methyl-hept-2-ene with yields between 20 and 50% .[51] Similarly, a 60% yield of epoxidation of the reactive alkene tetramethylethylene is obtained with Mn(TPP)Cl and *tert*-BuOOH in the presence of imidazole.[49] These results point to two important improvements of the Fe(porphyrin)Cl-ROOH or Mn(porphyrin)Cl-ROOH systems upon imidazole addition:

1. a large increase of the rate of ROOH reaction with Mn(III)-porphyrins;[49,50]
2. a very important increase of oxygen atom transfer to alkenes.

The latter improvement could be the result of a more favored heterolytic cleavage of the O-O bond increasing metal(V)-oxo over RO• (see Eqs. 3 and 4) formation, and/or to a relative reactivity of the imidazole-metal-oxo species lower for ROOH than for alkenes when compared to the Cl-metal-oxo species formed in the absence of imidazole. Another possible explanation for the beneficial role of imidazole in such systems is derived from a kinetic study of the reaction of *tert*-BuOOH with Mn(TPP)Cl in the presence of imidazole.[49] This study provided evidence for a Mn-catalyzed transfer of an oxygen atom of *tert*-BuOOH to imidazole itself, the corresponding oxidation product of imidazole being able then to act as an oxygen atom donor. From this mechanism, a possible beneficial role of imidazole would be to store, more or less reversibly, the active oxygen atom from ROOH allowing the ROOH concentration and the importance of its oxidation to ROO• to be reduced.

2. Model Systems Using H$_2$O$_2$

Among the various possible oxygen atom donors susceptible to transfer their oxygen atom to hydrocarbons, H$_2$O$_2$ is particularly interesting since it is a readily available and cheap oxidant and since it gives only water as secondary product. Iron-porphyrins such as the very robust Fe(TDCPP)Cl lead to a fast decomposition of H$_2$O$_2$ without significant oxygen atom transfer to a mixture of cyclooctane and cyclooctene under the conditions given in Table I (P. Battioni, M. Fort, and D. Mansuy, unpublished results, see also refs. 55, 56). Presumably, this is the result of the ability of iron-porphyrins to dismutate H$_2$O$_2$.[53,57] The same experiment performed in the presence of catalytic amounts of imidazole leads to a 17% yield of cyclooctene oxide but failed to give cyclooctanol (Table I). Under identical conditions, replacement of

Table I. Oxidation of a Mixture of Cyclooctane and Cyclooctene by H$_2$O$_2$ or PhIO Catalyzed be Fe- or Mn-Porphyrins and Imidazole[a]

Catalyst	Oxidant	Presence of imidazole	Products [b]		
			Epoxide	Cyclooctanol	Cyclooctanone
Fe(TDCPP)Cl	H$_2$O$_2$	-	2	tr	tr
Fe(TDCPP)Cl	PhIO	-	76	2	1
Fe(TDCPP)Cl	H$_2$O$_2$	+	17	tr	tr
Fe(TDCPP)Cl	PhIO	+	26	0.5	tr
Mn(TDCPP)Cl	H$_2$O$_2$	+	72	10	2
Mn(TDCPP)Cl	PhIO	+	57	9	1.5

[a]Conditions: catalyst 1.6 mM, oxidant 48 mM, imidazole (if present) 38 mM, cyclooctane 0.5 M, cyclooctane 0.2 M in CH$_3$CN:CH$_2$Cl$_2$ (1:1) at room temperature under argon.
[b]Yields based on the oxidant, after 24 hr; tr, trace.

H$_2$O$_2$ by PhIO leads also to cyclooctene epoxide and to very low amounts of cyclooctanol, indicating that the imidazole-Fe(V)=O intermediate reacts much more rapidly with cyclooctene than with cyclooctane. These results suggest that, in the presence of imidazole, the reactivity of the Fe(TDCPP)Cl-H$_2$O$_2$ system is closer to that of the Fe(TDCPP)Cl-PhIO system. However, the epoxide yield remains low. When identical experiments are performed with Mn(TDCPP)Cl instead of Fe(TDCPP)Cl, a good transfer of oxygen atom toward hydrocarbons is observed in the presence of imidazole (Table I).[55,56] Epoxidation of cyclooctene is the preferred reaction but significant amounts of products of cyclooctane hydroxylation are also observed.   Interestingly, an identical reaction using PhIO instead of H_2O_2 leads to similar results and particularly to a very similar cyclooctanol:cyclooctene oxide ratio (Table I). This indicates that the H_2O_2 - and PhIO-dependent systems involve   an   identical   active   species,   presumably   a (porphyrin)(imidazole)Mn(V)=O complex. In fact, we have found almost identical results for the Mn(TDCPP)Cl-H_2O_2-imidazole and Mn(TDCPP-Cl)PhIO-imidazole systems for all the substrates that we have studied so far. Accordingly the Mn(TDCPP)Cl-H_2O_2-imidazole system was found able to oxidize various alkanes such as cyclooctane, cyclohexane, ethylbenzene, indane, and tetralin leading to the corresponding alcohols and ketones with yields based on H_2O_2 between 45 and 65% when the alkane was used in excess.[56] For cyclooctene and cyclooctane used separately and in excess the yields of oxygen transfer from H_2O_2 were 95% and 60% respectively, the rest of H_2O_2 in the case of the less reactive cyclooctane having been transformed mainly   into O_2 and H_2O by   dismutation.   The H_2O_2-Mn(TDCPP)Cl-imidazole system could be used for conversion of alkanes by using a 5-fold excess of H_2O_2.[56] Within 2hr at room temperature, 50-80% conversion of cyclohexane and of the alkanes quoted previously was obtained, the corresponding alcohols and ketones being formed with yields between 40 and 75% (Fig. 10). Under these conditions, adamantane is almost completely converted

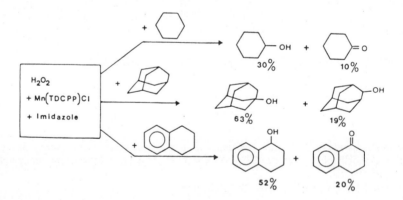

Figure 10. Hydroxylation of alkanes by H_2O_2 catalyzed by Mn-porphyrins and imidazole.

(95%) with preferential formation of 1-adamantanol (63%) and minor formation of 2-adamantanol and 2-adamantanone (19 and 3%). The particular ability of the Mn(TDCPP)Cl-H_2O_2-imidazole system to transfer oxygen atoms to hydrocarbons is presumably the result of two important effects of imidazole:

   1. It enables a fast heterolytic cleavage of the O-O bond of H_2O_2 by Mn(III)-porphyrins leading to an imidazole-Mn(V)-oxo intermediate, because of its double role as a Mn ligand and as an acid-base catalyst [49,58] (Fig. 11).
   2. In its presence, the relative importance of the dismutation of H_2O_2 is considerably reduced.

Iron-porphyrins are less interesting catalysts in such systems, since they retain a high tendency to dismute H_2O_2, even in the presence of imidazole.

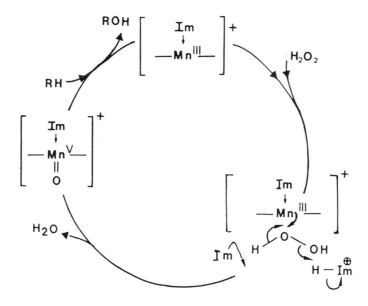

Figure 11. Possible mechanism for hydrocarbon oxidation by H_2O_2 catalyzed by Mn-porphyrins: possible roles of imidazole.

IV.   Model Systems Using O_2 Itself and Reducing Agents

Borohydrides were the first reducing agents used in O_2-dependent oxidations of hydrocarbons catalyzed by Fe- or Mn-porphyrins.[59-62] With $NaBH_4$ and Mn(TPP)Cl, cyclohexene is oxidized to cyclohexanol, which is believed to derive from reduction of cyclohexene oxide by $NaBH_4$, and to cyclohexen-3-ol, which derives from an allylic hydroxylation of cyclohexene.[59] With $(n\text{-}Bu_4N)BH_4$, which is soluble in CH_2Cl_2, and Mn(TPP)Cl, terminal olefins such as styrene are oxidized to the corresponding methyl ketones and to the alcohol formed by their reduction by the borohydride in excess.[60]

Sodium *ascorbate* was also used as a reducing agent in a biphasic system with the hydrocarbon and the Mn(TPP)Cl catalyst in benzene and ascorbate in a buffer pH 8.5. In the presence of catalytic amounts of a phase transfer agent, this system epoxidizes alkenes selectively and oxidizes alkanes into alcohols and ketones.[63] Contrary to the Mn(TPP)Cl-PhIO system that oxidizes alkanes with predominant formation of alcohols, the Mn(TPP)Cl-ascorbate-O_2 system oxidizes alkanes with major formation of the corresponding ketones[64] (Eq. 5). Moreover, the former system oxidizes cyclohexene to the corresponding epoxide and to minor amounts of cyclohexen-3-ol whereas the latter system leads exclusively to the epoxide. Thus, it was suggested that these two systems involved different active oxygen species.[64]

$$\text{(5)}$$

Hydrogen in the presence of catalytic amounts of colloidal platinum was used as a reducing agent for hydrocarbon monooxygenation by O_2 catalyzed by Mn- and Fe-porphyrins. A Mn(TPP)Cl-O_2-H_2-Pt system was found active for the epoxidation of various olefins.[65,66] This system hydroxylates adamantane with predominant formation of the tertiary alcohol[65] (Eq. 6). With the

$$\text{Adamantane} \xrightarrow[\text{Mn(TPP)Cl}]{\text{O}_2 + \text{H}_2 \text{ (Pt)}} \text{1-ol} + \text{2-ol}$$

$$\quad\quad\quad\quad\quad\quad\quad\quad 0.22 \quad 0.02 \text{ mol/mol cat./hr} \quad\quad \text{(6)}$$

iron-"picket-fence"-porphyrin, Fe(TpivPP)Cl, as a catalyst, the different steps of the epoxidation of an alkene by an O_2-H_2-Pt system have been studied.[67] It was found that a crucial step in O_2 activation by these systems was the heterolytic cleavage of the O-O bond of porphyrin-Fe(III)-O-O⁻ intermediate formed after O_2 binding to Fe(II) and one-electron reduction of the Fe(II)-O_2 complex. This cleavage is greatly facilitated by protonation or acylation of the Fe-O-O⁻ intermediate. Accordingly, it was recently shown that the heterolytic cleavage of the O-O bond of porphyrin-Fe(III)-O-O-CO-R complexes involves a very low activation enthalpy of 4 kcal/mol and was acid catalyzed.[68] The use of stoichiometric amounts of acid chlorides or acid anhydrides led to successful results in oxygenating systems using $O_2^{\cdot-}$ as oxygen atom donor and Mn[69]- or Fe[70]-porphyrins as catalysts. Carboxylic acid anhydrides were also used in electrochemical systems reported to hydroxylate alkanes and to epoxidize alkenes with O_2 as the oxygen atom donor, Fe^{70}- or Mn^{71}-porphyrins as catalysts, and electrons coming from an electrode. The electrochemical $Fe(TPP)Cl-Ac_2O-O_2$ system hydroxylates alkanes with formation of alcohols as major products.[70] This $Fe(TPP)Cl-Ac_2O-O_2$ system demethylates anisole with a $k_H:k_D$ isotopic effect of 7.[70]

All these O_2-dependent systems reproduce in a qualitative manner the main reactions of cytochrome P-450 and, particularly, the hydroxylation of alkanes, which is the subject of this chapter. However, their catalytic activities (number of turnovers per minute) and yields based on the reducing agent remain very low when compared to those of cytochrome P-450. The low yields based on the reducing agent, which are between 0.1 and 5% for alkene epoxidation and between 0.01 and 0.5% for alkane hydroxylation by the borohydride-, ascorbate- and H_2-dependent systems,[8] are due to a competition between the hydrocarbon substrate and the reducing agent in excess for reaction with the high-valent metal-oxo intermediates. As there is no separation between the active oxygen species and the reducing agent in excess in the model systems contrary to the enzymatic system, the reduction of the metal-oxo species by the reducing agent is faster than its reaction with the hydrocarbons.[64,67] The best yield based on the reducing agent reported so far, 56%, was obtained for cyclooctene epoxidation by an electrochemical system using $Mn(TPP)(Cl)$ as a catalyst and $(PhCO)_2O$ as a stoichiometric acylating agent. However, because of the slow arrival of the electrons in that system, the rate remained low (about 2 turnovers per hour).[71]

Very recently, three systems were reported to give at the same time good yields based on the reducing agent (up to 50%) and rates (up to 9 turnovers per minute) that are not too far from those observed with cytochromes P-450. The first one is based on a *dihydropyridine* as a reducing agent in the presence of a flavin mononucleotide as an electron transfer catalyst. It also uses a water-soluble anionic Mn-porphyrin and *N*-methylimidazole as catalysts and epoxidizes nerol and cyclohexene with respective rates of 9 and 3.6 mol of epoxide formed per mol of catalyst per minute and yields based on the reducing agent up to 33%.[72]

The second system also uses a Mn-porphyrin [$Mn(TPP)Cl$] and *N*-methylimidazole as catalysts but *Zn powder* as a reducing agent and CH_3COOH as a proton source.[73] This heterogeneous system epoxidizes cyclooctene with a turnover frequency of 3.3 mol of epoxide formed per mol of catalyst per minute and with a yield based on Zn of 50%. It hydroxylates cyclooctane with formation of cyclooctanol and cyclooctanone with total rates (0.5 turnovers per minute) and yields based on Zn (15%) that remain satisfactory. Adamantane is oxidized with major formation of the tertiary alcohol adamantan-1-ol (Eq. 7).[73] The third system performs also the reductive activation of O_2 by Zn.[74]

$$\text{Adamantane} \xrightarrow[\text{+ Mn(TPP)Cl + N-CH}_3\text{Imid.}]{O_2 + Zn + CH_3COOH} \text{1-ol + 2-ol}$$

$$26 \quad 4 \quad \text{mol/mol cat./hr} \quad (7)$$

However, *Zn amalgam* is used instead of Zn powder and an iron-porphyrin instead of a Mn-porphyrin. Moreover, this last system uses in addition methylviologen as an electron-transfer catalyst and stoichiometric amounts of acetic anhydride as an acylating agent. It hydroxylates cyclohexane with yields based on O_2 consumed up to 30% and rates up to 1.1 turnover per minute (Eq. 8). As expected for a monooxygenase-like system, $C_6H_{11}{}^{18}OH$ is formed when using $^{18}O_2$. Moreover, hydroxylation of cyclohexane occurs with a high kinetic isotope effect $k_H{:}k_D{=}7,$[74] similar to those reported in the case of cytochrome P-450.

$$\bigcirc \xrightarrow[\text{Fe(TPP)Cl + methylviologen}]{O_2 + Zn\,(Hg) + Ac_2O} \bigcirc\!\!-OH \quad (8)$$

$$30\% \;/\; O_2$$

V.   Conclusion

The mechanisms generally admitted now for alkane hydroxylation by the various cytochromes P-450 involve a hydrogen atom abstraction by a high valent iron-oxo intermediate followed by the transfer of an OH residue to the alkane-derived free radical. This mechanism is the result of the intrinsic reactivity of the iron-oxo active species but is independent of specific apoprotein structure. It is thus common to the various cytochrome P-450 isozymes. On the contrary, product regio- and stereoselectivities are mainly dependent on the apoprotein structure. The intrinsic characteristics of cytochrome P-450-dependent alkane hydroxylations

1. large internal isotopic effects $k_H{:}k_D$ (around 10);
2. preferential hydroxylation of tertiary C-H bonds in the absence of particular steric constraints in the active site;
3. rates between 1 and 100 turnovers per minute;

have been obtained in biomimetic systems using Fe- or Mn-porphyrins as catalysts and ArIO as oxygen atom donors.

Particular problems have been encountered with the use of alkylhydroperoxides or H_2O_2 as oxygen atom donors in such systems. They appear to be related to the fast reaction of these oxidants with the high-valent metal-oxo species leading to dismutation reactions. However, by using Mn-polyhalogenoporphyrin catalysts that are resistant toward oxidative degradation, very efficient alkane hydroxylations by H_2O_2 have been performed in the presence of imidazole cocatalysts. Model systems using ArIO or H_2O_2 are now efficient enough to be used in preparative organic chemistry. Based on these systems, the development of new catalysts for regioselective oxidations should be obtained in the near future. Some initial encouraging results have been described with the use of hindered Mn-porphyrins as catalysts for the ω-hydroxylation of linear alkanes.

Recently, systems that hydroxylate alkanes by dioxygen in the presence of a reducing agent have been reported. They exhibit qualitative characteristics similar to those of cytochrome P-450. Their catalytic activities and yields based on the reducing agent are still lower than those found for cytochromes P-450, though they now reach values of 9 and 1 turnovers of the catalyst per minute respectively, for alkene epoxidation and alkane hydroxylation, and 50% (electronic yields), which are not too far from those reported for cytochrome P-450. Major advances on such O_2-dependent model systems should occur in the next few years.

References

1. Leak, D. J. and Dalton, H., *Biocatalysis* **1987**, *1*, 23.
2. Ruettinger, R. T., Griffith, G. R., and Coon, M. J., *Arch. Biochem. Biophys.* **1977**, *183*, 528.
3. Ortiz de Montellano, P. R., ed., *Cytochrome P-450, Structure, Mechanism and Biochemistry*. Plenum Press, New York and London, 1986.
4. Guengerich, F. P., MacDonald, T. L., *Acc. Chem. Res.* **1984**, *17*, 9.
5. Ortiz de Montellano, P. R., in *Cytochrome P-450, Structure, Mechanism and Biochemistry*, (P. R. Ortiz de Montellano, ed.) p. 217. Plenum Press, New York and London, 1986.
6. McMurry, T. J., Groves, J. T., in *Cytochrome P-450, Structure, Mechanism and Biochemistry*, (P. R. Ortiz de Montellano, ed.) p. 1. Plenum Press, New York and London, 1986.
7. Meunier, B., *Bull. Soc. Chim. Fr.* **1986**, *4*, 578.
8. Mansuy, D., *Pure and Appl. Chem.* **1987**, *59*, 759.
9. Mansuy, D., Battioni, P., in *Frontiers of Biotransformation*, (K. Ruckpaul, H. Rein, eds.). Akademie-Verlag, Berlin, in press.
10. Poulos, T. L., Finzel, B. C., Gunzalus, I. C., Wagner, G. C., Kraut, J., *J. Biol. Chem.* **1985**, *260*, 16122.
11. Poulos, T. L., Finzel, B. C., Howard, A. J., *Biochemistry* **1986**, *25*, 5314.
12. Collman, J. P., Sorrell, T. N., in *Drug Metabolism Concepts,* (D. M. Jerina ed.), p. 27. J. Am. Chem. Soc., Symp. Series, Washington D.C., 1977.
13. Mansuy, D., in*The Coordination Chemistry of Metalloenzymes*, (I. Bertini, R. S. Drago, C. Luchinat, eds.), p. 343. D. Reidel, Dordrecht, 1983.

14. Schappacher, M., Weiss, R., Montiel-Montoya, R., Trautwein, A., Tabard, A., *J. Am. Chem. Soc.* **1985**, *107*, 3736 and references cited therein.
15. Groves, J. T., Haushalter, R. C., Nakamura, M., Nemo, T. E., Evans, B. J., *J. Am. Chem. Soc.* **1981**, *103*, 2884.
16. Boso, B., Lang, G., McMurry, T. J., Groves, J. T., *J. Chem. Phys.* **1983**, *79*, 1122.
17. Penner-Hahn, J. E., McMurry, T. J., Renner, M., Latos-Grazynsky, L., Eble, K. S., Davis, I. M., Balch, A. L., Groves, J. T., Dawson, J. H., Hodgson, K. O., *J. Biol. Chem.* **1983**, *258*, 12761.
18. Groves, J. T., McClusky, G. A., White, R. E., Coon, M. J., *Biochem. Biophys. Res. Commun.* **1978**, *81*, 154.
19. Gelb, M. H., Heimbrook, D. C., Malkonen, P., Sligar, S. G. *Biochemistry* **1982**, *21*, 370.
20. White, R. E., Miller, J. P., Favreau, L. V., Bhattacharyya, A., *J. Am. Chem. Soc.* **1986**, *108*, 6024.
21. Tanaka, K., Kurihara, N., Nakajima, M., *Pestic. Biochem. Biophys. Res. Commun.* **1979**, *76*, 541.
22. Groves, J. T., Subramanian, D. V., *J. Am. Chem. Soc.* **1984**, *106*, 2177.
23. Ortiz de Montellano, P. R., Stearns, R . A., *J. Am. Chem. Soc.* **1987**, *109*, 3415.
24. Hjelmeland, L. M., Aronow, L., Trudell, J. R., *Biochem. Biophys. Res. Commun.* **1977**, *76*, 541.
25. Shapiro, S., Piper., J. U., Caspi, E. *J. Am. Chem. Soc.* **1982**, *104*, 2301.
26. Jones, J. P., Trager, W. F., *J. Am. Chem. Soc.* **1987**, *109*, 2171.
27. Sugiyama, K., Trager, W. F., *Biochemistry* **1986**, *25*, 7336.
28. Miwa, G. T., Walsh, J. S., Lu, A. Y. H., *J. Biol. Chem.* **1984**, *259*, 3000.
29. Mansuy, D., Chottard, J. C., Lange, M., Battioni, J. P., *J. Mol. Catal.* **1980**, *7*, 215.
30. (a) Mansuy, D., in *Reviews in Biochemical Toxicology*, (E. Hodgson, J. R. Bend, and R. M. Philpot, eds.), Vol. 3, p. 283. Elsevier, New York, 1981; (b) Brothers, P. J., and Collman, J. P., *Acc. Chem. Res.* **1986**, *19*, 209.
31. Anders, M. W., Sunram, J. M., and Wilkinson, C. F., *Biochem. Pharmacol.* **1984**, *33*, 577.
32. Mansuy, D., Battioni, J. P., Chottard, J. C., and Ullrich, V., *J. Am. Chem. Soc.* **1979**, *101*, 3971.
33. Groves, J. T., Avaria-Neisser, G. E., Fish, K. M., Imachi, M., and Kuczkowski, R. L., *J. Am. Chem. Soc.* **1986**, *108*, 3837.
34. Ortiz de Montellano, P. R. and Grab, L. A., *J. Am. Chem. Soc.,* **1986**, *108*, 5584.
35. (a) Björkhem, I., and Danielsson, H., *Eur. J. Biochem.* **1970**, *17*, 450; (b) Sligar, S. G., and Murray, R. I., in *Cytochrome P-450, Structure, Mechanism and Biochemistry* (P. R. Ortiz de Montellano, ed.), p. 429 Plenum Press, New York and London, 1986.
36. Ullrich, V., *Topics Curr. Chem.* **1979**, *83*, 67.
37. Atkins, W. M. and Sligar, S. G., *J. Am. Chem. Soc.* **1987**, *109*, 3754 and references cited therein.
38. Groves, J. T., and Nemo, T. E., *J. Am. Chem. Soc.* **1983**, *105*, 6243.
39. Groves, J. T., Nemo, T. E., and Myers, R. S., *J. Am. Chem. Soc.* **1979**, *101*, 1032.

40. Lindsay Smith, J. R., and Sleath, P. R., *J. Chem. Soc. Perkin Trans II* **1983**, 1165.
41. Chang, C. K. and Ebina, F. *J. Chem. Soc., Chem. Commun.* **1981**, 778.
42. Traylor, P. S., Dolphin, D., and Traylor, T. G., *J. Chem. Soc., Chem. Commun.* **1984**, 279.
43. Traylor, T. G., Marsters, J. C., Nakano, T., and Dunlap, B. E., *J. Am. Chem. Soc.* **1985**, 107, 5537.
44. Traylor, T. G., Tsuchiya, S., *Inorg. Chem.* **1987**, 26, 1338.
45. (a) Hill, C. L., Schardt, B. C., *J. Am. Chem. Soc.* **1980**, *102*, 6374; (b) Groves, J. T., Kruper, W. J., Haushalter, R. C., *J. Am. Chem. Soc.* **1980**, *102*, 6375; (c) Schardt, B. C., Hollander, F. J., Hill, C. L., *J. Am. Chem. Soc.* **1982**, *104*, 3964; (d) Smegal, J. A., Hill, C. L., *J. Am. Chem. Soc.* **1983**, *105*, 3515.; (e) Hill, C. L., Smegal, J. A., Henly, T. J., *J. Am. Chem. Soc.* **1983**, *48*, 3277; (f) Lindsay Smith, J. R., Sleath, P. R., *J. Chem. Soc. Perkin Trans II* **1983**, 621; (g) Nappa, M. J., Tolman, C. A., *Inorg. Chem.* **1985**, *24*, 4711; (h) Collman, J. P., Kodadek, T., Raybuck, S. A., Brauman, J. I., Papazian, L. M., *J. Am. Chem. Soc.* **1985**, *107*, 4343.
46. Cook, B. R., Reinert, T. J., Suslick, K. S., *J. Am. Chem. Soc.* **1986**, *108*, 7281.
47. Mansuy, D., Battioni, P., Renaud, J. P., Guérin P., *J. Chem. Soc., Chem. Commun.* **1985**, 155.
48. Mansuy, D., Bartoli, J. F., Chottard, J. C., Lange, M., *Angew. Chem.* Int. Ed. **1980**, *19*, 909.
49. Balasubramanian, P. N. Sinha, A., Bruice, T. C., *J. Am. Chem. Soc.* **1987**, *109*, 1456.
50. Mansuy, D., Bartoli, J. F., Momenteau, M., *Tetrahedron Lett.* **1982**, *23*, 2781.
51. Mansuy, D., Battioni, P., Renaud, J. P., *J. Chem. Soc., Chem. Commun.* **1984**, 1255.
52. Lee, W. A., Bruice, T. C., *J. Am. Chem. Soc.* **1985**, *107*, 513.
53. Traylor, T. G., Xu, F., *J. Am. Chem. Soc.,* **1987**, *109*, 6202.
54. (a) Kalyanaraman, B., Mottley, C., Mason, R.P., *J. Biol. Chem.* **1983**, *258*, 3855; (b) Sheldon, R. A., Kochi, J. K., *Metal-Catalyzed Oxidations of Organic Compounds*, p. 34. Academic Press, New York, 1981.
55. Renaud, J. P., Battioni, P., Bartoli, J. F., Mansuy, D., *J. Chem. Soc., Chem. Commun.* **1985**, 888.
56. Battioni, P., Renaud, J. P., Bartoli, J. F., Mansuy, D. *J. Chem. Soc., Chem. Commun.* **1986**, 341.
57. Bruice, T. C., Zipplies, M. F., Lee, W.A., *Proc. Natl. Acad. Sci. U.S.A.* **1986**, *83*, 4646.
58. (a) Battioni, P., Renaud, J. P., Bartoli, J. F., Momenteau, M., Mansuy, D., *Rec. Trav. Chim. Pays Bas.* **1987**, *106*, 332; (b) Yuan, L.-C., Bruice, T. C., *J. Am. Chem. Soc.* **1986**, *108*, 1643; (c) Traylor, T .G., Lee, W. A., Stynes, D. V., *J. Am. Chem. Soc.* **1984**, *106*, 755.
59. Tabushi, I., Koga, N., *J. Am. Chem. Soc.* **1979**, *101*, 6456.
60. Perrée-Fauvet, M. Gaudemer, A., *J. Chem. Soc., Chem. Comm.* **1981**, 874.
61. Santa, T., Mori, T., Hirobe, M., *Chem. Pharm. Bull.* **1985**, *33*, 2175.
62. Mori, T., Santa, T., Hirobe, M., *Tetrahedron Lett.* **1985**, *26*, 5555.
63. Mansuy, D., Fontecave, M., Bartoli, J. F., *J. Chem. Soc., Chem. Commun.* **1983**, *26*, 253.

64. Fontecave, M., Mansuy, D., *Tetrahedron* **1984**, *40*, 4297.
65. Tabushi, I., Yazaki, A., *J. Am. Chem. Soc.* **1981**, *103*, 7371.
66. Tabushi, I., Morimitsu, K., *J. Am. Chem. Soc.* **1984**, *106*, 6871.
67. Tabushi, I., Kodera, M., Yokoyama, M., *J. Am. Chem. Soc.* **1985**, *107*, 4466.
68. Groves, J. T., Watanabe Y., *J. Am. Chem. Soc.* **1986**, *108*, 7834.
69. Groves, J. T., Watanabe, Y., McMurray, T. J., *J. Am. Chem. Soc.* **1983**, *105*, 4489.
70. Khenkin, A. M., Shteinman, A. A., *J. Chem. Soc., Chem. Commun.* **1984**, 1219.
71. Creager, S. E., Raybuck, S. A., Murray, R. W., *J. Am. Chem. Soc.* **1986**, *108*, 4225.
72. Tabushi, I., Kodera, M., *J. Am. Chem. Soc.* **1986**, *108*, 1101.
73. Battioni, P., Bartoli, J. F., Leduc, P., Fontecave, M., Mansuy, D., *J. Chem. Soc., Chem. Commun.* **1987**, 791.
74. Karasevich, E. I., Khenkin, A. M., Shilov, A. E., *J. Am. Chem. Soc.,Chem. Commun.* 1987, 731.

Chapter VII

SHAPE-SELECTIVE OXIDATION OF HYDROCARBONS

Kenneth S. Suslick

School of Chemical Sciences
University of Illinois at Urbana-Champaign
Urbana, Illinois 61801

I.   Introduction

The uniqueness of enzymes originates from their ability *to alter* properties of a *specific* substrate, including both spatial location (i.e., transport) and chemical structure. Many guest-host systems, such as the cavitands,[1] calixarenes,[2] crown ethers,[3] functionalized cyclodextrins,[4] and lacunar macrocycles,[5] have been synthesized, which exhibit remarkable substrate recognition.[6,7] None, however, is capable of catalyzing chemical reactions with the kind of regioselectivity typical of enzymatic reactions.

Such regioselectivity often originates from discrimination based on the size and shape of substrate molecules, i.e., shape selectivity. Thus, molecular recognition is not enough. *Two* goals exist for those interested in mimicking enzymatic function: first, the design and synthesis of chemical species capable of molecular shape recognition; and second, the inclusion in such synthetic hosts of a reactive center capable of catalyzing a regiospecific or enantioselective chemical transformation on the guest substrate. In microporous extended solids, most notably zeolites, shape-selective catalysis of hydrocarbon reforming has been successful and formed the basis for important industrial processes.[8] Nonetheless, the creation of shape-selective catalysts for hydrocarbon oxidations remains largely unexplored.

The nature of oxidative processes inherently suggests an important role for shape selectivity. In the absence of steric restraints, regioselectivity in oxidations are determined by thermodynamic control.

For example, in the hydroxylation of alkanes by metalloporphyrins, product distribution analyses are consistent with hydrogen atom abstraction and recombination, involving a radical intermediate.[9,10] Hydroxylation selectivities are therefore dominated by C-H bond strengths: 3º>>2º>>1º. In epoxidations, the active oxidant is extremely electrophilic and therefore preferentially attacks the most electron-rich (i.e., most substituted) double bond, as do essentially all oxidants, porphyrinic or otherwise.[11] In the presence of specific substrate binding sites, however, regioselectivities of this sort can be altered, and various metalloproteins can yield dramatically different products. Most notable are some of the isozymes of cytochrome P-450 and the nonheme, iron-containing proteins known as ω-hydroxylases, which are responsible for primary alcohol synthesis via terminal hydroxylation of alkyl chains (e.g., cholesterol, fatty acids, and *n*-alkanes). The enzymatic cycle of cytochrome P-450 has been particularly well delineated[12] (as shown in Fig. 1), in part because of the presence of an iron-porphyrin active site, which provides a relatively easily definable coordination geometry.

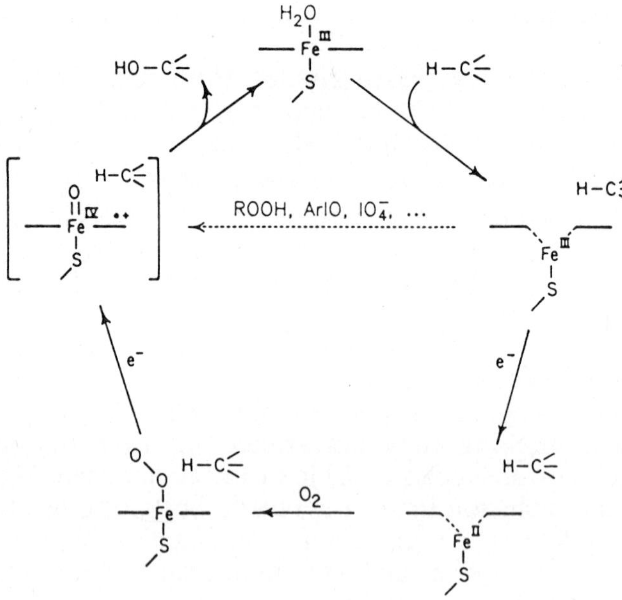

Figure 1. The catalytic cycle of cytochrome P-450.

In order to elucidate the reaction mechanisms for the oxidation of hydrocarbons by cytochrome P-450, intense efforts have been made to generate similar catalytic behavior with synthetic iron-, manganese-, and chromium-porphyrin complexes.[13-17] The synthetic versatility of these macrocycles make them attractive tools with which to construct enzymatic analogs.[18]

Suslick and co-workers[19-22] and others[23-34] have attempted to use steric protection of the metalloporphyrin faces to reverse these selectivities for both hydroxylation and epoxidation. In this chapter, we shall review the development of such metalloporphyrins as shape-selective oxidation catalysts (with special emphasis on alkane functionalization), their use as mechanistic probes of the nature of the active catalyst, and the extension of these concepts to the synthesis of heterogeneous shape-selective catalysts.

II. Shape Selectivity

These discussions will focus first on the regioselectivity for alkane hydroxylation shown by various metalloporphyrin catalysts whose functionalization produces a wide range of steric constraint over the face of the macrocycle; second, on similar studies on the shape-selective epoxidation of alkenes; and third, on the use of shape-selective catalysts as mechanistic probes to demonstrate the nature of the catalytic species and the site of oxidation. Finally, comparisons to enzymatic selectivities will be made.

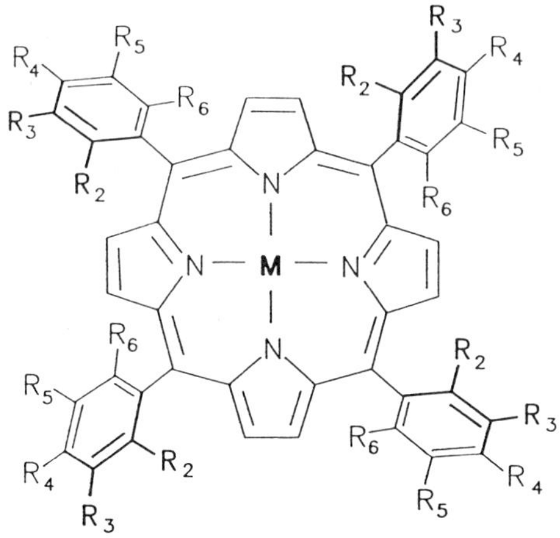

TPP: 5, 10, 15, 20-tetraphenylporphyrin; $R_2 = R_3 = R_4 = R_5 = R_6 = H$
TF$_5$PP: 5, 10, 15, 20-tetrakis(pentaflurophenyl)porphyrin; $R_2 = R_3 = R_4 = R_5 = R_6 = F$
T(2-MeP)P: 5, 10, 15, 20-tetra(2'-methylphenyl)porphyrin; $R_2 = CH_3$
T(4-HDP)P: 5, 10, 15, 20-tetra(4'-hexadecyloxyphenyl)porphyrin; $R_4 = OC_{16}H_{33}$
TDCPP: 5, 10, 15, 20-tetrakis-(2', 6'-dichlorophenyl)porphyrin; $R_2 = R_6 = Cl$
TMP: 5, 10, 15, 20-tetramesitylporphyrin; $R_2 = R_4 = R_6 = CH_3$
TTMPP: 5, 10, 15, 20-tetrakis(2',4',6'-trimethoxyphenyl)porphyrin; $R_2 = R_4 = R_6 = OCH_3$
TTPPP: 5, 10, 15, 20-tetrakis(2',4',6'-triphenylphenyl)porphyrin; $R_2 = R_4 = R_6 = C_6H_5$

Figure 2. Chemical structure of a series of sterically hindered porphyrins.

A number of different metalloporphyrins will be discussed in this chapter. As an aid to clarity, Figure 2 gives the abbreviations that will be used. Special emphasis will be placed on the detailed studies[19-22] made with a series of three porphyrins that span the steric constraints attainable with such complexes: the unhindered TPP, the shallow-pocketed TTMPP, and our highly hindered, bis-pocket porphyrin, TTPPP. Figure 3 shows computer-generated space-filled models of this series.

Figure 3. Computer-generated molecular models of sterically hindered porphyrins; for scale, heptane is also shown. The porphyrins represented are the unhindered TPP, the moderately hindered TTMPP, and the deeply pocketed TTPPP. For the sake of clarity, the atomic radii shown are only 0.8 of the van der Waals radii. These models were generated with the ChemX program suite, Chemical Design, Ltd., Oxford, U.K.

1.   Homogeneous Catalysts

In order to use any catalyst for oxidation of hydrocarbons, it is important that the catalyst be oxidatively robust. This is generally *not* the case with simple metalloporphyrins. A semiquantitative comparison of the rates of porphyrin degradation in these systems is enlightening. In highly dilute manganese porphyrin solutions (10 μM), the half-lives of the metalloporphyrins upon addition of a large excess of oxidant are independent of oxidant concentration, but strongly dependent on steric encumbrance at the *meso*-position of the porphyrin.   For iodosylbenzene, pentafluoroiodosylbenzene, or *m*-chloroperbenzoic acid, MnTPP(OAc) has a half-life of 5 min, MnTTMPP(OAc) of 10 min, and MnTTPPP(OAc) of 25 hr.[20] Thus, steric protection of the periphery of metalloporphyrins *dramatically enhances* the oxidative robustness of the catalyst. Electron-withdrawing substituents can have similar effects,[29] as with, for example, the particularly robust TDCPP.

A. Hydroxylation

The involvement of hydrogen atom abstraction, followed by rapid recombination of a metal bound OH, has been demonstrated for at least some of the alkane hydroxylations catalyzed by metalloporphyrins.[14] In the absence of steric considerations, regioselectivity will therefore be determined by relative bond dissociation energies. With a sterically undemanding metalloporphyrin (such as MnTPP), one would expect (and observes) that the selectivity would be 3°>>2°>>1°, and that there would be no statistically significant preference for one 2° site over another 2° site, etc. With sterically bulky catalysts, however, access of the alkane to the metal should be restricted to the more exposed C-H bonds, giving rise to shape selectivity. Similarly, epoxidation appears to involve the same electrophilic oxidant (i.e., probably a terminal metal oxo species), which means that the preferred site of epoxidation is the most electron-rich (i.e., most substituted) double bond, in the absence of other steric constraints. The possibility for reversal of selectivities by steric means is therefore established for both hydroxylation and epoxidation.

The *most difficult* substrate for the demonstration of regioselectivity by any catalyst must be the *n*-alkanes, which lack any functionality or polarity with which the catalyst may differentiate one site from another. Only relatively modest, anisotropic differences in the shape of the *n*-alkanes may allow for the distinction of one methylene from another, or, even more challenging, for the selection of terminal methyl groups.

In all cases involving modestly hindered metalloporphyrins[19,20,23-28] (such as complexes of 5,10,15,20-tetrakis(2',4',6'-trimethoxyphenyl)porphyrin, TTMPP, as shown in Figs. 2 and 3), only very modest shape selectivity has proved possible: for example, slight regioselectivity for hydroxylation of one secondary carbon versus another was observed, but no significant production of primary alcohols was seen. These subtle differences in shape selectivity may be examined by noting the production of 2-ols relative to 3-ols from *n*-alkane hydroxylation, as shown in Figure 4. In this plot, the normalized [2-ol]/[3-ol] ratio would be 1.0 if there is no discrimination between one secondary site and another. As expected, unhindered metalloporphyrins give a ratio of about 1. As the level of steric restraint increases, however, this ratio can become considerably greater than 1. Terminal hydroxylation, to be discussed later, is observed only with the most sterically hindered catalysts.

The selectivity for the most accessible secondary site is greatest for our bis-pocket porphyrin, MnTTPPP(OAc), and is even quite significant for substrates as small as pentane.[20] Results from molecular modeling suggest that smaller alkanes can enter the deep pocket of MnTTPPP(OAc) sideways. However, the observed product distribution proves that even small alkanes enter in an oriented fashion.

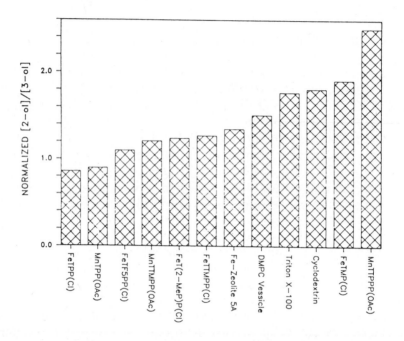

Figure 4. Secondary vs secondary hydroxylation: ω-1 selectivity as a function of steric hindrance of the catalyst. In most cases, the ratio of 0.5*[2-pentanol]/[3-pentanol] from hydroxylation of *n*-pentane with iodosylbenzene is given. The ratio [2-hexanol]/[3-hexanol] is shown for the hydroxylation of *n*-hexane with iodosylbenzene for DMPC vesicle [dimyristoylphosphatidylcholine vesicles containing Fe(T(4-HDP)P)(Cl)], for Triton *X*-100 [micelles containing Fe(T(4-HDP)P)(Cl)], and for cyclodextrin [aqueous solution containing β-cyclodextrin and Fe(T(4-HDP)P)(Cl)]. References given in text.

These experimental observations are consistent with both space-filling models (Fig. 3) and with computer molecular modeling (ChemX Program Suite, Chemical Design, Ltd., Oxford, U.K.). The calculated pocket dimensions from such molecular modelling are 4.0 Å across by 5.0 Å deep, measured from the van der Waals surfaces.[20,22]

Regioselectivity among primary sites of different steric environments may be exhibited in the hydroxylation of branched alkanes. 2,2-Dimethylbutane is a particularly interesting case, since it has both a 2° site and two quite different 1° sites. This substrate is a bit like a fist with a sore thumb, and as such is well suited as a probe for shape selectivity! As shown in Figure 5, the secondary alcohol is the predominant product (≈90%) for catalysis by either MnTPP(OAc) or MnTTMPP(OAc). With the deeply pocketed MnTTPPP(OAc), however, the steric inaccessibility of the 2° site is quite pronounced and 3,3-dimethylbutan-2-ol becomes only a minor product. Even more striking is the selectivity shown in favor of the most exposed methyl group (which gives 3,3-dimethylbutan-1-ol) and against the hindered *tert*-butyl group methyls (which give 2,2-dimethylbutan-1-ol). The ratio of the primary alcohols, weighted for their total number of hydrogens, increases from 0.3 to 0.89 to 34 for MnTPP(OAc), MnTTMPP(OAc), and MnTTPPP(OAc), respectively. Thus, 2,2-dimethylbutane may enter the pocket of MnTTPPP(OAc) in only one way to minimize the steric interaction between the bulky *tert*-butyl group and the triphenylphenyl substituents on the porphyrin's periphery. This very specific, enzyme-like, shape selection of the substrate by the catalyst gives rise to the impressive preference for hydroxylation of the sterically most accessible methyl group.

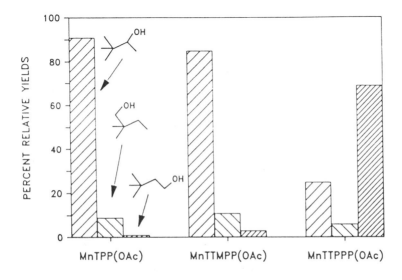

Figure 5. Primary vs primary hydroxylation: 2,2-dimethylbutane hydroxylation as a function of steric hindrance on the metalloporphyrin catalyst. Reactions were done in benzene under Ar using iodosylbenzene as oxidant.[20]

The *most difficult* goal in this research is clearly the hydroxylation of terminal methyl groups in *n*-alkanes. This task has at least two different motivations. First, the terminal hydroxylation of alkyl chains is an important metabolic pathway in mammals as well as in bacteria. Second, alkanes are inexpensive (so much so that we routinely burn them!) and potentially attractive chemical feedstocks. As shall be seen, a few synthetic systems now exist which are capable of some selectivity for terminal hydroxylation.

In order to make meaningful comparisons between various alkane substrates in Figure 6, normalized selectivity indices have been used. The primary selectivity index was defined as the ratio of the concentration of primary alcohol to secondary alcohols, normalized for the respective number of hydrogens. The following general formula was used: $N*[1°$ alcohol]/[2° alcohols], where N is the ratio of the number of primary to secondary H atoms (1.00 for n-pentane, 1.33 for n-hexane, 1.67 for n-heptane, 2.00 for n-octane, 2.67 for n-decane, and 4.00 for n-tetradecane).

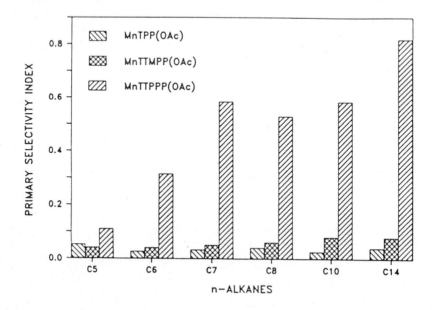

Figure 6. Primary selectivity index vs n-alkane chain length: primary selectivity index is the ratio of primary to total secondary alcohols, normalized for the relative number of hydrogens. Reactions were done in benzene under Ar using iodosylbenzene as oxidant.[20]

As expected, MnTPP(OAc) catalysis of iodosylbenzene hydroxylation of n-alkanes gives nearly exclusively 2° alcohols; the shallow pocketed metalloporphyrin, MnTTMPP(OAc), is little better for the production of 1° alcohol (Fig. 6). With the deeply pocketed MnTTPPP(OAc), however, regioselectivity for the hydroxylation of the more sterically accessible methyl groups is observed, using iodosylbenzene as an oxidant (Fig. 6). MnTTPPP(OAc) also shows increasing primary selectivity for n-alkanes with increasing chain length (Fig. 6). Relatively large increases in the primary selectivity index are observed in going from pentane to hexane to heptane, after which much smaller increases are observed with each added methylene unit. Thus, hexane marks the size cutoff for the "sideways" entry of n-alkanes into the pocket of this porphyrin, and alkanes of larger size are restricted in their entry to an end-on approach, consistent with results from computer molecular modeling.

The selection for primary over *tertiary* hydroxylation is extremely difficult as a result of the large difference in bond dissociation energies. It is not surprising, then, that even with MnTTPPP(OAc), the principal product from hydroxylation of 2,3-dimethylbutane remains 2,3-dimethylbutan-2-ol, as seen in Figure 7. Nonetheless, there is a 40-fold increase in primary selectivity for MnTTPPP(OAc) compared to the less hindered porphyrin complexes, and the primary alcohol (2,3-dimethylbutan-1-ol) becomes a substantial, rather than trace, product.

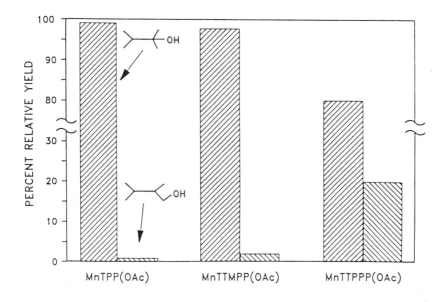

Figure 7. Primary vs tertiary hydroxylation: 2,3-dimethylbutane hydroxylation as a function of steric hindrance on the metalloporphyrin catalyst. Reactions were done in benzene under Ar using iodosylbenzene as oxidant.[20]

B.   Epoxidation

This same sort of shape selectivity can be extended to alkene epoxidation, as well. A number of researchers have examined the effect of small steric restraints and electronegative substitution on the regioselectivity of alkene epoxidation[29-33] of, for example, norbornene (in which the *exo* to *endo* ratio is exquisitely sensitive to steric and electronic factors) or the sterically demanding *trans*-stilbene. For example, an *ortho*-chloro substitution (as in TDCPP) will give a 10-fold decrease in the production of *exo*-epoxynorbornane.[29] Some success has also been made by Groves and Myers on catalytic asymmetric

epoxidation with chiral iron-porphyrins,[34] yielding enantiomeric
excesses as high as 48%.

    The effects of severe steric protection of epoxidation catalysts
have only recently been examined. We used a series of nonconjugated
dienes of varying shapes and sizes as intramolecular probes of shape
selectivity in metalloporphyrins[21,22] in analogy to the shape-selective
hydroxylation of alkanes. These dienes were epoxidized using aqueous
NaOCl in the presence imidazole and three manganese-porphyrins
with varying steric demands. In all cases, MnTTPPP(OAc) shows
enhanced selectivity for epoxidation of the most exposed double bond of
the substrate (Fig. 8). Even with nonconjugated straight-chain dienes
(e.g., *trans*-1,4-hexadiene or *trans*-1,4-octadiene), MnTTPPP(OAc)
shows quite a remarkable preference for epoxidation at the terminal
position, when compared to the unhindered MnTPP(OAc) or even the
modestly hindered MnTTMPP(OAc). As expected, this selectivity for
terminal epoxidation increases as the steric bulk of the diene increases,
similar to the trend in the hydroxylation of *n*-alkanes by these same
catalysts. The shape selectivity originates from the steric demands of
the metalloporphyrins' superstructures. Consistent with this, the ratio

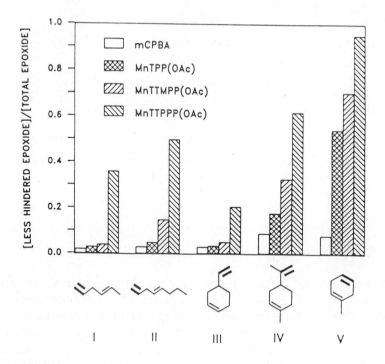

Figure 8. Shape selectivity for nonconjugated diene epoxidation. The less hindered
double bond is highlighted in the drawings of the diene substrates. *m*-CPBA
is *meta*-chloroperbenzoic acid, uncatalyzed; all other cases are metalloporphyrin-
catalyzed epoxidations in benzene with aqueous NaOCl under Ar.[21]

of enhancement for terminal epoxidation in going from C-6 to C-8 is very similar to that observed for hydroxylation of *n*-hexane versus *n*-octane.

Limonene (**IV** in Fig. 8) and its structural analog 4-vinyl-1-cyclohexene (**III**) are very useful chiral starting materials for many organic syntheses.[35] In all previous cases, however, epoxidation of these molecules gave exclusively ring epoxidation rather than external epoxidation. This is still the case with the unhindered MnTPP(OAc) and the modestly hindered MnTTMPP(OAc). The extremely hindered MnTTPPP(OAc), however, enhances the epoxidation of the external double bond, and, for the more sterically demanding limonene, external epoxide is the major product formed. Indeed, the selectivity observed for MnTTPPP(OAc)-catalyzed epoxidation is the largest ever reported for epoxidation with any synthetic oxidant system.[30]

The epoxidation of limonene was examined with representative oxidant systems of both iron- and manganese porphyrin catalysts. The differences in selectivity between iron and manganese are quite dramatic. The ratio of external epoxide to ring epoxide is 0.22 for MnTPP(OAc), 0.49 for MnTTMPP(OAc), and 1.63 for MnTTPPP(OAc), but only 0.16 for FeTPP(OAc), 0.18 for FeTTMPP(OAc), and 0.35 for FeTTPPP(OAc).

Extremely high selectivity is observed in the epoxidation of 1-methyl-1,4-cyclohexadiene (**V** in Fig. 8) for the sterically less-hindered double bond. Even the unencumbered MnTPP(OAc) gives a large enhancement in selectivity relative to uncatalyzed epoxidation with *m*-chloroperoxybenzoic acid, which involves the sterically less sensitive 3-member intermediate [MnTPP(OAc) shows modest enhancements compared to *m*-chloroperoxybenzoic acid for all substrates examined]. MnTTPPP(OAc), with its very hindered pockets, gives almost *exclusive* epoxidation of the more exposed, disubstituted double bond.

2.   Heterogeneous and Microheterogeneous Catalysts

Considerably less work has been done with heterogeneous catalysis for shape-selective oxidation, although it is an approach with much promise. There are two classes of catalysts in this area: microheterogeneous systems that utilize micelles or lipid bilayers to achieve a steric isolation of the catalytic center (which is usually still a metalloporphyrin) and microporous solids with interior catalytic sites (usually zeolites). Each will be considered in turn.

A number of researchers have incorporated porphyrins into micellar or bilayer structures, and there have been two recent reports of the use of such systems to induce shape-selective oxidation. The first is an intriguing, but often overlooked, note by Shilov and co-workers.[36] These researchers used a long-tailed iron porphyrin complex [tetrakis(4'-hexadecyloxyphenyl)porphyrin, T(4-HDP)P] incorporated into Triton *X*-100 micelles or phospholipid vesicles to hydroxylate hexane. The preference for ω-1 hydroxylation to give 2-hexanol in both cases was good, as shown in Figure 4. Terminal hydroxylation occurs, albeit to a limited extent. The primary selectivity index was 0.04 for the micellar system and 0.10 for the vesicles, as compared to 0.31 for the bis-pocket porphyrin, MnTTPPP(OAc). Even more remarkable, however, is the effect of added β-cyclodextrin to Fe[T(4-HDP)P](Cl); in this system, the [2-ol]/[3-ol] ratio is reasonably good, but the terminal hydroxylation is amazing, with a primary selectivity index of 0.47! The origin of this effect has not been explained, and the effect of substrate or oxidant also requires further examination.

Several other groups have also incorporated metalloporphyrins into lipid bilayer vesicles for both O_2 binding and oxidation.[37,38] Recently, however, Groves and Neumann have succeeded in using the system represented in Figure 9 to epoxidize regioselectively sterols and fatty acids.[39] Given a sufficiently large sterol, the selectivity for the more extended double bond can be significant. In the case of simple diene fatty acids, however, the selectivities are relatively modest. Alkanes are apparently not effective substrates. Removal of products from the active site has also proved to be difficult, which limits the catalytic utility of this system.

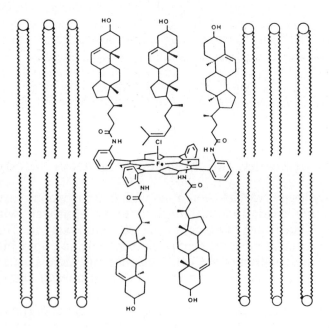

Figure 9. Idealized molecular assembly of 5,10,15,20-tetra(*o*-3β-ol-5-cholenylamidophenyl)porphyrinatoriron(III) chloride and desmosterol in a phospholipid bilayer.[39]

Various workers have recently begun an exploration of the use of modified zeolites as shape-selective heterogeneous catalysts for hydrocarbon oxidation.[40-43] In one study, an iron phthalocyanine was synthesized *in situ* within a large pore zeolite and examined as an alkane hydroxylation system.[42] The number of turnovers observed was quite low, but considerably increased compared to iron phthalocyanine, which is quite oxidatively sensitive. No terminal hydroxylation was observed. For octane with the most selective zeolite (type Y with 0.13 wt% Fe), the ω-1 selectivity was 1.33, which may be compared to moderately hindered porphyrins as shown in Figure 4.

Of more interest is the recent use by Herron and Tolman of a small pore zeolite, Fe(II)/Pd(II) exchanged zeolite 5A (Si/Al ≈ 1.2%). This catalyst was used to hydroxylate alkanes in the presence of H_2/O_2 mixtures.[43] Good shape selectivity was observed for product extracted from within the zeolite: [2-ol]/[3-ol] ratios were comparable to moderately hindered porphyrins (Fig. 4), and the primary selectivity index for terminal hydroxylation of octane was extremely high [0.54 to 0.67 vs 0.53 for MnTTPPP(OAc)]. As noted by the authors, there are four points that significantly limit the utility of this system. First, in the absence of a surface poison (such as 2,2'-bipyridine), most oxidation occurs in a *nonselective* fashion on the surface of the zeolite, with little terminal hydroxylation. This is hidden in the analysis technique which examines *only* product caught in the interior of the zeolite. Second, product formed in the interior of the zeolite is permanently trapped there, and must be extracted by *dissolution of the zeolite in concentrated sulfuric acid!* Third, the catalyst is self-poisoning; since product is entrapped, substrate cannot continue to have access to the catalytic sites (turnovers are 0.3/Fe). Finally, >95% of the H_2 is utilized in the formation of H_2O rather than substrate oxidation. If a larger pore zeolite host is used (such as ZSM-5), the products can be removed without zeolite dissolution, but the selectivities are reduced. Nonetheless, this approach has tremendous potential for future industrial application.

Finally, a brief mention will be made of the so-called Gif process,[44-46] which uses an iron salt, zinc, and O_2 in a pyridine/acetic acid suspension to form ketones from nonactivated methylene groups. The system will not attack either primary or tertiary carbons. Although the selectivity is quite high, it does not appear that this is a case of shape selectivity, but must represent an unusual mechanistic requirement. The pathways of this reaction are under continued investigation.

III.   Shape Selectivity as a Mechanistic Probe

1.   Alkane Hydroxylation

The presence and degree of shape selectivity can be used as a mechanistic probe of various porphyrin-catalyzed oxidations. The detailed mechanisms of hydroxylation and of epoxidation are different, and the discussion will be divided accordingly.

The primary selectivity index in the hydroxylation of *n*-heptane with various oxidants in the presence of MnTTPPP(OAc) indicates whether the hydroxylation reaction occurs at the metal center or in the bulk solution (Fig. 10). When pentafluoroiodosylbenzene or *m*-chloroperbenzoic acid is used, the primary alcohol selectivities are very similar to those of iodosylbenzene, suggesting that these three oxidants generate the same catalytically active species at the metal. A high spin d^2, Mn(V)-oxo complex has been suggested as one possible intermediate.[47] It had also been suggested in the case of iodosylbenzene oxidation that the active hydroxylating agent might be a metalloporphyrin-iodosylbenzene complex;[48] this hypothesis is inconsistent with the *constancy* of the primary selectivity index observed with iodosylbenzene, pentafluoroiodosylbenzene, and *m*-chloroperoxybenzoic acid. Hill and co-workers have isolated and characterized[49,50] two other species formed during these reactions, both of which are capable of oxidizing alkanes in solution: $[XMn(IV)TPP(OIPh)]_2O$ ($X = Cl^-$, Br^-) and $(N_3Mn(IV)TPP)_2O$.

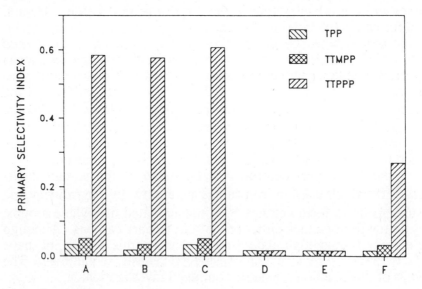

Figure 10. Primary selectivity index for various hydroxylating systems.[20] (*A*) Mn-porphyrin catalyst + iodosylbenzene as oxidant. (*B*) Mn-porphyrin catalyst + pentafluoroiodosylbenzene as oxidant. (*C*) Mn-porphyrin catalyst + *m*-chloroperoxybenzoic acid as oxidant. (*D*) Mn-porphyrin as catalyst + trifluoroethanol-solubilized pentafluoroiodosylbenzene as oxidant [reactions done in 1 ml benzene/heptane with 100 μl trifluoroethanol; also done with CH_2CCl_2/F_3-EtOH/H_2O (80:18:2 v/v) as cosolvent.). (*E*) Mn-porphyrin catalyst + *tert*-butylhydroperoxide as oxidant. (*F*) Fe-porphyrin catalyst + iodosylbenzene as oxidant.

Dimerization to form species similar to Hill's μ-oxo complexes can also be ruled out as the catalytic species, because MnTTMPP and particularly MnTTPPP cannot form dimers. Thus, the active oxidant in all likelihood is the terminal oxo complex.

      The large kinetic isotope effect observed shows that significant C-H bond breaking is taking place in the activated complex for manganese-catalyzed hydroxylation. This is consistent with the radical-like pathway proposed by Hill and Schardt[9] and Groves et al.[10] Thus, the mechanism of hydroxylation in these systems is well established: complete oxygen atom transfer from the iodosylbenzene takes place to form a monomeric catalytic species that is capable of H-atom abstraction. Consistent with these results is the following partial reaction scheme, in which X=O is C_6H_5IO, C_6F_5IO, or m-$ClC_6H_4CO_3H$. No direct evidence for the coordination of the acetate

$$Mn(Porph)\,(OAc) + X{=}O \longrightarrow O{=}Mn(Porph)\,(OAc) + X$$

$$RH + O{=}Mn(Porph)\,(OAc) \longrightarrow R{\bullet} + HO{-}Mn(Porph)\,(OAc)$$

$$R{\bullet} + HO{-}Mn(Porph)\,(OAc) \longrightarrow ROH + Mn(Porph)\,(OAc)$$

is available; however, its dissociation is unlikely in nonpolar solvents. This scheme is not comprehensive; some involvement of carbenium ions, as an alternate path to radical abstraction, has also been suggested.[27,34,50]

      Addition of 2,2,2-trifluoroethanol or other alcohols solubilizes iodosylbenzene or pentafluoroiodosylbenzene, by alcoholysis, and enhances the rate of olefin epoxidation using iron-porphyrins.[51] When F_3CCH_2OH is added to the manganese-porphyrin-pentafluoro-iodosylbenzene system, however, hydroxylation rates are diminished and shape selectivity is no longer observed. Thus, the species responsible for hydroxylation in this system is not the same as in the unsolubilized iodosylbenzene and is *not* localized at the metal center. In contrast, however, epoxidation with this oxidant yields selectivity very similar to other oxidants, as discussed shortly. This suggests that the initial oxidant formed in this system *is* the same metal oxo species, but that in the case of poor substrates (i.e., alkanes) secondary reactions occur that generate a radical chain oxidation.

      Using *tert*-butylhydroperoxide as an oxidant, rapid alkane hydroxylation occurs, but no change in selectivity is observed as the manganese-porphyrin catalyst is varied. Therefore, with this oxidant, hydroxylation is *not* taking place at the metal center; rather, it must be the result of a free radical chain pathway, initiated by the metalloporphyrin. The regioselectivity observed by this radical chain shows little primary product and the predominant secondary product is at the ω-1 methylene position similar to that reported for the radical chlorination of *n*-heptane.[52]

When *iron* porphyrin complexes are used as catalysts, with iodosylbenzene as oxidant, shape selectivity is still observed, as shown in Figure 10. Thus, for both Mn- and Fe-porphyrins, substrate oxidation is taking place in close proximity to the metal center. Primary selectivity, however, is diminished for iron- relative to manganese-porphyrin complexes. The diminution in selectivity may be caused by either electronic differences between the two different oxometalloporphyrin intermediates or to differences in the steric constraints present during the transition state of H-atom abstraction. The diminished selectivity does suggest that in the transition state more C-H bond breaking is occurring in the iron system than in the manganese. This conclusion is confirmed by the relative isotope effects observed for iron[53] ($k_H/k_D = 11.5$ for cytochrome P-450 and 12.9 for FeTPP(Cl) with iodosylbenzene) as compared to manganese[20] ($k_H/k_D = 3.5$).

The amount of steric contact generated between substrate and metalloporphyrin during hydroxylation may be determined from the relative rates of 1° vs 2° alcohol production for TTPPP relative to TPP with either Mn or Fe. This change is the energy associated with the greater steric contact of TTPPP over TPP with the substrate. For *n*-heptane hydroxylation, this calculated energy is 1.6-1.7 kcal/mole for *both* metal systems.

Electronic effects of substituents on the porphyrin periphery have also been examined by Nappa and Tolman.[27] Such substituents can change the redox potentials for porphyrin ring oxidation by as much as 0.4 V. The effect on regioselectivity for one secondary site versus another (in the absence of steric hindrance), however, is small. For example, the difference in redox potentials for FeTPP(Cl) and FeTF$_5$PP(Cl) is 0.2 V, but as seen in Figure 3 this has little effect on the selectivity observed for alkane hydroxylation. Likewise, in the hydroxylation of pentane, both FeTPP(Cl) and FeTF$_5$PP(Cl) have very small primary selectivity indices (≈0.02). In the case of sterically demanding substrates, such as *tert*-butylcyclohexane, greater influence of electronic effects has been seen in unhindered metalloporphyrins, but are still overwhelmed by the steric demands of the catalyst.

2.   Alkene Epoxidation

For alkene epoxidation catalyzed by metalloporphyrins, several reaction intermediates have been proposed. These include long-lived metallooxetane intermediates,[54-56] electron transfer mechanism resulting in either a metal-bound radical or carbocation intermediate,[17,57] or direct oxygen atom transfer. In addition, radical autoxidation (mostly at allylic positions) via radical chain pathways can also be initiated by metalloporphyrins.[58] The metallooxetane and radical or carbocation intermediates are shown below.

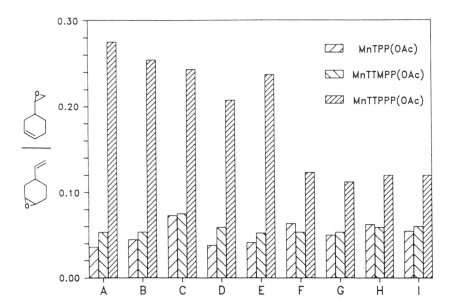

When the three manganese porphyrin complexes of increasing steric demands are used as catalysts for the epoxidation of 4-vinyl-1-cyclohexene, shape selectivity for the external epoxide is shown and increases as the metalloporphyrin catalyst becomes more encumbered (Fig. 11). The substantial selectivity observed with

Figure 11. Shape selectivity for epoxidation as a function of oxidant. Ratio of external epoxide to ring epoxide for 4-vinyl-1-cyclohexene is shown for various oxidant systems.[22] (*A*) MnPorph(OAc) + NaOCl$_{(aq)}$ + 20 m*M* 4'-(imidazol-1-yl)-acetophenone (ImAcP). (*B*) MnPorph(OAc) + iodosylbenzene + 20 m*M* ImAcP. (*C*) MnPorph(OAc) + iodosylbenzene. (*D*) MnPorph(OAc) + 1,1,1-trifluoroethanol-solublized iodosylbenzene. (*E*) MnPorph(OAc) + O$_2$/H$_2$ + Pt colloid. (*F*) FePorph(OAc) + pentafluoroiodosylbenzene. (*G*) FePorph(OAc) + iodosylbenzene. (*H*) FePorph(OAc) + 1,1,1-trifluoroethanol solublized iodosylbenzene. (*I*) FePorph(OAc) + iodosylbenzene + 20 m*M* ImAcP.

MnTTPPP(OAc) is essentially independent of oxidant used and is insensitive to the addition of nitrogenous bases. This is consistent with initial oxygen atom transfer from the oxidant to form a monomeric O=Mn(V) and demonstrates that epoxidation proceeds through very similar transition states for *all* of these epoxidation systems. Interestingly, selectivity is not significantly altered by addition of ligands (such as imidazoles). Although these ligands hasten the reaction of the metalloporphyrin with the oxidant in forming the active oxidizing species,[33,59,60] surprisingly, they do *not* significantly alter the geometry or energetics of the transition state. This is not the result of lack of imidazole coordination, however, since moderate sized ligands (such as imidazoles) can bind even to TTPPP complexes as easily as to the sterically undemanding TPP complexes.[61]

The iron(III) complexes of these porphyrins also show shape selectivity for epoxidation, albeit less so than manganese. This indicates a much less sterically demanding intermediate for the iron-catalyzed epoxidations relative to manganese, such as the ring open radical or carbocation intermediate expected from an electron transfer mechanism.[17] Epoxidation *cannot exclusively* involve a metallooxetane *intermediate* (as opposed to transition state) for both iron and manganese, since very similar selectivities would then be expected for both metals. Either there must exist at least two pathways, one sterically demanding and the second less so, or there is *no preequilibrium* to a discrete intermediate, and the differences in selectivity reflect changes in steric demands of a late versus early transition state. In fact, evidence continues to demonstrate that under different conditions several different mechanisms are probably operable.[21,47]

IV.   Comparisons to Enzymatic Hydroxylation

There are two classes of ω-hydroxylases (i.e., enzymes that hydroxylate terminal methyl groups of alkyl chains): a nonheme iron monooxygenase found in bacteria,[62] and specific isozymes of cytochrome P-450 found, for example, in bacteria, yeast, and mammalian mitochondria.[63-65] With *in vivo* substrates (e.g., fatty acids and cholesterol steroids), the regioselectivities can be quite striking:[66] the ratio of $\omega/(\omega-1)$ hydroxylation of capric acid by kidney cytochrome P-450, for example, can be as high as 20.

Direct comparisons to enzyme regioselectivity in *alkane* hydroxylation are difficult, as a result of the variations of one isozyme to another and to the weakness of alkane binding in the enzyme active site. Although several forms of P-450 have been isolated (particularly from bacterial sources) that give terminal hydroxylation of *n*-alkanes, data are seldom given for the amounts of secondary alcohols that may also be produced. Nonetheless, data are available[66-69] for the hydroxylation of hexane and heptane by rat liver microsomal cytochrome P-450. The regioselectivities shown for these *n*-alkanes, however, are not nearly as high as those shown for fatty acids. The primary selectivity, as defined earlier, of rat liver microsomal P-450 (uninduced) is 0.16 for hexane and 0.26 for heptane, as compared to 0.32 and 0.59, respectively, for MnTTPPP(OAc) and C_6H_5IO. On treatment with phenobarbital, different and less selective isozymes of cytochrome P-450 are induced, with primary selectivities of 0.03 for hexane and 0.10 for heptane. These data, together with those of the synthetic porphyrins, are shown in Figure 12.

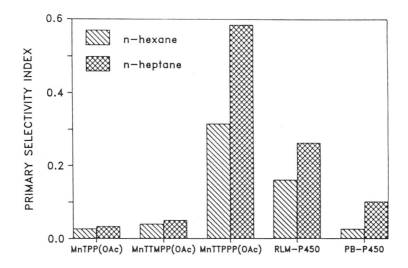

Figure 12. Comparison of shape selectivity of synthetic porphyrins vs cytochrome P-450 for *n*-hexane and *n*-heptane hydroxylation.[20] Oxidizing systems represented are Mn-porphyrins with C_6H_5IO, uninduced rat liver microsomal cytochrome P-450 (RLM-P450), and phenobarbital-induced rat liver microsomal cytochrome P-450 (PB-P450).

Certain isozymes of the heme protein cytochrome P-450 are capable of dramatic regioselectivity epoxidation,[70] as well. A comparison of the selectivities for epoxidation of limonene and of 4-vinyl-1-cyclohexene can be made between our synthetic metalloporphyrins and limonene-induced rat liver microsomal P-450 (Fig. 13). As the steric demands of the active site of the oxidizing system increase from uncatalyzed *m*-chloroperoxybenzoic acid to the very bulky MnTTPPP(OAc), the selectivity for external epoxidation dramatically increases, and MnTTPPP(OAc) is nearly as selective as the native enzyme.

It is clear, then, that our most hindered porphyrin, TTPPP, is comparable in its shape selectivity to native cytochrome P-450.

Although the details of the steric interactions undoubtedly differ between the enzyme and our synthetic analogs, the total steric interaction must be quite similar in magnitude for both MnTTPPP(OAc) and various isozymes of cytochrome P-450. It will prove interesting to compare the relative size and shape of binding sites in the enzymes to the synthetic analogs, as structures become available.

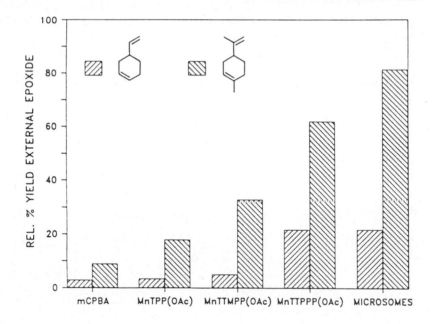

Figure 13. Comparison of shape selectivity of synthetic porphyrins vs cytochrome P-450 for epoxidations. Plotted are the relative yields of external (less hindered) to ring (more hindered) epoxides for 4-vinyl-1-cyclohexene and limonene. The epoxidations were run as before with uncatalyzed *m*-chloroperoxybenzoic acid, MnPorph(OAc) + NaOCl$_{(aq)}$ + 20 mM ImAcP, or limonene-induced rat liver microsomes as oxidants. For the enzyme comparison, the corresponding epoxide and diol (from epoxide hydrolysis) products were summed.

V.   Conclusions

Shape selectivity for the hydroxylation of alkanes and the epoxidation of dienes has been demonstrated using very sterically hindered metalloporphyrins as catalysts. The selectivities depend on the choice of metal and substrate. A few catalysts, both homogeneous and heterogeneous, have been discovered recently that show remarkable enhancements for primary hydroxylation of branched and *n*-alkanes, unprecedented in nonbiological catalysis, and comparable to enzymatic ω-hydroxylase activity. Nonetheless, the absolute level of selectivity for *n*-alkane hydroxylation remains modest. In addition, the turnover rates for such systems remain relatively low (in both synthetic and enzymatic systems). It is clear that industrial application of this work remains only a future possibility.

The presence and degree of shape selectivity are offered as conclusive proof of direct metalloporphyrin involvement during the actual hydroxylation and epoxidation of substrates with some, but not all, oxidants under conditions of homogeneous catalysis. Several different oxidants yield similar selectivity for both hydroxylation and epoxidation, thus demonstrating metal-based oxidation via a common monomeric intermediate (probably a terminal oxo metal complex) for these systems. In the case of epoxidation of nonconjugated dienes, shape selectivity is substantially less for Fe than Mn complexes, which means that both metals cannot be generating a long-lived metallooxetane intermediate and suggests that iron-porphyrin-catalyzed epoxidations probably proceed through an electron transfer pathway. Thus, shape selectivity is proving a useful tool for mechanistic studies of new metalloporphyrin oxidizing systems.[71]

Acknowledgments

The special efforts of Dr. Bruce R. Cook and Dr. Thomas J. Reinert are greatly appreciated. We thank the National Institutes of Health (Heart, Lung, and Blood Institute) and the American Heart Association for their generous support. K.S.S. gratefully acknowledges the receipt of an N.I.H. Research Career Development Award and of a Sloan Foundation Research Fellowship.

References

1. Cram, D. J., *Science* **1983**, *219*, 1177.
2. Gutsche, C. D., *Acc. Chem. Res.* **1983**, *16*, 161.
3. Wipff, G., Kollman, P. A., Lehn, J. M., *Theochem.* **1983**, *10*, 153.
4. D'Souza, V. T., Bender, M. L., *Acc. Chem. Res.* **1987**, *20*, 146.
5. Busch, D. H, Cairns, C., *Prog. Macrocyclic Chem.* **1987**, *3*, 1.
6. Rebek, J., Jr., *Science* **1987**, *235*, 1478.
7. Sutherland, I. O., *Chem. Soc. Rev.* **1986**, *15*, 63.
8. Newsam, J. M., *Science* **1986**, *231*, 1093.
9. Hill, C. L., Schardt, B. C., *J. Am. Chem. Soc.* **1980**, *102*, 6374.
10. Groves, J. T., Kruper, W. J., Haushalter, R. C., *J. Am. Chem. Soc.* **1980**, *102*, 6375.
11. March, J., *Advanced Organic Chemistry: Reactions, Mechanisms, and Structure*, 3rd Ed. pp. 733-736. Wiley-Interscience, New York, 1985.
12. Ortiz de Montellano, P. R., ed., *Cytochrome P-450: Structure, Mechanism, and Biochemistry*, Plenum Press, New York, 1986.
13. Collman, J. P., Hampton, P. D., Brauman, J. I., *J. Am. Chem. Soc.* **1986**, *108*, 7861.

14.   McMurry, T. J., Groves, J. T., in *Cytochrome P-450: Structure, Mechanism, and Biochemistry* (P. R. Ortiz de Montellano, ed.), p. 1. Plenum Press, New York, 1986.

15.   Mansuy, D., Battioni, P., *Bull. Soc. Chim. Belg.* **1986**, *95*, 959.

16.   Meunier, B., *Bull. Soc. Chim. Fr.* **1986**, 578.

17.   Traylor, T. G., Miksztal, A. R., *J. Am. Chem. Soc.* **1987**, *109*, 2770.

18.   Suslick, K. S., Reinert, T. J., *J. Chem. Ed.* **1985**, *62*, 974.

19.   Suslick, K. S., Cook, B. R., Fox, M. M., *J. Chem. Soc., Chem. Commun.* **1985**, 580.

20.   Cook, B. R., Reinert, T. J., Suslick, K. S., *J. Am. Chem., Soc.* **1986**, *108*, 7281.

21.   Suslick, K. S., Cook, B. R., *J. Chem. Soc., Chem. Commun.* **1987**, 200.

22.   Suslick, K. S., Cook, B. R., in *Inclusion Phenomena and Molecular Recognition* (J. L. Atwood, ed.), Plenum Press, New York, 1989.

24.   Traylor, T. G., Tsuchiya, S., *Inorg. Chem.* **1987**, *26*, 1338.

25.   Khenkin, A. M., Shteinman, A. A., *J. Chem. Soc., Chem. Commun.* **1984**, 1219.

26.   Mansuy, D., Bartoli, J. F., Momenteau, M., *Tetrahedron. Lett.* **1982**, *23*, 2781.

27.   Nappa, M. J., Tolman, C. A., *Inorg. Chem.* **1985**, *24*, 4711.

28.   Khenkin, A. M., Koifman, O., Semeikin, A., Shilov, A., Shteinman, A. A., *Tetrahedron Lett.* **1985**, 4247.

29.   Traylor, T. G., Nakano, T., Dunlap, B. E., Traylor, P. S., Dolphin, D., *J. Am. Chem. Soc.* **1986**, *108*, 2782.

30.   De Carvalho, M. E., Meunier, B., *New J. Chem.* **1986**, *10*, 223.

31.   Groves, J. T., Nemo, T. E., *J. Am. Chem. Soc.* **1983**, *105*, 5786.

32.   Lindsay Smith, J. R., Sleath, P. R., *J. Chem. Soc. Perkin Trans. II* **1982**, 1009.

33.   Collman, J. P., Brauman, J. I., Meunier, B., Hayashi, T, Kodadek T., Raybuck, S. A., *J. Am. Chem. Soc.* **1985**, *107*, 2000.

34.   Groves, J. T., Myers, R. S., *J. Am. Chem. Soc.* **1983**, *105*, 5791, 6243.

35.   Szabo, W. A., Lee, H. T., *Aldrichimi. Acta* **1980**, *13*, 13.

36.   Sorokin, A. B., Khenkin, A. M., Marakushev, S. A., Shilov, A. E., Shteinman, A. A., *Dokl. Akad. Nauk. SSSR.* **1984**, *279*, 939.

37.   Tsuchida, E., Kaneko, M., Nishide, H., Hoshino, M., *J. Phys. Chem.*, **1986**, *90*, 2283.

38.   van Esch, J., Roks, M. F. M., Nolte, R. J. M., *J. Am. Chem. Soc.* **1986**, *108*, 6093.

39.   Groves, J. T., Neumann, R., *J. Am. Chem. Soc.* **1987**, *109*, 5045.

40.   Chang, C. D., Hellring, S. D., "Shape Selective Catalytic Oxidation of Phenol," U.S. Patent 4,578,521, 25 March 1986.

41.   Dessau, R. M., *J. Catal.* **1982**, *77*, 304.

42.   Herron, N., Stucky, G. D., Tolman, C. A., *J. Chem. Soc. Chem. Commun.* **1986**, 1521.

43.   Herron, N., Tolman, C. A., *J. Am. Chem. Soc.* **1987**, *109*, 2837.

44.   Barton, D. H. R., Boivin, J., Gastiger, M., Morzycki, J., Hay-Motherwell, R. S., Motherwell, W. B., Ozbalik, N., Schwartzentruber, K. M., *J. Chem. Soc., Perkin Trans. I* **1986**, 947.

45.   Barton, D. H. R., Boivin, J., Crich, D., Hill, C. H., *J. Chem. Soc., Perkin Trans. I* **1986**, 1811.

46.   Barton, D. H. R., Boivin, J., Motherwell, W. B., Ozbalik, N., Schwartzentruber, K. M., Jankowski, K., *Nouv. J. Chim.* **1986**, *10*, 387.

47. Groves, J. T., Stern, M. K., *J. Am. Chem. Soc.* **1987**, *109*, 3812.
48. Birchall, T., Smegal, J. A., Hill, C. L., *Inorg. Chem.* **1984**, *23*, 1910.
49. Smegal, J. A., Schardt, B. C., Hill, C. L., *J. Am. Chem. Soc.* **1982**, *104*, 3964.
50. Smegal, J. A., Hill, C. L., *J. Am. Chem. Soc.* **1983**, *105*, 3515.
51. Traylor, T. G., Marsters, J. C., Jr., Nakano, T., Dunlap, B. E., *J. Am. Chem. Soc.* **1985**, *107*, 5537.
52. Bernardi, R., Galli, R., Minisci, F., *J. Chem. Soc. (B)*, **1968**, 324.
53. Groves, J. T., *J. Chem. Ed.* **1985**, *62*, 928.
54. Collman, J. P., Kodadek, T., Raybuck, S. A., Brauman, J. I., *J. Am. Chem. Soc.* **1985**, *107*, 4343.
55. Collman, J. P., Kodadek, T., Brauman, J. I., *J. Am. Chem. Soc.* **1986**, *108*, 2588.
56. Groves, J. T. Watanabe, Y., *J. Am. Chem. Soc.* **1986**, *108*, 507.
57. Traylor, T. G., Iamamoto, Y., Nakano, T., *J. Am. Chem. Soc.* **1986**, *108*, 3529.
58. Mlodnicka, T., *J. Mol. Catal.* **1986**, *36*, 205.
59. Guilmet, E., Meunier, B., *Nouv. J. Chim.* **1982**, *6*, 511.
60. Meunier, B., Guilmet, E., de Carvalho, M. E., Poilblanc, R., *J. Am. Chem. Soc.* **1984**, *106*, 6668.
61. Suslick, K. S., Fox, M. M., Reinert, T. J., *J. Am. Chem. Soc.* **1984**, *106*, 4522.
62. McKenna, E. J., Coon, M. J., *J. Biol. Chem.* **1970**, *245*, 3882.
63. Jefcoate, C. F., in *Cytochrome P-450: Structure, Mechanism, and Biochemistry* (P. R., Ortiz de Montellano, ed.), p. 387. Plenum Press, New York 1986.
64. Sligar, S. G., Murray, R. I., in *Cytochrome P-450: Structure, Mechanism, and Biochemistry* (P. R., Ortiz de Montellano, ed.), p. 429. Plenum Press, New York, 1986.
65. Kappeli, O., *Microbiol. Rev.* **1986**, *50*, 244.
66. Ellin, A., Orrenius, S., *Mol. Cell Biochem.* **1975**, *8*, 69.
67. Frommer, U., Ullrich, V., Standinder, H., Orrenius, S., *Biochim. Biophys. Acta* **1972**, *280*, 487.
68. Morohashi, K., Sadano, H., Okada, Y., Omura, T., *J. Biochem.* **1983**, *93*, 413.
69. Karasevich, E. I., Khenkin, A. M., *Biokhimiya (Moscow)* **1986**, *51*, 1454.
70. Watabe, T., Hiratsuka, A., Ozawa, M., Isobe, M., *Xenobiotica* **1980**, *11*, 333.
71. Suslick, K. S., Acholla, F. V., Cook, B. R., *J. Am. Chem. Soc.* **1987**, *109*, 2818.

Chapter VIII

CATALYTIC OXYGENATION OF UNACTIVATED C-H BONDS: SUPERIOR OXO TRANSFER CATALYSTS AND THE "INORGANIC METALLOPORPHYRIN"

Craig L. Hill

Department of Chemistry
Emory University
Atlanta, Georgia 30322

I.   Introduction

Although the oxidation and isomerization of alkanes have been and are at present major features of the chemical economy, the low reactivity of these most abundant of organic compounds has restricted their use to a handful of processes. One of the major problems in catalysis and in the reactivity of organic materials remains the development of catalytic methods for the selective functionalization of unactivated carbon-hydrogen bonds. The hundreds of papers on this subject generated on the part of the scientific community in the decade of the 1980s testify to this fact. It should be pointed out, however, that although the principal thrust of recent research on the cleavage of unactivated carbon-hydrogen bonds has focused on alkanes, the ability to systematically alter unactivated carbon-hydrogen bonds in functional group bearing molecules also constitutes a major and general problem with a limited number of promising experimental leads at the present time. Although this chapter addresses new approaches under development in our laboratory to resolve some of the well-defined and long standing problems in the existing chemistry of alkanes, the principal thrust of this chapter is to review the chemistry dealing with oxo transfer to alkanes.

   The few commercial processes involving alkanes that exist are as messy and difficult to control as they are large in scale.[1-3] A low level of selectivity is characteristic of the catalytic reforming of hydrocarbons, as well as processes done under net oxidizing conditions: the halogenation and catalytic air oxidation of hydrocarbons. The latter processes are beset with all the problems entailed by free radical chain reactions. It would be hard to find many commercial processes that produce as many products as catalytic hydrocarbon autoxidation.[1-3] The catalyzed air oxidation of cyclohexane carried out by DuPont in conjunction with the nylon process, for example, discussed in some detail in Chapter X, produces over 100 detectable products by chromatography! It should be noted parenthetically that the free radical chain autoxidations of methyl arenes, such as the oxidation of *p*-xylene to terephthalic acid, exhibit reasonable selectivity at high conversions in large part by virtue of the fact the benzylic methyl group(s) are the only ones on the substrate(s) of high reactivity; the limited possible products, aryl carboxylic acids, are not particularly reactive to further oxidation.

Since 1980, investigators in nearly all subdisciplines of chemistry have addressed the problems associated with the reactivity of alkanes and the selective conversion of these abundant materials into more useful compounds. Among the most noteworthy accomplishments in alkane chemistry since this time have been the development of alkane dehydrogenation by organometallic polyhydride systems,[4,5] the oxidative addition of unactivated alkane C-H bonds to mid-transition metal cyclopentadienyl complexes,[6] alkane activation by lanthanide and other d° systems[7,8] and the research areas that each of these initial discoveries has spawned (Chapters III, IV, and V, respectively).[9-15] Over the period of time that the alkane chemistry of these organometallic systems has developed, the functionalization of hydrocarbons including methane by superacid and electrophilic systems, largely by Olah and his co-workers, continued to develop (see Chapter II),[16,17] as has the homogeneous catalytic functionalization of alkanes by metalloporphyrins,[18-27] Schiff base complexes,[28] and related species. The latter systems functionalize alkanes in large part by C-H bond hydroxylation through the intermediacy of oxometal species in a fashion analogous to C-H bond hydroxylation by cytochrome P-450, the most abundant oxidant of hydrocarbons in the biosphere (Chapter VI).[29-34] These oxo transfer systems have been sterically modified in elegant ways synthetically and by inclusion in inorganic support hosts to facilitate modification of the substrate selectivities and regioselectivities observed in alkane oxidation (see Chapters VII and X). Also in the last few years new systems that functionalize alkanes in heretofore unprecedented ways that do not fit into any of the above categories have been discovered and are under current investigation. Primary among these are the Gif and Gif-Orsay systems of the D. H. R. Barton group and the polyoxometalate systems of the Hill group (Chapters IX and this chapter, respectively).

As indicated in the preface, the focus of this book is principally on homogeneous systems investigated by academic and industrial groups for the modification of saturated hydrocarbons. The importance of garnering an in-depth knowledge of the chemistry of alkanes is buttressed by the *substantial* work that has been done on the chemistry of these molecules by the chemical physics and heterogeneous catalytic communities. All the gas phase ion molecule work with metal ions and alkanes in the last few years,[35-38] has contributed markedly to our knowledge of the energetic features involved in the cleavage of alkane bonds, while the extensive and ongoing investigations of methane oxidation and dehydrogenation over heterogeneous catalysis remain an area of acute investigation by many petrochemical firms and academic researchers alike.[39,40] In addition to the gas phase and heterogeneous studies, theoretical studies have provided yet more valuable insight into the nature of alkane reactivity.[41] Only lack of space precludes these gas phase and heterogeneous catalytic experimental studies and theoretical calculations from being included in this volume. Some of the results of these studies have influenced much of the chemistry that is discussed in this volume and in this chapter, however, and allusions have been made to some of the key work in these latter areas. It is also from the collective efforts of chemists in all subdisciplines, that existing challenges in the chemistry of saturated hydrocarbons are best defined.

The work in this decade has seen some problems associated with the chemistry of alkanes solved, but many major problems remain. At the present time the relationships between selectivity (substrate selectivity, regioselectivity, and stereoselectivity) and the electronic and structural properties of most alkane-activating species are reasonably well understood. At the same time, rendering the most desirable processes catalytic, such as the thermodynamically and kinetically preferred terminal C-H bond cleavage observed in several organometallic systems, has been difficult. The stability of the system, which usually but not always entails the stability of the catalyst, is often a problem as is developing catalytic systems for the selective functionalization of alkanes in which the selectivity can be systematically altered in simple ways. For alkane functionalization processes that are net oxidations (processes that consume an oxidant versus some dehydrogenation and isomerization processes that do not), the use of the least expensive oxidants is desirable -- *tert*-butyl hydroperoxide (TBHP)[42] if oxygen donors are required, and ideally, dioxygen. Unfortunately the use of dioxygen for selective organic oxidations of any kind, and particularly those involving alkanes, is less than straightforward. A selective and effective catalytic method of oxidizing alkanes using O_2 as the terminal oxidant, in which the typical radical chain processes and their dictated selectivities are not dominant, has yet to be achieved.

It was in response to these and other general or substantial problems associated with the chemistry of alkanes that we began to explore the use of the early transition metal polyoxometalates and their derivatives as catalysts for various processes. Polyoxometalates are aggregates of d° transition metal and oxide ions held together by metal oxygen bonds.[43] This basic property renders these complexes thermodynamically stable to oxidation under most if not all the conditions that would ever be encountered in the selective oxidation of organic compounds including alkanes. A principal theme in our recent research, and substantial work from our laboartory will be published in the near future, is that the polyoxometalate surface represents an oxidatively resistant yet soluble and synthetically modifiable entity. *Effectively polyoxometalate-based catalysts combine the selectivity and control advantages of homogeneous catalysts with the stability advantages of metal oxide and other heterogeneous catalysts.*

II. General Features of Carbon-Hydrogen Bond Oxygenation

The oxidation of hydrocarbon substrates, particularly alkanes and alkenes, and other organic substrates, by the transfer of oxygen atoms from an oxometal or other oxo-containing species (Eq. 1) is of widespread importance in both biological and industrial oxidation processes[1-3,27,29] and will likely feature prominently in the future

$$DO \quad + \quad S \quad \xrightarrow{\text{catalyst}} \quad SO \quad + \quad D \qquad (1)$$

DO = oxygen atom donor, S = organic substrate (e.g., alkane)

development of selective catalysts for oxidation of these substrates. In Eq. 1 the oxygen atom donor, or "oxo donor," DO is not limited to transition metal species, although for the oxidation of hydrocarbon substrates in the absence of a catalyst it is primarily transition metal oxo species that are both kinetically competent to react with these substrates and also potentially amenable to manipulation to modulate the selectivities (substrate selectivity, regioselectivity, and stereoselectivity) that are possible. The organic substrate, S, can be a heteroatom-containing molecule, which is oxidized to a product containing one (or more) oxo groups, or, as in a multitude of papers from industrial and academic laboratories recently, an alkene or alkane.[18-34] This chapter focuses on alkane oxidation processes, S = alkane, SO = alcohol.

A good review of processes involving oxo transfer from oxometal and related species to organic substrates is needed, particularly one that addresses all the studies involving alkene and alkane substrates that have dominated much of the literature in the last few years. Fortunately a comprehensive and excellent review by Holm on oxo transfer not involving these substrates recently appeared.[44] I recommend that those interested in the problems associated with the catalytic oxo transfer oxidation of hydrocarbons and other organic substrates and the development of new and innovative catalysts for these types of processes, in addition to those interested by oxygen atom transfer in general, read this review.

Many authors, particularly in the last few years, have referred to the oxidation of organic substrates by oxygen atom transfer as "oxygenation." The name follows from function and stoichiometry and can simplify some discussions of complex oxidation mechanisms. For sake of convenience only, the discussion of alkane oxidation by the formal transfer of oxygen atoms in this chapter will be referred to as "oxygenation." It should be pointed out that a substrate oxidation by oxygenation (Eq. 1) really implies little about mechanism. Oxygenation is more a bookkeeping device and a thermodynamic concept as opposed to a mechanistic one. As discussed in Sec. V, oxygenation of C-H bonds including the reactions of alkanes, by oxo transition metal species, can take place by one of several mechanisms. Some of these mechanisms in turn are constituted by more than one elementary process that is well defined based on experimental work with other or simpler reactions. Another point that concerns organic oxygenation processes that applies perhaps with most pertinence to alkane oxidation is that oxygenation (Eq. 1) may not be the only oxidation process involved. Often oxidation pathways that are not oxygenations are operable in processes that are nominally viewed as the latter. On the other extreme, processes that are traditionally viewed as dominantly free radical chain in character such as industrial hydrocarbon autoxidations (for the case of cyclohexane oxidation by DuPont see Chapter X) may entail oxygenation processes as minor components.

The oxygenation of hydrocarbons can be viewed in context with a general class of group transfer reactions (Table I) in which formal neutral divalent 6-electron moieties are the species being transferred. In context with alkane oxidation, however, only for the case of

Table I. Transfer of Divalent Six-Electron Groups[a]:

$$DX \quad + \quad S \quad \xrightarrow{\text{catalyst}} \quad SX \quad + \quad D$$

X (6-e)	Name of group	Name of transfer	Comments
O	Oxene (oxo)	Oxygenation	Common, well studied
NR	Nitrene	Nitrogenation (alkanes) Aziridination (alkenes)	Documented, but not well studied
CR$_2$	Carbene	Carbenation?	Not documented for alkanes
Others (S, PR, etc.)	-		Not yet explored

[a]DX = donor of 6-electron group "X," S = organic substrate (alkane, alkene, heteroatom substrates, etc.).

oxygenations, X = O, has there been extensive development, and in-depth investigations of mechanism carried out. The catalytic transfer of nitrene equivalents from donor species DX, X = NR, catalyzed by metalloporphyrins has been documented by the groups of Breslow,[19] Mansuy,[45,46] and, in an indirect manner, by Groves and Takahashi.[47] The transfer of carbenes to hydrocarbons has not yet been achieved. In the one case in which this goal was sought using metalloporphyrin catalysts, the iodosyl carbene equivalent, DX, X = a dimedone derived moiety, functioned as a suicide inactivator of the catalyst.[48,49] Sharpless and others pointed out the similarity of oxo (oxene) transfer to nitrene and other transfer processes some time ago.[100] A problem not only with the genesis of catalytic carbene transfer processes, but also with these divalent neutral 6-electron group transfers to organic substrates in general, is the stability of the catalytic system under the desired conditions. In addition, a problem that plagues the development of new catalytic processes in general, and one that will be a factor here, is the availability of facile alternate irreversible processes leading to undesired products.

A cycle for the oxo transfer oxidation or oxygenation of an organic substrate, by an oxygen donor, DO, catalyzed by a transition metal complex, L_nM, can be written in the most general manner as in Figure 1. The processes as depicted in the figure do not imply mechanism. Several different oxygen donors, DO, and catalysts, L_nM,

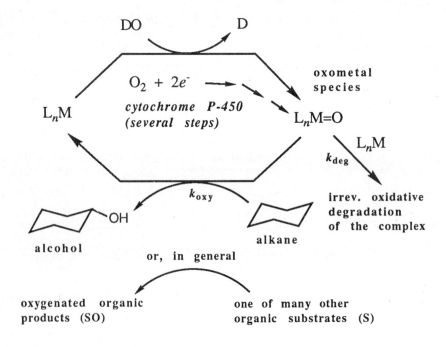

Figure 1. Catalytic oxometal oxygenation cycles.

have been used successfully in such processes. Among the oxygen donors that have been used are iodosylarenes, peracids, amine *N*-oxides, alkyl hydroperoxides, persulfates, hypochlorite, periodate, hydrogen peroxide, oxaziridines, as well as molecular oxygen. Among the catalysts that have been used are metalloporphyrins,[19-27,31-34] Schiff bases complexes,[28] metallophthalocyanines,[50] soluble transition metal salts without macrocylic or chelating ligands such as the triflates and nitrates,[51,52] and transition metal-substituted polyoxometalates.[53-57]

III.   Reactivity and Selectivity in C-H Bond Oxygenation

The ideal oxo transfer system would involve a catalyst, L_nM in Figure 1, that could be oxygenated by inexpensive oxidants, for example, *tert*-butyl hydroperoxide (TBHP), hypochlorite (OCl⁻), or, ideally, O_2, forming an oxometal species, $L_nM=O$, that could rapidly transfer its oxygen to the organic substrate in a completely selective manner. The ideal oxo transfer systems would be compatible with several different reaction media and would be effective for oxygenation of a range of substrates, not just low-molecular weight ones or some other particular category. In addition, the selectivity should be amenable to systematic alteration. This ideal oxo transfer system would also be highly resistant to the inactivation that all catalysts, homogeneous and heterogeneous alike, inevitably face. The initial approach to these desirable attributes involves optimization of the reactivity, selectivity, and stability characteristics of the system.

Although little of the recent extensive research on hydrocarbon oxygenation by metalloporphyrins and other transition metal complexes has focused on the rates of reaction, all these studies collectively have begun to define the relationships between the structural and electronic features of the oxygen donor and catalyst, DO and L_nM in Figure 1, respectively, and the observed catalytic turnover rates. As shall be discussed in detail later, the mechanisms of hydrocarbon oxygenation processes not only can and do vary with both the oxygen donor and the catalyst, but also multiple modes of oxygenation may be operable in any given system.

It is clear, referring to the idealized and oversimplified cycle for oxygenation in Figure 1, that the rate of the overall process can be dictated either by components of the catalyst oxygenation by DO (top reaction), or by transfer of the oxygen atom from the oxometal intermediate to the substrate (bottom reaction). Three well-documented reactions in which the formation of the oxometal intermediate is rate determining are the oxygenation of $Fe^{III}(TPP)Cl$ by *p*-cyano-*N,N*-dimethylaniline *N*-oxide,[20b] $(ImH)Mn^{III}(TPP)Cl$ by TBHP,[42,58] and $[(TPP)Fe^{III}]_2O$ by *p*-cyano-*N,N*-dimethylaniline

N-oxide.[42,59]   In contrast, two well-investigated examples of hydrocarbon oxygenation in which a component of the substrate oxygenation process is rate determining are the epoxidation of olefins by hypochlorite catalyzed by Mn-porphyrins,[33a,60] and the epoxidation of olefins by pentafluoroiodosylarenes catalyzed by Fe-porphyrins.[33b,61]

The dependence of the rate of the top process, oxometal formation, is largely covered by the research addressed in Holm's authoritative review of oxo transfer processes.[44] The formation of the oxometal species involves in principle either homolytic or heterolytic cleavages of the D-O bonds, O-O single bonds in the case of the peroxide compounds, (Eqs. 2 and 3, respectively).   In context with selective organic oxygenation reactions, the heterolytic process to generate an oxometal complex with the metal in a formal oxidation state two equivalents higher than that in the reactant complex, $L_nM^{(n+2)+}=O$, would usually be preferred over the homolytic process to generate an oxometal complex with the metal in a formal oxidation state one equivalent higher than that in the reactant compound, $L_nM^{(n+1)+}=O$. The principal reason for this is that homolytic D-O cleavage generates radical species, D·, that are usually capable of reacting with the organic substrates, including alkanes, and that can lead to several undesirable

$$L_nM^{n+} + O\text{-}D \xrightarrow{\text{homolytic}} L_nM^{(n+1)+}=O \;+\; D\cdot \qquad (2)$$

$$L_nM^{n+} + O\text{-}D \xrightarrow{\text{heterolytic}} L_nM^{(n+2)+}=O \;+\; D^- \qquad (3)$$

side reactions.   Generation of $L_nM^{(n+1)+}=O$ and D· from homolysis of DO, for example, is often of importance in metal-catalyzed autoxidation processes.   Indeed, generation of alkoxy radicals by Haber-Weiss decomposition of hydroperoxides (Eqs. 4 and 5) and other homolytic processes during the industrial air oxidation of cyclohexane and other

$$L_nM^{n+} + ROOH \longrightarrow L_nM^{(n+1)+} + RO\cdot + OH^- \qquad (4)$$

$$L_nM^{(n+1)+} + ROOH \longrightarrow L_nM^{n+} + ROO\cdot + H^- \qquad (5)$$

hydrocarbons constitutes important propagating steps in the radical chain chemistry that dominates these processes (see footnotes cited in Chapter X).

Recently intense efforts have been focused on the mechanism of cleavage of the peroxide oxygen donors, both alkyl hydroperoxides and peracids, by transition metal complexes, and, in particular, by first row transition metal metalloporphyrins. Although the principal impetus for a detailed examination of such a process is a greater understanding of similar molecular processes that are operable in the heme oxidases, and, in particular, in the cytochrome P-450 enzymes, these studies also provide information of relevance to several catalytic oxidation processes, and generally provide information pertinent to the manipulation of dioxygen.

Key initial work on the formation of oxometal species of relevance to the oxygenation of hydrocarbons from reaction of a variety of peroxide species, both peracids and alkyl hydroperoxides, with Fe-,[62] Cr-,[63] Mn-,[64] and Co[65]-porphyrins was reported by Bruice and co-workers.[58] The groups of Traylor[66] and Groves[67] have also examined the kinetics of formation spectroscopically characterized intermediates and products of metalloporphyrin/peroxide reactions as the peroxide oxygen donor and/or the reaction conditions are varied. This work collectively indicates that a reaction dominated by homolysis of the peroxide O-O bond can change to a reaction dominated by heterolysis of this bond with changes only in the leaving group of the peroxide or with changes in the reactions conditions.[67] It is apparent that in some of these systems, minor changes in conditions can alter the mechanism, and, as a consequence, the reactivity toward hydrocarbon substrates.

Although both homolytic and heterolytic reactions forming oxometal complexes (Eqs. 2 and 3) with several different oxygen donors, DO, and transition metal complexes, L_nM^{n+}, appear to be thermodynamically favorable (see, for example, the thermodynamic oxo transfer scale in Holm's review), predicting the kinetic behavior of all L_nM^{n+} + DO reactions is less than straightforward. The potentials for oxidizing L_nM^{n+} by one versus two electrons, the polarity and hydrogen-bonding ability of the media, the susceptibility to acid catalysis, and other factors could in principle affect the partitioning between homolytic and heterolytic processes. Further data on nonmetalloporphyrin systems and with nonperoxide oxygen donors would be of great value whether or not they lead to the selective or controllable oxygenation of hydrocarbons.[44]

A working model for oxometal formation by cleavage of the oxygen donor, D-O bond would involve, at a minimum, the d-orbitals on the transition metal, d_{xz} and d_{yz} using the usual convention, and the σ^*-antibonding orbital of the D-O bond as indicated below. However, the molecular orbitals involving the lone pairs on the D-O unit, π-orbitals on atoms proximal and potentially overlapping in some low-energy conformation with molecular orbitals principally localized on D, as well as the overlapping frontier orbitals of the other ligands on M, and, in particular, the π-symmetry orbitals of the axial ligand trans to DO, cannot be ignored. In addition, the perturbations of these frontier orbitals by hydrogen bonding and other interactions with the solvent, which can be of considerable significance based on

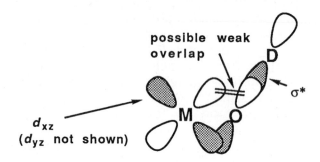

metalloporphyrin/peroxide studies, must also be accommodated to generate a more complete and predictively competent electronic model of the metal complex-induced O-D cleavage process.

The reactivity and selectivity exhibited in hydrocarbon oxygenation by oxometal species should be principally dependent on the bottom reaction in Figure 1, the formal transfer of oxygen from the oxometal intermediate to the substrate. In principle, and to some extent in practice, these reactivities and selectivities are dictated by two effects that can to a large extent be differentiated from each other, the steric constraints in the local environment of the oxometal complex, and the electronic features of the oxometal complex. The voluminous data acquired to date on hydrocarbon oxygenation by the cytochrome P-450 enzymes and many synthetic metalloporphyrin model compounds of the active site of these heme monooxygenases indicate that the selectivities in generation of organic products (substrate selectivities, regioselectivities, and stereoselectivities) are dictated principally by the former and not the latter. In fact, oxometalloporphyrins or oxometal transition metal complexes, in general, that hydroxylate alkanes appear to show similar selectivities resulting from similar electronic structure-reactivity characteristics. Namely, C-H bond cleavage processes in most if not all oxometal systems devoid of proximal steric congestion that have been studied in some depth to date are largely radical hydrogen atom abstraction in character, and all, as a consequence, exhibit the highest reactivities for the weakest C-H bonds. Typically allylic and benzylic C-H bonds (~85 kcal/mol) are most reactive > tertiary C-H bonds (~91 kcal/mol) > secondary C-H bonds (~94 kcal/mol) > primary C-H bonds (~98 kcal/mol) > methane C-H bonds (104 kcal/mol), which are the least reactive.[68] No examples exist of methane oxygenation with any demonstrable selectivity, that is, without major oxidation of initial products, methanol, etc., by oxometal species whether they be in a homogeneous or heterogeneous molecular environment. This is unfortunate in that development of methods for the selective high-conversion oxidation, oxidative dimerization, or otherwise transformation of methane still remains the single greatest unsolved problem in oxidation chemistry.[69] Furthermore, no examples exist of even moderately selective functionalization of primary C-H bonds in the presence of secondary or tertiary C-H bonds by any sterically unconstrained oxometal species. The inherent selectivity of the likely active form of the active site of cytochrome P-450, an oxoiron(IV) porphyrin π-cation radical, is likewise for the weakest (or most substituted) C-H bonds. Only the steric environment of the active site in these enzymes, elaborately tailored through evolution and environmental induction, facilitates the preferential hydroxylation of methyl groups or terminal C-H bonds.[19-27,30-34] Some of this chemistry is discussed further in references cited in Chapters VI and VII.

There are now sufficient data available on sterically unconstrained oxometal complexes, $L_nM^{(n+1)+}=O$ and $L_nM^{(n+2)+}=O$, that trends are emerging concerning their relative thermodynamic stabilities as well as their relative reactivities and selectivities in hydrocarbon oxygenation, provided both alkane hydroxylation and alkene epoxidation are considered together. There are not sufficient data exclusively on alkane C-H bond hydroxylation by any one structurally and electronically related family of oxometal complexes to allow much to be said at this time regarding electronic fine tuning of the C-H cleavage processes. When both alkane hydroxylation and olefin epoxidation are included, however, it is clear that reactivities and selectivities depend greatly on the metal. Although it would normally be expected that the more highly oxidized $L_nM^{(n+2)+}=O$ complexes resulting from heterolytic transfer of oxo to the metal would be higher in free energy, more reactive, and, as a consequence of the Hammond postulate, less selective in hydrocarbon oxygenation than the $L_nM^{(n+1)+}=O$ complexes resulting from homolytic transfer of oxygen to the metal, this is not always the case.

From the reactivities and selectivities seen in hydrocarbon oxygenation by the best documented family of oxometal complexes by far, the $L_nM^{(n+2)+}=O$ and $L_nM^{(n+1)+}=O$ complexes generated from the metallotetraphenylporphyrin reactants, $M^{III}TPPX$ (M = Cr, Mn, and Fe, X = halogen, acetate, or other monodentate monovalent axial ligand), by reaction with iodosylarenes, hypochlorite, etc. as oxygen donors, the following experimental observations have been noted and reproduced:

1. The reactivity is higher for the $O=M^VTPPX$ complexes than for the corresponding $O=M^{IV}TPP(L)$ complexes, which may or may not have a neutral sixth ligand, L.

2. Only the O=Mn^{V}TPPX and O=Fe^{IV}(TPP+·)X have been documented by work to date to be competent to oxidize unactivated alkane C-H bonds rapidly at or below room temperature.

3. The reactivity toward hydrocarbons of the high-valent oxometalloporphyrins generally follows the order Cr complexes (least reactive) < Mn complexes < Fe complexes (most reactive).

4. The spin state of the high-valent metal complex can be crucially important. The most dramatic case of the latter involves the XMn(V) porphyrins. The oxomanganese(V), X = O,[19-24,26,33] and the alkyl nitrido(nitrene) manganese(V), X = NCOR,[47] species are the hydrocarbon-oxidizing forms in these Mn^{III}(porphyrin)X + oxidant systems. These extremely reactive species are more than likely high spin and they thermally decompose rapidly at temperatures well below 0°C. In contrast, the nitridomanganese(V) porphyrins are low-spin d^2 systems. These nitrido complexes are not only completely unreactive to alkanes and alkenes but they can be recovered intact after refluxing for hours in inert solvents at ~140°C![70]

Table II summarizes reactivity and selectivity data for both alkane hydroxylation and olefin epoxidation by these O=M^{IV}(TPP) and O=M^V(TPP)$^+$ systems.

Table II.   Reactivity   of   O=M^{n+}(TPP)   Complexes   Toward   Hydrocarbons

Complex	Thermal stability at 25°C	Reactivity, 25°C Alkanes	Alkenes	Selectivity[a]	Reference
O=Cr(IV)	Stable[b]	None	None	NA	73,74
O=Cr(V)	min-hr[c]	None	Yes	E, quite S	75
O=Mn(IV)	sec-min	None	Yes	E, NS	76-78
O=Mn(V)	<1 sec	Yes	Yes	H, NS; E, S[d]	79,80
O=Fe(IV)	<1 sec	No	Yes	E, S[e]	81,82
"O=Fe(V)"[f]	<1 sec	Yes	Yes	H, NS; E, S	83

[a]E, epoxidation; H, hydroxylation; S, stereospecific; NS, nonstereospecific; NA, not applicable.
[b]Stable provided highly purified hydrocarbon solvents are used.
[c]Depends on type of solvent and its purity.
[d]Stereospecificity depends on conditions including axial ligand.
[e]Appears to be capable of epoxidation albeit nonstereospecifically under some conditions,[81] but not active under others.[82]
[f]The formal "O=Fe(V)" porphyrins characterized to date are O=Fe(IV)(porphyrin π-cation radicals), with low-spin d^4 Fe(IV) ions.[83,84]

It is reasonable that the reactivity of transition oxometal species for C-H bond hydroxylation should correlate with the degree of radical character in the ground electronic state of the oxometal complex in addition to the oxidation potential of the complex. Oxometal species with high unpaired electron density in frontier orbitals on oxygen and high electrophilicities would be expected to be the most reactive, provided that radical hydrogen abstraction is the mode of C-H bond cleavage and that the so-called "oxygen atom rebound" is the mechanism for C-H bond oxygenation. As is discussed in Sec. V, radical oxygen atom rebound still appears to be the most likely mechanism operable in most C-H hydroxylating oxometal systems (see also the middle portion of Chapter X), although the overall situation appears to be a more complex that this. In other words, canonical forms **1** and **2** of the five illustrated in Figure 2 correlate best with kinetic competence for C-H hydroxylation by oxometal compounds.

$$L_nM^-O^+ \longleftrightarrow L_nM^.O^. \longleftrightarrow L_nM{=}O$$
$$\quad\quad\mathbf{1} \quad\quad\quad\quad\quad\quad \mathbf{2} \quad\quad\quad\quad\quad\quad \mathbf{3}$$

$$\longleftrightarrow L_nM^+O^- \longleftrightarrow L_nM^{2+}O^{2-}$$
$$\quad\quad\quad\quad \mathbf{4} \quad\quad\quad\quad\quad\quad \mathbf{5}$$

Figure 2. Possible canonical forms depicting the electronic character of oxometal cores, $O{=}M^{n+}$, in the porphyrin and most conventional ligand environments.

In accord with the points above, the diamagnetic ground electronic state of the $O{=}Cr^{IV}(TPP)$ complex [as well as the corresponding $O{=}Ti^{IV}(TPP)$ and $O{=}V^V(TPP)$ complexes] explains the negligible hydroxylation activity shown by these compounds. The high thermodynamic stability and low electrophilicity of the d^1 oxometal complexes, $O{=}V^{IV}$ and $O{=}Cr^V$, explain the low or negligible reactivities of these compounds. Whereas $O{=}V^{IV}(TPP)$ and, for that matter, all vanadyl complexes show no ability to hydroxylate alkanes or epoxidize olefins under *mild conditions*, however, the substantially higher oxidation potential of the $O{=}Cr^V(TPP)X$ complexes dictates the thermal instability and limited oxygenation activity of the latter species. The next most active oxometal complex in Table II, the $O{=}Mn^{IV}(TPP)$ complex, is clearly a t_{2g}^3 species. The EPR spectrum indicates three unpaired electrons and the resonance Raman and infrared data point to an anomalously weak $O{=}M$ stretching frequency and bond,[77,85,86] in accord with a thermodynamically higher energy and more reactive oxometal complex that those previously mentioned. The most reactive complexes in this series, $O{=}Mn^VTPPX$ and $O{=}Fe^{IV}(TPP+\cdot)X$, have both unpaired electrons and very high electrophilicities.

The minimal theoretical model for the oxometal cores, $O=M^{n+}$, in the porphyrin and most conventional ligand environments involves the donation of both σ and π electron density from the formal O^{2-} ion to the $d_{xz/yz}$ and d_{z^2} orbitals of the $M^{(IV\ or\ V)}$ ions, respectively, rendering the latter antibonding and oxygen atom slightly cationic. Although the ground electronic states of the $O=M^{n+}$ complexes, M^{n+} = V, Cr, Mn in the IV or V oxidation states and Fe in the IV oxidation state, can be rationalized by the relative energies of the largely nonbonding d_{xy}, the variably antibonding d_{xz}- and d_{yz}-orbitals, and the electron pairing interaction, a more sophisticated picture of the electronic structure of these oxometal species has been afforded by recent ab initio molecular orbital calculations. One set of computations by Yamaguchi and co-workers points out that at least for the oxometal complexes with weak or long M-O π bonds, and based on experimental data the $O=Mn^{IV}$ and $O=Mn^{V}$ species are among the best candidates exhibiting such bonds, the description of the ground electronic state of the oxometal core involves a finite contribution of a pseudo $\pi^*\pi^*$ double excitation from configuration interaction (CI) theory.[87] In conjunction with this latter character, the bonding closed shell molecular orbitals of π symmetry bifurcate into the different-orbital-for-different spin (DODS) π-m.o.'s. These elongated and reactive oxometal species then display partial closed shell and DODS shell character. From Yamaguchi's calculations, the up- and down-spin m.o.'s are "more or less localized" on the transition metal and oxygen atoms, respectively, and, in this sense, the DODS molecular orbitals for these unstable oxometal species are similar to the generalized valence bond (GVB) orbitals of Goddard et al.[88]

These theoretical studies are better at confirming some of the experimentally observed electronic features of reactive transition metal oxo complexes than predicting subtle features of reactivity that have yet to be observed. In other ab initio calculations, Rappé and Goddard have provided an insightful view of the energetics of reactions between alkanes and high-valent (d°) Group 6 dioxo compounds, L_nMO_2, M = Cr, and Mo.[89] It is clear from these computations that the presence of the second oxo group greatly improves the energetics for cleavage of the strong alkane C-H bonds relative to the monooxometal complexes, L_nMO, that are operable in most alkane hydroxylating oxometal systems and more than likely operable in cytochrome P-450. An in-depth theoretical investigation of C-H cleavage processes by monooxometal complexes addressing high-valent and hypercoordinate organometallic species and other intermediates would be most worthwhile.

IV. Stability of the Catalyst

As indicated earlier, it is superior stability in addition to reactivity and selectivity that is requisite in the development of better oxygenation catalysts. It is not just the loss of catalytic activity, either complete or partial, as a consequence of catalyst degradation that should be considered, but also the loss of selectivity as a function of the number of turnovers of the catalyst. Alteration of the physical integrity of a catalyst with time under operating conditions can lead to multiple discrete species catalyzing various processes and hence an overall lowered selectivity with time. This can take place long before the total catalytic activity of the system has reached a point that demands fresh catalyst be used. In many homogeneous oxidation processes, and not just those of the oxygenation type, the catalysts often contain organic ligands. A principal function of these ligands is not only to provide the redox potentials and coordination environment that are required for the oxygenation, substrate binding, and other features essential for catalytic turnover, but also to protect the transition metal ions from precipitating as insoluble and catalytically far less active or inactive metal hydroxides or oxides. Oxygenation of organic substrates by oxygen donors facilitated by transition metal complexes that contain only weak ligands, such as the nitrate or triflate salts of Cu, Fe, and Mn, stops after a few turnovers, even in the best case. The rapid precipitation of the transition metal ions as insoluble hydroxides or oxides appears to explain the rapid and irreversible inactivation of these systems.[52a,90]

Irreversible oxidative degradation of the organic ligands of transition metal oxidation catalysts in general results in the loss of catalytic activity. In a great majority of the homogeneous oxometal-based catalytic hydrocarbon oxidation systems investigated to date including the metalloporphyrin systems, there is appreciable decomposition of the catalyst ligands and an appreciable loss in activity after a few turnovers. Similarly, in a large-scale industrial hydrocarbon autoxidation process, the transition metal complexes used for initiation of the autoxidation and catalytic breakdown of the initial hydroperoxide products are susceptible to inactivation when the organic ligands, often hydrophobic long-chain carboxylates, are oxidatively degraded.[91]

Consider again the simplified scheme for organic substrate oxygenation by oxometal complexes in Figure 1. In no system does substrate oxygenation proceed with 100% selectivity. Oxidative degradation of the organic ligand(s) of the catalyst (processes collectively indicated as k_{deg} in Fig. 1) competes with the desired substrate oxygenation process(es) (k_{oxy} in Fig. 1). In simplest terms, control of partitioning of the reactive oxometal intermediate between the desirable substrate oxidation process(es) and the undesirable degradation process(es) is of fundamental importance. Not only must the ratio k_{oxy}/k_{deg} be optimized, but ideally, k_{deg} should be rendered effectively zero. Unfortunately with a great majority of the catalysts used in the homogeneous catalytic functionalization of hydrocarbons, including the metalloporphyrins, Schiff base complexes, and metallophthalocyanines, this is a very difficult goal indeed. Major efforts have been made to stabilize the porphyrin ligand with respect to oxidative degradation either from intramolecular attack by the oxometal group on its own ligand, or from intermolecular attack by one of several possible freely diffusing intermediate oxidizing species. There has been much discussion of the enhanced oxidative stability of the tetraarylmetalloporphyrins with sterically encumbering substituents on the *ortho* position of the *meso* aryl rings that preclude μ-oxo dimer formation and decrease the ease of intermolecular attack by one oxometal complex on another porphyrin ligand. The tetrakis-2,6-dichlorophenylporphyrin (TDCPP), whose superior oxidative stability in the form of its iron complexes was originally reported by Traylor and Dolphin, imparts a resistance to oxidative degradation in the presence of iodosylarene oxygen donors that is about two orders of magnitude superior to that seen in the sterically unencumbered iron complexes of the parent tetraphenylporphyrin (TPP).[34,90] TDCPP not only has large substituents on all *ortho* positions of the *meso* aryl rings, but these substituents, unlike the methyl groups in the popular tetramesitylporphyrin, are oxidatively resistant. The formation of TDCPP under Rothemund conditions has been examined in some depth.[92]

In oxygenations of the less reactive substrates, including all alkanes, oxidative degradation of the catalyst and proportionate premature loss of the catalytic activity become real problems even with M(TDCPP)X complexes. More recently, Traylor and Tsuchiya reported the preparation and partial characterization of an octabrominated derivative of TDCPP.[93] This octabromooctachlorotetraarylporphyrin apparently has bromides on all the oxidatively susceptible β-pyrrole positions of the ring. The Fe complex of this porphyrin is apparently substantially more stable than the corresponding TDCPP complex in hydrocarbon oxygenation.[93] The efforts in rendering the porphyrin and other organic ligands ever more halogenated and oxidatively resistant will doubtless continue.

The underlying problem in this approach, however, stems from that fact that all organic ligands are thermodynamically unstable with respect to oxidative degradation either by oxygen donors or by dioxygen itself. Inspite of the great kinetic resistance of some recently prepared metalloporphyrins to oxidative destruction, this underlying thermodynamic instability with respect to CO_2, H_2O, and other oxidation products should give one pause. It is precisely in response to the thermodynamic problem represented by organic ligands that our group introduced the concept of using early transition metal polyoxometalates substituted with d^n transition metal ions in surface sites as soluble and oxidatively resistant analogs of metalloporphyrins.[53-57] Figure 3 illustrates conventional organic

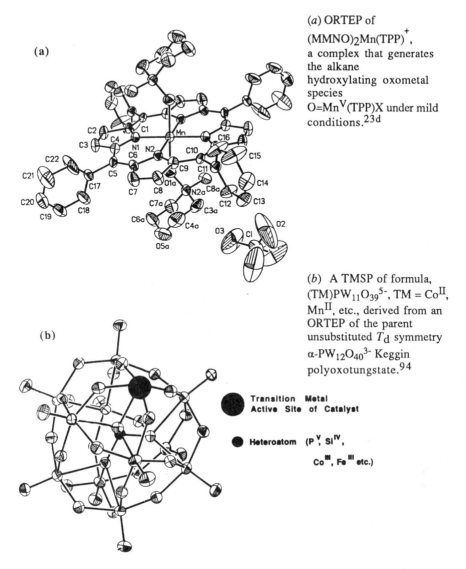

(a) ORTEP of $(MMNO)_2Mn(TPP)^+$, a complex that generates the alkane hydroxylating oxometal species $O=Mn^V(TPP)X$ under mild conditions.[23d]

(b) A TMSP of formula, $(TM)PW_{11}O_{39}^{5-}$, TM = Co^{II}, Mn^{II}, etc., derived from an ORTEP of the parent unsubstituted T_d symmetry α-$PW_{12}O_{40}^{3-}$ Keggin polyoxotungstate.[94]

● **Transition Metal Active Site of Catalyst**

● Heteroatom (P^V, Si^{IV}, Co^{III}, Fe^{III} etc.)

    <u>Features of both the metalloporphyrin and the TMSP:</u> (1) soluble, (2) thermally stable (TMSP, in principle, more so), (3) d^n transition metal ion "active site" firmly held (affecting tendency to deligate and precipitate), (4) ligand has a redox capacity that can be used in context with oxygenation and other catalytic redox processes, (5) vacant coordination positions available on d^n transition metal ion "active site," and (6) properties of "active site" can be manipulated by alteration of axial ligand (heteroatom(s) in case of the TMSP complexes).
    <u>Additional features unique to TMSP:</u> (1) ligand environment thermodynamically resistant to oxidation, and, as one consequence, (2) resistance to deligation of active site under catalytic oxidation conditions.

Figure 3. The conventional organic metalloporphyrin versus the transition metal-substituted polyoxometalate (TMSP), an "inorganic metalloporphyrin."

metalloporphyrin and TMSP ("inorganic porphyrin") structures based on ORTEP plots of recently published work from our group and summarizes some of the features of both types of oxygenation catalysts. There are two critical advantages exhibited by the TMSP complexes: the d^n transition metal ion active site is protected against deligation and subsequent precipitation, and the ligand framework in the TMSP complexes, as a consequence of its composition (d^0 W^{VI} and oxide ions), is thermodynamically, not just kinetically, resistant to oxidative degradation under the conditions of oxygenation catalysis.[53-57]

The TMSP complexes can catalyze a number of difficult oxidations including the epoxidation of olefins and the hydroxylation of alkanes. The rates and selectivities using metalloporphyrin, triflate salts, and TMSP complexes as catalysts for these hydrocarbon oxygenation reactions with iodosylarene oxygen donors have been compared.[53] In olefin epoxidation, the rates, selectivities, and stabilities compare favorably with any system based on metalloporphyrins, Schiff base complexes, or triflate salts. These results can be summarized as follows:

Comparison of Oxygenation Catalysts for Epoxidation Using Iodosylarenes[42,53,90]

Reactivity: $Co^{II}PW_{11}O_{39}^{5-}$ (most active) > $Mn^{II}PW_{11}O_{39}^{5-} \geq Fe^{III}(TDCPP)Cl$ > $Fe^{III}(TPP)Cl$ > $M(OTf)_2$ (least active)

Selectivity for Epoxide: $Co^{II}PW_{11}O_{39}^{5-} \sim Mn^{II}PW_{11}O_{39}^{5-}$ (most selective) > $M(OTf)_2$ > $Fe^{III}(TDCPP)Cl$ > $Fe^{III}(TPP)Cl$ (least selective)

Stability: $M(OTf)_2$ (least stable) < $Fe^{III}(TPP)Cl$ < $Fe^{III}(TDCPP)Cl$ < the TMSPs = $(TM)PW_{11}O_{39}^{5-}$, TM = Co^{II}, Mn^{II}, etc. (the only oxygenation catalysts stable in the presence of oxygen donor, DO in Figure 1, and in the absence of substrate)

For alkane substrates using the inexpensive reagent *tert*-butyl hydroperoxide (TBHP) as the oxygen donor, the yields of alkane oxidation products based on the oxygen donor consumed are higher, in the best solvents for this chemistry (benzene and other quite nonpolar media) than for any alkane hydroxylation reaction by oxygen donors catalyzed by metalloporphyrins, Schiff base complexes, triflate or nitrate salts, or other soluble transition metal complex. Perhaps most significantly, however, the stabilities of the TMSP complexes exceed those of all other homogeneous complexes.[53-57] The TMSP complexes are the only oxygenation catalysts that are stable in the absence of a kinetically protecting substrate. That is, k_{deg} in the TMSP/DO systems is very close to zero, and k_{oxy}/k_{deg} is then very large. The TMSP complexes are effectively "inorganic metalloporphyrins". Figure 4 summarizes the stabilities of a variety of oxo transfer catalysts. The numbers specifically cited here (turnovers before catalyst destruction) are for alkane hydroxylation.

Although the TMSP complexes offer substantial promise as catalysts for oxygenation as well as other difficult oxidation processes,[95,96] detailed delineation of the energetic and mechanistic features of these processes is just beginning. Some shortcomings of TMSP and other polyoxometalate-derived catalysts, as all catalysts, homogeneous and heterogeneous alike, are sure to be found in the course of further research. Establishing the origins of the high selectivities observed in some TMSP/DO oxygenation processes and the detailed mechanism of these oxygenations is a priority at the moment. The general directions of future work in this area are enumerated in Sec. VI.

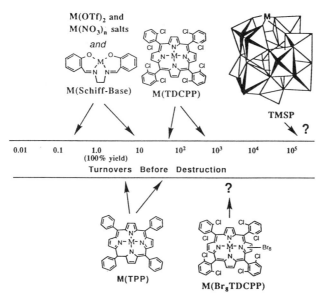

Figure 4. Stability of catalysts for oxo transfer to hydrocarbons. (The numbers given are for alkane oxygenations.)

V. Mechanism of Oxo Transfer to Unactivated C-H Bonds

Although most of the detailed information we have on the mechanism of oxo transfer from transition metal oxo species to the unactivated C-H bonds of alkanes comes from experiments executed since 1980, substantial pertinent literature on the subject was published prior to this time. A majority of the recent papers involve metalloporphyrin systems whereas a majority of the papers prior to 1975 involve simple transition metal oxo compounds such as Mn^{VII} and Cr^{VI} reagents. The following discussion, like the previous ones, does not attempt to cover the pertinent literature comprehensively. It is a chronology of the most significant papers that, in the view of this author, have shaped our understanding of the mechanism of C-H hydroxylation by oxometal species (Eq. 6). Allusion will be made to some studies that involve hydroxylation of activated (allylic, benzylic, etc.) C-H bonds in

$$L_nM{=}O \ + \ R\text{-}H \ \longrightarrow \ ROH \ + \ L_nM \qquad (6)$$

R-H = alkane, or other organic substrate with C-H bond

addition to those involving alkanes because the experiments on the former substrates address some general points regarding oxo transfer to C-H bonds and because the appropriate experiments with alkanes have yet to be carried out.

The principal mechanisms for Eq. 6 are summarized in Figure 5. The electrophilic process (Eq. 7) that entails simultaneous formation of the carbon-metal bond and the oxygen-hydrogen bond in the C-H cleavage process generating an organometallic complex has been omitted

from Figure 5 for the sake of clarity. Although an electrophilic cleavage of alkane C-H bonds appears to be operable in superacid systems (see Chapter II) and some low-valent or organometallic transition metal systems,[97] there is no compelling experimental evidence that such a mechanism is involved in C-H oxygenation by transition metal oxo species. Probably the strongest evidence against the electrophilic mechanism for alkane oxygenation comes from early studies on alkane oxidation by dioxochromium(VI) compounds in which the bridgehead position of norbornane was reported to be very unreactive, suggesting a tricoordinate rather than a pentacoordinate transition state for the C-H cleavage process.[98,99]

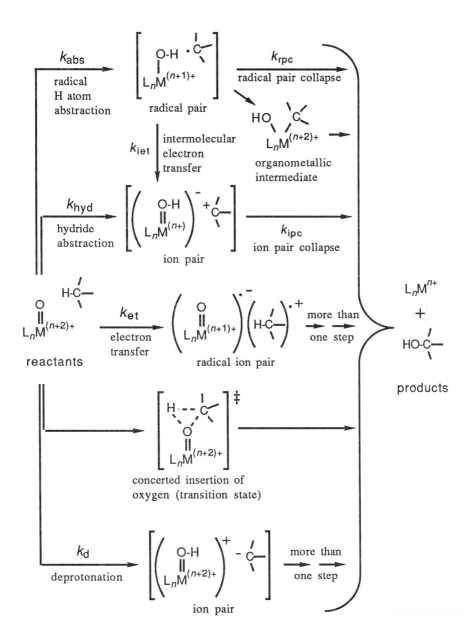

Figure 5. Pathways for the hydroxylation of C-H bonds by oxometal species, Eq. 6.

Sharpless et al. first pointed out that interaction of the organic substrate with the metal ion of a transition metal oxo complex could be of major significance in hydrocarbon oxygenation and related oxidation processes.[100] Although precomplexation of substrate to the transition metal is more applicable and experimentally precedented for reactions of

alkenes and arenes than for reactions of alkanes, the potential
significance of alkane-metal interactions in alkane oxidation by oxometal
species should not be ruled out. There is now, after all, some evidence
for alkane complexes with transition metal ions (see Chapter IV).

In the early days prior to the publication of any substantive
mechanistic investigations of hydrocarbon oxidation by transition metal
oxo complexes, there was some feeling that such processes, particularly
in the enzymatic systems, might be concerted, proceeding by a
mechanism similar to the second one from the bottom in Figure 5.
Although such a mechanism is analogous to one of those proposed for
epoxidation of olefins by peracids, there is no experimental evidence the
author is aware of that strongly suggests such a mechanism is applicable
for alkane hydroxylation by oxometal species in the liquid phase under
mild conditions. This mechanism will doubtless be the most difficult
one for which hard evidence can be obtained inasmuch as it involves the
least number of intermediates of any of the likely mechanisms (none).

Although the direct oxidation of alkanes by hydride abstraction
has been suggested by Benson and Nangia as possibly operable in
combustion processes,[101] there is no evidence for such a process
involving alkane oxidation by transition metal oxo compounds. Indeed
alkane C-H cleavage by hydride abstraction would entail very high
selectivities with respect to the type of C-H bond. Although there is no
direct data on the relative rates of hydride abstraction from primary (1º),
secondary (2º), and tertiary (3º) C-H bonds for any one class of organic
compounds, Wiberg, Hamilton, and other investigators have pointed
out similarities between the transition states seen in hydride abstraction
and solvolysis of alkyl derivatives, RX, with good leaving groups, X,
such as bromide, iodide, and tosylate. Inasmuch as both types of
processes generate a high degree of carbonium ion character on the
carbon, and the relative rates of solvolysis of 1º, 2º, and 3º halides are
~ $1:10^5:10^7$, it is not unreasonable to propose that hydride abstraction
from alkanes should occur with such high selectivities. Very few if any
products resulting from oxidative functionalization at 1º or 2º centers
should be observed if hydride abstraction is operable.[102,103] In nearly
all the alkane hydroxylations by oxometal complexes, the 1º:2º:3º
cleavage selectivities seen are far less pronounced than those above and
in the range of radical hydrogen abstraction reactions, albeit very
selective ones. The other argument against hydride abstraction is that
generation of carbonium ions, at least at secondary and primary
positions, should lead to isomerization of the carbon-carbon bonded
skeletons of the alkanes. The counter argument that renders this point
of minimal value is that, as just pointed out, direct generation of
carbonium ions by hydride abstraction at these less substituted positions
should be rarely if ever observed in the nonenzymatic systems under
mild reaction conditions.

There is currently no evidence for thermal alkane oxidation by an
oxometal species under mild conditions in the liquid phase that proceeds

by an initial electron transfer to the oxometal moiety, the middle equation in Figure 5. Furthermore, the few processes that do generate alkane cation radicals require conditions that do not permit ready comparison with those likely to be encountered in any homogeneous (or heterogeneous) alkane oxygenation reaction. Nevertheless, the reactions that may or doubtless do involve oxidation of alkanes by initial electron transfer are discussed briefly below.

The principal evidence that argues against a rate-determining electron transfer mechanism for alkane oxidation is one based on kinetic isotope effect data. In nearly all oxometal-based alkane oxidation processes in which primary kinetic isotope effects, k_H/k_D, have been measured, the values are substantial ranging from a low of 1.6 to over 10. Since there should be a small perturbation of the C-H(D) bonds by removal of one electron from an alkane with several carbon atoms, a small isotope effect would be expected. More importantly, it should be pointed out, as far as this author is aware, however, that no data exist regarding the kinetic isotope effect of alkane oxidation by electron transfer by any chemical (nonphotochemical and nonelectrochemical) means in the liquid phase. There simply are no bona fide examples of the thermal and chemical electron transfer oxidation of unstrained alkanes in solution. Thus, no data exist for purposes of comparison.

There are several reports in the literature addressing the electrooxidation of alkanes, with most of them appearing in the 1970s.[104-106] Although there is ample evidence that strained alkanes can be electrooxidized by an initial electron transfer mechanism to generate alkane cation radicals under mild conditions in neutral solution,[104] the corresponding electrooxidations of common unstrained cyclic and acyclic alkanes generally do not go as well. The few reports of anodic oxidation of unstrained alkanes in the literature involve extremely acidic media,[105] or acetonitrile.[106] For the few alkanes that have been successfully electrooxidized in the neutral medium, acetonitrile, the products are largely the acetamides, suggesting that at the extremely anodic potentials required to see any reaction at all, radical cation or radical species generated initially are rapidly oxidized to the corresponding carbonium ions. An important point in this interesting electrochemistry is that although the anodic oxidations of unstrained alkanes probably proceed by an initial electron transfer, there is still little proof for this mechanism.

The electron transfer oxidation of strained alkanes by photoinduced electron transfer to a stoichiometric oxidant is a fairly general reaction and has allowed Roth,[107] Gassman and Hay,[108] and others to characterize several polycyclic alkane cation radical species with some rigor. These methods have not been broadly applied to unstrained alkanes or to neutral alkane solutions at room temperature.

Recently our group reported a family of reactions that involves the oxidation of unstrained alkanes in neutral solutions at room temperature by a form of oxometal species. These reactions entail the photochemical dehydrogenation or functionalization of alkanes and other organic substrates, SH_2, catalyzed by the early transition metal polyoxometalates, and, in particular, the polyoxotungstates, P (Eq. 8).[95] Two processes sum to Eq. 8, reduction of the excited state of the polyoxotungstate accessible with near UV or blue light, P*, by the organic substrate forming the reduced polyoxotungstate, P_r, the oxidized organic compound, "S," and protons (Eq. 9) and reoxidation of the P_r by hydrogen evolution (Eq. 10). The oxidized organic products, S , in the case of the alkane oxidation can be either alkenes or

$$SH_2 \xrightarrow[\text{light } (\lambda > 300 \text{ nm})]{\text{P (catalyst)}} S + H_2 \qquad (8)$$

$$P^* + SH_2 \longrightarrow P_r + S + 2 H^+ \qquad (9)$$

$$P_r + 2 H^+ \longrightarrow P + H_2 \qquad (10)$$

other products. Although experimental evidence, including kinetics, product distribution, laser flash photolysis, and spectroscopic studies, indicate that the alkane oxidation step in these reactions (Eq. 9) may involve electron transfer,[95] more work needs to be done on these highly complex but potentially useful redox processes before the nature of this step is understood satisfactorily. The polyoxometalate-catalyzed photochemical dehydrogenation and functionalization of alkanes will not be discussed further here for three reasons. (1) These reactions do not produce the corresponding alcohols or other oxygenated products in most cases, thus they do not strictly fall under the purview of this review on alkane oxygenation. (2) These processes have a number of other features that distinguish them from any other type of alkane process in the literature, and more work needs to be done before several key points can be definitively addressed. (3) This chemistry will be reviewed elsewhere in the near future.

    The bottom mechanism in Figure 5, deprotonation followed by at least two steps to generate the products, has never been seriously forwarded as the mechanism of C-H cleavage in any alkane + oxometal reaction. It has been included for the sake of completeness. The reason for this derives less from experimental evidence that rules out such a mechanism, than from the physical attributes of the two reactants. Transition metal oxo complexes are usually very weak bases whereas alkanes are some of the least acidic compounds known.

The mechanism for C-H hydroxylation by oxometal species that involves initial hydrogen atom abstraction by the oxygen atom of the oxometal species followed by radical pair collapse, k_{abs} and k_{rpc} in Figure 5, is one of the first mechanisms proposed that was based on experimental facts. This mechanism does explain best the experimental data on many C-H oxygenations by oxometal complexes. It also constitutes the presently popular "oxygen atom rebound mechanism," a phrase used by the sizable group of investigators still actively engaged in probing the mechanism of C-H bond hydroxylation catalyzed by cytochrome P-450 and numerous synthetic model metalloporphyrins.

The following lines of evidence led investigators to propose k_{abs} with formation of the radical pair as the principal path for alkane oxidation by aqueous acidic chromic acid. No one line of evidence alone constitutes strong evidence for this mechanism at the exclusion of other potential mechanisms, but taken collectively all the experimental observations constitute a strong case for k_{abs} and radical pair formation. First, the rate law is first order in both alkane and Cr^{VI}. This result taken alone says little inasmuch as a similar rate law should be observed with several alternative mechanisms. Second, the relative rates of attack at $1°/2°/3°$ C-H bonds (1:110:7000) are in the range expected for hydrogen atom abstraction by fairly selective radicals.[98] This point is discussed further subsequently. Third, there is a demonstrable primary hydrogen kinetic isotope effect, $k_H/k_D = 2.5$.[109] Fourth, steric acceleration is fairly generally observed.[98] Fifth, no anchimeric assistance is found in the oxidations of camphane, cyclobutane and isocamphane.[110]

Behavior similar to that in Cr^{VI} + alkane reactions was observed in other oxometal alkane oxygenations studied subsequently. Additional features of the k_{abs}, k_{rpc} mechanism documented in subsequently investigated reactions are the partial loss of stereochemistry at the hydroxylated carbon and carbon-carbon skeleton rearrangements characteristic of radical intermediates. The partial loss of stereochemistry has been seen, for example, in oxidation of the unactivated $3°$ C-H bond of 4-methylhexanoic acid by MnO_4^- in early work by Wiberg and Fox,[111] and in the hydroxylation of alkanes by various oxometalloporphyrins in more recent studies.[18] More substantial internal kinetic isotope effects upon hydrocarbon C-H bond hydroxylation by oxometalloporphyrins by k_{abs}, k_{rpc} mechanisms were observed by our group and the groups of Groves, Bruice, Mansuy, and our investigators (see Chapter VI).[18]

The rearrangements of the neophyl radical generated during hydroxylation of *tert*-butylbenzene by oxomanganese(V) porphyrins,[23] the norcaranyl radical generated by reaction with oxoiron(IV) porphyrin π-cation radicals[22b] and oxomanganese(V) porphyrins[22a] with norcarane, and allylic radicals generated on abstraction of allylic hydrogens by oxoiron(IV) porphyrin π-cation radicals[83c] are all in accord with radical intermediates. The products observed in the neophyl and norcaranyl systems rule out the presence of carbonium ion intermediates in at least the reactions of these substrates (primary and secondary centers, respectively). These and other rearrangement studies collectively indicate that the fate of the radical pair, $[L_nM^{(n+1)+}\text{-OH}$ $\cdot CR_3]$, is highly dependent on the electronic structure of the hydroxymetal intermediate, $L_nM^{(n+1)+}$-OH in Figure 5.

Two recent interesting studies constitute a final note on alkane C-H cleavage by k_{abs} forming a radical pair. Unpublished work by Nappa and McKinney, which is covered in part in the middle portion of Chapter X, indicates that with appropriate electronic destabilization and increased reactivity of oxomanganese(V) porphyrins, most easily effected by incorporation of electron-withdrawing substituents into the porphyrin ring, alkane oxygenation by these reactive complexes begins to look concerted in nature. Alternatively, however, these interesting reactions can be viewed, in the opinion of this author, as still stepwise processes but with a markedly earlier and lower transition state. The alteration of the selectivities in these metalloporphyrin-induced alkane oxygenation reactions investigated by Nappa is noteworthy. The other interesting recent study involves the oxygenation not of alkanes but of cyclohexene-3,3,6,6-d_4 by $[(bpy)_2(bpy)Ru^{IV}=O]^{2+}$ reported by Meyer and co-workers.[112] This reaction proceeds without allylic rearrangement. This is also a significant and unusual result inasmuch as this allylic rearrangement is far faster than diffusion and faster even than some rapid intramolecular processes. In this case too, we also favor the interpretation that this oxygenation is proceeding by an early k_{abs} transition state with a minimal activation energy followed by an extremely rapid capture of the resulting radical as opposed to a concerted insertion of oxygen into the C-H bond.

With a fairly comprehensive delineation of C-H cleavage by oxometal species culminating with a discussion of what is likely the most prevalent mechanism, hydrogen atom abstraction, we now turn to subsequent processes in this mechanism and the fate of the radical pair. Figure 6 summarizes the possible fates of the radical pair resulting from alkane hydrogen abstraction by an oxometal complex, $L_nM^{(n+2)+}=O$. In this figure we illustrate the proposed processes derived from mechanistic studies on many different oxometal-based oxygenation systems. Although these processes usually contribute little to the overall regioselectivities observed for alkane oxygenation, they nevertheless can be critical with respect to the stereochemical and product selectivities. By product selectivities we mean the type of product generated; in oxygenations, the quantity of alcohol(s) versus other products such as alkenes, ketones, and dimeric species. It is thus desirable to understand the features that influence the relative rates of the various processes in Figure 6 such that the partitioning of the radical pair between the various paths can be systematically controlled by experimental conditions.

The first fate of the radical pair in Figure 6 to be considered is attack of the radical at the metal to form an organometallic intermediate (Eq. 11). Although there is little direct evidence for this pathway in

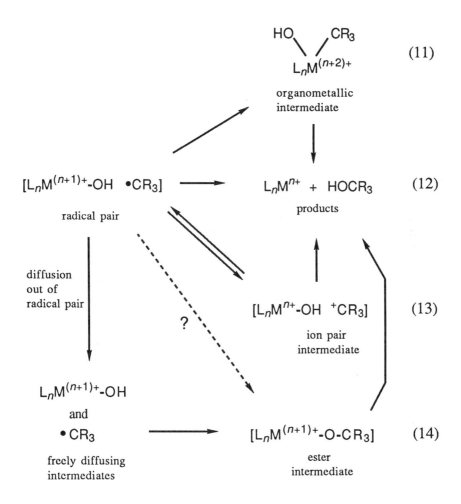

Figure 6. Possible fates of the radical pair derived from alkane hydrogen abstraction by an oxometal species.

hydroxylation reactions, it is not an unreasonable mechanism. Numerous studies, principally by Kochi and co-workers, have indicated the imminent feasibility of attack by freely diffusing radicals at metal centers in reactions involving ligand transfer oxidation and radical oxidation by transition metal complexes.[113] Although the principal fate of a σ-hydroxy, σ-organometallic intermediate that has been suggested by various authors is concerted reductive elimination to form the products, the extensive literature addressing the mechanisms of oxidative addition and reductive elimination in organometallic complexes suggests that alternative paths for breakdown of this species involving intermediates cannot be excluded.

There is substantial evidence for electron transfer oxidation of the radical within the radical pair producing an ion pair prior to rapid formation of the products (Eq. 13). First, investigations, not of the behavior of the caged radical in the case of alkane hydroxylations by the usual radical cage mechanism, but of the behavior of *freely diffusing* radicals in the presence of transition metal oxidants indicate that radical oxidation is very common. Where it is thermodynamically favorable it generally appears to be a very rapid process.[113] Some time ago, Rocek invoked hydride-carbonium ion character in the radical abstraction transition state itself to explain the fairly high selectivity manifested in the relative rates of attack at 1°/2°/3° C-H bonds (1:110:7000) in the alkane/Cr^{VI} systems.[114] It is clear that more research needs to be done to understand further the relationships between the radical pair and ion pair in Figure 6. Based on the previous points, however, we have connected these two intermediates in Figure 6 with double arrows. A single double ended arrow indicating different canonical forms of the same species may be a better depiction of the mechanism of some oxometal hydrogen atom abstraction processes.

Extensive   work   by   our   group   on   the Mn^{III}(TPP)X/ iodosylarene, X = halide or pseudohalide, catalytic alkane functionalization systems clearly indicates multiple fates for the intermediate radical pair resulting from substrate hydrogen atom abstraction by oxomanganese(V) porphyrin intermediates. In these systems, diffusion of the radicals out of the radical pair, intramolecular electron transfer to give an ion pair followed by ion pair collapse, and likely a third mode of product determining breakdown of the radical pair in Figure 6 are operable.[23c] Furthermore, the partitioning of radical pair between these paths depends on the complex and the type of C-H bond being hydroxylated. The tertiary centers appear to undergo oxidation to the carbonium ion oxidation state, Eq. 13, whereas the primary centers do not. A majority of the primary radicals resist oxidation and other processes long enough to diffuse out of the radical cage. This latter property, and the extremely high rates of ligand transfer oxidation of alkyl radicals by the tetragonally distorted high-spin d^4 Mn^{III} porphyrin complexes that are present in these reactions, enabled the development of methods for converting catalytic alkane oxygenation to catalytic alkane halogenation and nitrogenation processes (incorporation of azide and nitro groups).[115]

As in the case for the concerted oxo insertion mechanism in the C-H bond cleavage step discussed (Fig. 5), it is difficult to garner direct evidence for the concerted generation of products from the radical pair (Eq. 12) (Fig. 6). It is difficult to differentiate a mechanism involving the synchronous breaking of the M-O bond and making of the C-O bond

(Eq. 12) from mechanisms involving the attack of the caged radical at the metal (Eq. 11) and oxidation of the radical (Eq. 13).

The final point to be discussed pertaining to the fate of the radical pair in Figure 6 concerns with the possibility of radical attack at the pendant oxygen forming an intermediate ester, which then hydrolyzes to form the alcohol. Note that conversion of the ester depicted in Figure 6 to the products involves a further one-electron oxidation of the metal. No one has seriously discussed the direct formation of such as ester from the radical pair as this would involve

movement of two groups that are in principle independent, the $\cdot CR_3$ radical, and the proton initially on the hydroxymetal species in the radical pair. This type of ester intermediate has been proposed to occur in other alkane oxygenations, however, where freely diffusing alkyl radicals can encounter free oxo groups. The requirement of freely diffusing radicals and free oxo groups has limited the discussion of this type of process to those systems in which two or more oxo groups are present on the metal oxidant, most specifically in the Cr^{VI} oxidations.[102]

It is clear at the present time that there are multiple fates for the radical pair in Figure 6. Substantial research is needed to further define the structure-reactivity relationships in the behavior of such intermediates. A large number of fairly straightforward experiments with both metalloporphyrin and simpler transition metal oxo systems can be envisioned.

VI. Oxo Transfer Catalysis - Future Directions

There are several goals pertaining to the oxygenation of alkanes and the reactivity of oxometal species that need to be addressed experimentally. Five of the more important of these goals are enumerated here.

1. It is clear that control of the fate of radical pairs generated by hydrogen atom abstraction from alkanes (and other organic substrates) by transition metal oxo complexes (see Fig. 6) is of importance in the stereoselectivities and product selectivities observed in these reactions. Although it is clear from presently available data that the *d*-orbital occupancy in addition to the redox potentials of the hydroxymetal species in the intermediate radical pairs dictate in large part the partitioning of the radical pair intermediate between the various breakdown pathways, more information linking the structural and electronic features of the transition metal center and the proximal ligands with these breakdown processes is needed.

2. A second goal is to develop catalysts for oxo transfer to alkanes that are not only more reactive and selective but also more stable. The transition metal-substituted polyoxometalate (TMSP) complexes look quite promising in connection with this goal, but a diversity of systems needs to be examined. Both exploratory work

seeking new systems as well as in-depth studies on currently promising systems in order to render the latter more effective would be welcome.

3. A third goal is to design thermal alkane oxygenation catalysts that exhibit high selectivities for desired products at high substrate conversions. Nearly all existing thermal processes for the transformation of alkanes, including the industrial oxidations, fail to exhibit high selectivities at high substrate conversions. Some alkane transformations by organometallic complexes that convert alkanes in high selectivities need to be examined under conditions of high conversion. Although the organometallic-based alkane activation processes remain, in large part, stoichiometric, and production of functionalized alkane products versus organometallic products can be a problem, one of these fascinating systems may ultimately produce the first catalytic high-selectivity, high-conversion alkane modification process. The few alkane transformation reactions based on organometallic species that are catalytic in complex display irreversible degradation of the catalyst after a very limited number of turnovers.

Both our group[116] and that of Crabtree[117] have developed catalytic photochemical methods for the conversion of alkanes that exhibit reasonable selectivities at reasonable substrate conversions with minimal decomposition of the catalyst, but thermal processes that exhibit such features are lacking. Perhaps the further development of the provocative and unique Gif and Gif-Orsay liquid phase alkane functionalization systems of the Barton group may yield the first catalytic thermal high-selectivity, high-conversion alkane functionalization chemistry (see Chapter IX).

4. A fourth goal is to develop oxometal or other systems that can utilize dioxygen exclusively as the oxidant in non-radical-chain alkane oxidation reactions. In conjunction with this, processes are sought in which no reducing of any kind including the solvent is required and both atoms of the dioxygen are incorporated into the alkane. The alkane oxidations that use dioxygen are nearly all free radical chain reactions. Although oxometal species may be produced in some industrial autoxidation processes and function to cleave C-H bonds, they certainly constitute a minor percentage of the intermediates kinetically competent to react directly with the alkane substrates. In essence, systems should be sought in which Reactions 15 and 16 (S = alkane) proceed at rates faster than radical initiated autoxidation. Although the catalytic aerobic epoxidation of olefins has been achieved,[118] that is, Eqs. 15 and 16 (S = olefin) are faster than autoxidation processes, the corresponding reactions for alkanes have not been achieved. No reducing agent free dioxygen-based system is known that produces oxometal species, $L_nM^{(n+2)+}=O$, in Eqs. 15 and 16, that can react directly and readily with alkanes.

$$L_nM^{n+} \quad + \quad 1/2\ O_2 \quad \longrightarrow \quad L_nM^{(n+2)+}=O \tag{15}$$

$$L_nM^{(n+2)+}=O \quad + \quad S \quad \longrightarrow \quad L_nM^{n+} \quad + \quad SO \tag{16}$$

5.   A fifth goal in the oxygenation of alkanes is to develop chiral catalysts that effect enantioselective hydroxylation of unactivated aliphatic C-H bonds.

Acknowledgments

I am indebted to the able graduate students and postdoctoral fellows for the research results from our group that were presented in this review. Over the last 7 years the following co-workers of mine worked to establish the physical, chemical, photochemical or catalytic properties of various alkane functionalization systems. In alphabetical order: D. A. Bouchard, R. B. Brown, Jr., M. J. Camenzind, R. C. Chambers, L. A. Combs, M. Faraj, T. J. Henly, M. Kadkhodayan, C.-H. Lin, C. Prosser-McCartha, J. A. Smegal, B. S. Schardt, R. F. Renneke, and M. M. Williamson.   The National Science Foundation has been the principal source of support for this research. We also wish to thank ARCO Chemical Company, Chevron Corporation, and the Petroleum Research Fund administered by the American Chemical Society for direct support and the National Institutes of Health and the Department of Defense for partial or indirect support of these programs.

References

1.   Parshall, G. W., *Homogeneous Catalysis. The Applications and Chemistry of Catalysis by Soluble Transition Metal Complexes*, Wiley-Interscience, New York, 1980, Chapter 10 and references cited therein.

2.   Sheldon, R. A., Kochi, J. K., *Metal-Catalyzed Oxidations of Organic Compounds*. Academic Press, New York, 1981, Chap. 2 and references cited therein.

3.   Weissermel, K., Arpe, H.-J., *Industrial Organic Chemistry*, p. 212 Verlag Chimie, New York, 1978.

4.   (a) Crabtree, R. H., Mihelcic, J. M., Quirk, J. M., *J. Am. Chem. Soc.* **1979**, *101*, 7738;   (b) Crabtree, R. H., Mellea, M. F., Mihelcic, J. M., Quirk, J. M., *J. Am. Chem. Soc.* **1982**, *104*, 107.

5.   (a) Baudry, D., Ephritikhine, M., Felkin, H., *J. Chem Soc., Chem. Commun.* **1982**, 606; (b) Baudry, D., Ephritikhine, M., Felkin, H., Holmes-Smith, R., *J. Chem Soc., Chem. Commun.* **1983**, 788.

6.   Janowicz, A. H., Bergman, R. G., *J. Am. Chem. Soc.* **1981**, *103*, 2488.

7.   For example, see Fendrick, C. M., Marks, T. J., *J. Am. Chem. Soc.* **1986**, *108*, 425.

8.   Watson, P. L., *J. Am. Chem. Soc.* **1983**, *105*, 6491;   (b) Watson, P. L.,

J. Chem. Soc., Chem. Commun. **1983**, 276; (c) Watson, P. L., *J. Am. Chem. Soc.* **1982**, *104*, 337; (d) Watson, P. L., Roe, D. C., *J. Am. Chem. Soc.* **1982**, *104*, 5471.

9.   References 10-15 are recent reviews on organometallic alkane activation systems. Each review is written with a different orientation.

10.   Shilov, A. E., *Activation of Saturated Hydrocarbons Using Transition Metal Complexes*. R. Reidel, Dordrecht, 1984.

11.   Bergman, R. G., *Science (Washington D. C.)* **1984**, *223*, 902.

12.   Crabtree, R. H., *Chem. Rev.* **1985**, *85*, 245.

13.   (a) Halpern, J., *Inorg. Chim. Acta* **1985**, *100*, 41; (b) Halpern, J., in *Fundamental Research in Homogeneous Catalysts* (A. E., Shilov ed.), Vol I. p. 393. Gordon and Breach, New York, 1986.

14.   Rothwell, I. P., *Polyhedron* **1985**, *4*, 177.

15.   Ephritikhine, M., *Nouv. J. Chim.* **1986**, *10*, 9.

16.   (a) Olah, G. A., Schlosberg, R. H., *J. Am. Chem. Soc.* **1968**, *190*, 2726; (b) Olah, G. A., Halpern, Y., Shen, J., Mo, Y. K., *J. Am. Chem. Soc.* **1971**, *93*, 1251; (c) Olah, G. A., Olah, J. A., *ibid.* **1971**, *93*, 1256; (d) Olah, G. A.,Yoneda, N., Parker, D. G., *J. Am. Chem. Soc.* **1976**, *98*, 483 and many other papers by this group.[17]

17.   Olah, G., *Acc. Chem. Res.* **1987**, *11*, 422.

18.   References 19-27 are representative recent papers addressing the catalytic functionalization of *alkanes* by metalloporphyrins. They are listed in alphabetical order by the principal investigator's last name.

19.   (a) Breslow, R. S., Gellman, S. H., *J. Chem. Soc., Chem. Commun.*, **1982**, 1400; (b) Breslow, R., Gellman, S. H., *J. Am. Chem. Soc.*, **1983**, *105*, 6728; (c) Svastits, E. W., Dawson, J. H., Breslow, R., Gellman, S. H., *J. Am. Chem. Soc.* **1985**, *107*, 6427.

20.   (a) Nee, M. W., Bruice, T. C., *J. Am. Chem. Soc.* **1982**, *104*, 6123; (b) Dicken, C. M., Lu, F.-L., Nee, M. W., Bruice, T. C., *J. Am. Chem. Soc.* **1985**, *107*, 5776.

21.   Chang, C. K., Ebina, F., *J. Chem. Soc., Chem. Commun.* **1981**, 778.

22.   (a) Groves, J. T., Kruper, W. J., Jr., Haushalter, R. C., *J. Am. Chem. Soc.* **1980**, *102*, 6375; (b) Groves, J. T., Nemo, T. E., *J. Am. Chem. Soc.* **1983**, *105*, 6243.

23.   (a) Schardt, B. C., Hill, C. L., *J. Am. Chem. Soc.* **1980**, *102*, 6374; (b) Smegal, J. A., Hill, C. L., *J. Am. Chem. Soc.*, **1983**, *105*, 2920; (c) Smegal, J. A., Hill, C. L., *J. Am. Chem. Soc.* **1983**, *105*, 3515; (d) Brown, R. B., Jr., Williamson, M. M., Hill, C. L., *Inorg. Chem.* **1987**, *26*, 1602, and references cited in each.

24.   (a) Mansuy, D., Fontecave, M., Bartoli, J. -F., *J. Chem. Soc., Chem. Commun.* **1983**, 253; (b) Mansuy, D., Mahy, J.-P., Dureault, A., Bedi, G., Battioni, P., *J. Am. Chem. Soc.* **1984**, 1161; (c) Battioni, P., Renaud J.-P., Bartoli, J. F., Mansuy, D., *J. Chem. Soc., Chem. Commun.* **1986** 341; (d) Battioni, P., Bartoli, J. F., Leduc, P., Fontecave, M., Mansuy, D., *J. Chem. Soc., Chem. Commun.* **1987**, 791.

25.   (a) Nappa, M. J., Tolman, C. A., *Inorg. Chem.* **1985**, *24*, 4711; (b) Nappa, M. J., McKinney, R. J., *Inorg. Chem.* **1988**, *27*, 3740.

26.   (a) Suslick, K. S., Cook, B., Fox, M., *J. Chem. Soc., Chem. Commun.* **1985**, 580; (b) Cook, B. R., Reinert, T. J., Suslick, K. S., *J. Am. Chem. Soc.* **1986**, *108*, 7281; (c) Suslick, K. S., Acholla, F. V., Cook, B. R., *J. Am. Chem. Soc.*, **1987**, *109*, 2818.

27. Recent reviews on metalloporphyrin catalyzed oxygenation: (a) Meunier, B., *Bull. Soc. Chim. Fr.* **1986**, *4*, 578; (b) Hill, C. L., in *Advances in Oxygenated Processes*, (A. L. Baumstark, ed.), Vol. I, **1988**, p. 1.

28. For example: (a) Srinivasan, K., Michaud, P., Kochi, J. K., *J. Am. Chem. Soc.* **1986**, *108*, 2309; (b) Koola, J. D., Kochi, J. K., *Inorg. Chem.* **1987**, *26*, 908.

29. References 30-34 are leading references focusing on the modeling of reactivity features of cytochrome P-450. See also refs. 19-27.

30. Reviews: (a) McMurry, T. J., Groves, J. T., in *Cytochrome P-450: Structure, Mechanism, and Biochemistry* (P. R., Ortiz de Montellano, ed.) Chap. 1. Plenum, New York, 1986; Chap. 1. (b) Mansuy, D., *Pure Appl. Chem.* **1987**, *59*, 759. See also ref. 27.

31. (a) Castellino, A. J., Bruice, T.C., *J. Am. Chem. Soc.* **1988**, *110*, 1313; (b) Castellino, A. J., Bruice, T.C., *J. Am. Chem. Soc.* **1988**, *110*, 158; (c) Ostovic, D., Knobler, C. B., Bruice T. C., *J. Am. Chem.* Soc. **1987**, *109*, 3444; (d) Bruice, T. C., Dicken, C. M., Balasubramanian, P. N., Woon, T. C., Lu, F.-L., *J. Am. Chem. Soc.* **1987**, *109*, 3436; (e) Wong, W.-H., Ostovic, D., Bruice, T. C., *J. Am. Chem. Soc.* **1987**, *109*, 3428; (f) Balasubramanian, P. N., Sinha, A., Bruice, T. C., *J. Am. Chem. Soc.* **1987**, *109*, 1456, and references cited in each.

32. (a) Groves, J. T., Watanabe, Y., *Inorg. Chem.* **1987**, *26*, 785; (b) Groves J. T., Watanabe, Y., *Inorg. Chem.* **1986**, *25*, 4808; (c) Groves, J. T., Watanabe, Y., *J. Am. Chem. Soc.* **1986**, *108*, 7836; (d) Groves, J. T., Watanabe, Y., *J. Am. Chem. Soc.* **1986**, *108*, 7834; (e) Groves, J. T., Avaria-Neisser; G. E., Fish, K. M., Imachi, M., Kuczkowski, R. L., *J. Am. Chem. Soc.* **1986**, *108*, 3837; (f) Groves, J. T., Watanabe, Y., *J. Am. Chem. Soc.* **1986**, *108*, 507 and references cited therein.

33. (a) Collman, J. P., Hampton, P. D., Brauman, J. I., *J. Am. Chem. Soc.* **1986**, *108*, 7861; (b) Collman, J. P., Kodadek, T., Brauman, J. I., *J. Am. Chem. Soc.* **1986**, *108*, 2588; (c) Collman, J. P., Brauman, J. I., Meunier, B., Hayashi, T., Kodadek, T., Raybuck, S. A., *J. Am. Chem. Soc.* **1985**, *107*, 2000 and references cited therein.

34. (a) Traylor, P. S., Dolphin, D., Traylor, T. G., *J. Chem. Soc., Chem. Commun.* **1984**, 279; (b) Traylor, T. G., Nakano, T., Miksztal, A. R., Dunlap, B. E., *J. Am. Chem. Soc.* **1987**, *109*, 3625; (c) Traylor, T. G., Miksztal, A. R., *J. Am. Chem. Soc.* **1987**, *109*, 2770; (d) Traylor, T. G., Iamamoto, Y., Nakano, T., *J. Am. Chem. Soc.* **1986**, *108*, 3529; (e) Traylor, T. G., Nakano, T., Dunlap, B. E., Traylor, P. S., Dolphin, D., *J. Am. Chem. Soc.* **1986**, *108*, 2782 and references cited therein.

35. (a) Hanratty, M. A., Beauchamp, J. L., Illies, A. J., Bowers, M. T., *J. Am. Chem. Soc.* **1985**, *107*, 1788; (b) Tolbert, M. A., Mandich, M. L., Halle, L. F., Beauchamp, J. L., *J. Am. Chem. Soc.* **1986**, *108*, 5675; (c) Kang, H., Beauchamp, J. L., *J. Am. Chem. Soc.* **1986**, *108*, 7502; (d) Tolbert, M. A., Beauchamp, J. L., *J. Am. Chem. Soc.* **1986**, *108*, 7509;

36. (a) Jacobson, D. B., Freiser, B. S., *J. Am. Chem. Soc.*, **1985**, *107*, 7399; (b) Jacobson, D. B., Freiser, B. S., *J. Am. Chem. Soc.* **1985**, *107*, 7399; (c) Jackson, T. C., Carlin, T. J., Freiser, B. S., *J. Am. Chem. Soc.* **1986**, *108*, 1120; (d) Jacobson, D. B., Freiser, B. S., *J. Am. Chem. Soc.* **1985**, *107*, 7399; (e) Cassady, C. J., Freiser, B. S., *J. Am. Chem. Soc.* **1986**, *108*, 5690; (f) Tews, E. C.; Freiser, B. S., *J. Am. Chem. Soc.* **1987**, *109*, 4433 and references cited in each.

37. Baseman, R. J., Buss, R. J., Casavecchia, P., Lee, Y. T., *J. Am. Chem.*

Soc. **1984**, *106*, 4108.

38.   (a) Freas, R. B., Campana, J. E., *J. Am. Chem. Soc.* **1986**, *108*, 4659; (b) Mandich, M. L., Steigerwald, M. L., Reents, W. D., Jr., *J. Am. Chem. Soc.* **1986**, *108*, 6197; (c) Magnera, T. F., David, D. E., Michl, J., *J. Am. Chem. Soc.* **1987**, *109*, 936;

39.   Recent studies of methane dehydrogenation over metal oxides: (a) Liu, H.-F., Liu, R.-S., Liew, K. Y., Johnson, R. E., Lunsford, J. H., *J. Am. Chem. Soc.* **1984**, *106*, 4117; (b) Otsuka, K., Liu, Q., Morikawa, A., *J. Chem. Soc., Chem. Commun.* **1986**, 586; (c) Aika, K., Moriyama, T., Takasaki, N., Iwamatsu, E., *J. Chem. Soc., Chem. Commun.* **1986**, 1210; (d) Sofranko, J. A., Leonard, J. J., Jones, C. A., *J. Catal.* **1987**, *103*, 302; (e) Jones, C. A., Leonard, J. J., Sofranko, J. A., *J. Catal.* **1987**, *103*, 311; (f) Hutchings, G. J., Scurrell, M. S., Woodhouse, J. R., *J. Chem. Soc., Chem. Commun.* **1987**, 1388; (g) Martin, G.-A., Mirodatos, C., *J. Chem. Soc., Chem. Commun.* **1987**, 1393; (h) Anderson, J. R., Tsai, P., *J. Chem. Soc., Chem. Commun.* **1987**, 1435; (i) Lin, C.-H., Ito, T., Wang, J-X., Lunsford, J. H., *J. Am. Chem. Soc.* **1987**, *109*, 4808; (j) Kasztelan, S., Moffat, J. B., *J. Catal.* **1987**, *106*, 512; (k) Thomas, J. M., Kuan, X., Stachurski, J., *J. Chem. Soc., Chem. Commun.* **1988**, 162, and references cited in each.

40.   See also Cogen, J. M., Maier, W. F., *J. Am. Chem. Soc.* **1986**, *108*, 7752; (b) Lebrilla, C. B., Maier, W. F., *J. Am. Chem. Soc.* **1986**, *108*, 1606; (c) Maier, W. F., *Nature* London **1987**, *329*, 531.

41.   Saillard, J.-Y., Hoffmann, R., *J. Am. Chem. Soc.* **1984**, *106*, 2006.

42.   Abbreviations used in text: TBHP, *tert*-butyl hydroperoxide, TPP, *meso*-tetraphenylporphyrin dianion, TDCPP, tetrakis(2,6-dichlorophenylporphyrin), ImH, imidazole.

43.   Reviews of polyoxometalates: (a) Pope, M. T., *Heteropoly and Isopoly Oxometalates*. Springer-Verlag, Berlin, 1983; (b) Day, V. W., Klemperer, W. G., *Science* **1985**, *228*, 533.

44.   Holm, R. H., *Chem. Rev.* **1987**, *87*, 1401.

45.   (a) Mansuy, D., Mahy, J.-P., Dureault, A., Bedi, G., Battioni, P., *J. Chem. Soc., Chem. Commun.* **1984**, 1161; (b) Mahy, J.-P., Battioni, P., Mansuy, D., *J. Am. Chem. Soc.* **1986**, *108*, 1079. See also ref. 46.

46.   Mahy, J.-P., Battioni, P., Bedi, G., Mansuy, D., Fischer, J., Weiss, R., Morgenstern-Badarau, I., *Inorg. Chem.* **1988**, *27*, 353.

47.   Groves , J. T., Takahashi, T., *J. Am. Chem. Soc.* **1983**, *105*, 2073.

48.   Mansuy, D., Battioni, J.-P., Akhrem, I., Dupré. D., Fischer, J., Weiss, R., Morgenstern-Badarau, I., *J. Am. Chem. Soc.* **1984**, *106*, 6112. See also ref. 49.

49.   Battioni, J.-P., Artaud, I., Dupré, D., Leduc, P., Akrem, I., Mansuy, D., Fischer, J., Weiss, R., Morgenstern-Badarau, I., *J. Am. Chem. Soc.* **1986**, *108*, 5598.

50.   Herron, N., Stucky, G.D., Tolman, C. A., *J. Chem. Soc., Chem. Commun.* **1986**, 1521. See also Chapter X.

51.   Although alkanes were not examined as substrates per se, hydroxylation of cumene by the Valentine group has been reported, cf. VanAtta, R., Franklin, C. C., Valentine, J. S., *Inorg. Chem.* **1984**, *23*, 4121. For other papers by the Valentine group on the MX_2/PhIO (X = nitrate or triflate) systems, see reference 52.

52.   (a) Valentine, J. S., VanAtta, R., Margerum, L. D., Yang, Y., in *The Role of Oxygen in Chemistry and Biology* (W. Ando, ed.), p. 175. Elsevier,

New York, 1988; (b) Franklin, C. C., VanAtta, R., Tai, A. F., Valentine, J. S., *J. Am. Chem. Soc.* **1984**, *106*, 814; (c) Tai, A. F., Margerum, L. D., Valentine, J. S., *J. Am. Chem. Soc.* **1986**, *108*, 5008.

53. Hill, C. L., Brown , R. B., Jr., *J. Am. Chem. Soc.* **1986**, *108*, 536.
54. Brown, R. B., Renneke, R., Hill, C. L., *Prepr. Am. Chem. Soc. Div. Pet. Chem.* **1987**, *32*,(1), 205.
55. Faraj, M., Hill, C. L., *J. Chem. Soc., Chem. Commun.* **1987**, 1497.
56. Hill, C. L., Renneke, R. F., Faraj, M. K., Brown, R.B., Jr., in *The Role of Oxygen in Chemistry and Biology* (W. Ando, ed.), p. 185. Elsevier, New York, 1988.
57. Faraj, M., Lin, C.-H., Hill, C. L., *New J. Chem.* **1988**, *12*, 745.
58. Balasubramanian, P. N., Sinha, A., Bruice, T. C., *J. Am. Chem. Soc.* **1987**, *109*, 1456.
59. Dicken, C. M., Balasubramanian, P. N., Bruice, T. C., *Inorg. Chem.* **1988**, *27*, 197.
60. Collman, J. P., Brauman, J. I., Meunier, B., Raybuck, S. A., Kodadek, T., *Proc. Natl. Acad. Sci. U.S.A.* **1984**, *81*, 3245.
61. Collman, J. P., Kodadek, T., Raybuck, S. A., Brauman, J. I., Papazian, L. M., *J. Am. Chem. Soc.* **1985**, *107*, 4343.
62. Lee. W. A., Bruice, T. C., *J. Am. Chem. Soc.* **1985**, *107*, 513.
63. Yuan, L.-C., Bruice, T. C., *J. Am. Chem. Soc.* **1985**, *107*, 512.
64. Yuan, L.-C., Bruice, T. C., *Inorg. Chem.* **1985**, *24*, 987.
65. Lee., W. A., Bruice, T. C., *Inorg. Chem.* **1986**, *25*, 131.
66. (a) Traylor, T. G.; Lee, W. A.; Stynes, D., *J. Am. Chem. Soc.* **1984**, *106*, 755.
67. Groves, J. T., Watanabe, Y., *Inorg. Chem.* **1986**, *25*, 4808.
68. Kerr, J. A., *Chem. Rev.* **1966**, *66*, 465.
69. Review of bond strength data: See for example, Mimoun, H., *New J. Chem.* **1987**, *11*, 513. See also references in footnote 39.
70. Hill, C. L., Hollander, F. J., *J. Am. Chem. Soc.* **1982**, *104*, 7318. References 71 and 72 also address low-spin d^2 nitridomanganese(V) porphyrin complexes.
71. (a) Buchler, J. W., Dreher, C., Lay, K. L., *Z. Naturforsch., B:* **1982**, *37B*. 1155; (b) Buchler, J. W., Dreher, C., Lay, K.-L., Lee, Y. J. A., *Inorg. Chem.* **1983**, *22*, 888.
72. Bottomley, L. A., Neely, F. L., Gorce, J.-N., *Inorg. Chem.* **1988**, *27*, 1300.
73. (a) Groves, J. T., Kruper, W. J., Haushalter, R. C., Butler, W. M., *Inorg. Chem.* **1982**, *21*, 1363; (b) Penner-Hahn, J. E., Benfatto, M., Hedman, B., Takahashi, T., Doniach, S., Groves, J. T., Hodgson, K. O., *Inorg. Chem.* **1986**, *25*, 2255.
74. (a) Buchler, J. W., Lay, K. L., Castle, L., Ullrich. V., *Inorg. Chem.* **1982**, *21*, 842; (b) Budge, G. R., Gatehouse, B. M. K., Nesbit, M. C., West, B. O., *J. Chem. Soc., Chem. Commun.* **1981**, 370.
75. (a) Groves, J. T.; Kruper, W. J., *J. Am. Chem. Soc.* **1979**, *101*, 7613; (b) Groves, J. T., Haushalter, R. C., *J. Chem. Soc., Chem. Commun.* **1981**, 1163; (c) Creager, S. E., Murray, R. W., *Inorg. Chem.* **1985**, *24*, 3824.
76. (a) Groves, J. T., Stern, M. K., *J. Am. Chem. Soc.* **1987**, *109*, 3812.
77. Czernuszewicz, R. S., Su, Y. O., Stern, M. K., Macor, K. A., Kim, D., Groves, J. T., Spiro, T. G., *J. Am. Chem. Soc.* **1988**, *110*, 4158.
78. See also Schappacher, M., Weiss, R., *Inorg. Chem.* **1987**, *26*, 1190.
79. Many references cited in footnotes 19-33.

80.   Bortolini, O., Ricci, M., Meunier, B., Friant, P., Ascone, I., Goulon, J.,
       Nouv. J. Chim. **1986**, *10*, 39, and references cited therein.
81.   Groves, J. T., Stern, M. K., presented at the Third Chemical Congress of
       North American, Toronto, June 1988.
82.   (b) Chin, D. H., Balch, A. L., La Mar, G. N., *J. Am. Chem. Soc.* **1980**,
       102, 1446; (c) Chin, D. H., La Mar, G. N., Balch, A. L., *J. Am. Chem.
       Soc.* **1980**, *102*, 4344; (d) Balch, A. L., La Mar, G. N., Latos-Grazynski,
       L., Renner, M. W., Thanabal, V., *J. Am. Chem. Soc.* **1985**, *107*, 3003
       and references cited therein, (e) A. Balch, personal communication.
83.   (a) Groves, J. T., Nemo, T. E., *J. Am. Chem. Soc.* **1983**, *105*, 5786; (b)
       Groves, J. T., Nemo, T. E., *J. Am. Chem. Soc.* **1983**, *105*, 6243; (c) see
       also allylic rearrangement studies: Groves, J. T., Subramanian, D. V., *J.
       Am. Chem. Soc.* **1984**, *106*, 2177.
84.   (a) Groves, J. T., Haushalter, R. C., Nakamura, M., Nemo, T. E., Evans,
       B. J., *J. Am. Chem. Soc.* **1981**, *103*, 2884; (b) Penner-Hahn, J. E., Eble,
       K. S., McMurry, T. J., Renner, M., Balch, A. L., Groves, J. T., Dawson,
       J. R., Hodgson, K. O., *J. Am. Chem. Soc.* **1986**, *108*, 7819; see also (c)
       Gans, P., Buisson, G., Duée, E., Marchon, J.-C., Erler, B. S., Scholz, W.
       F., Reed, C. A., *J. Am. Chem. Soc.* **1986**, *108*, 1223 and references cited
       in each.
85.   Other thermally unstable monomeric Mn^{IV} porphyrins that are not
       porphyrin cation radicals[86] are S = 3/2 systems: (a) Camenzind, M. J.,
       Hollander, F. J., Hill, C. L., *Inorg. Chem.* **1982**, *21*, 4301; (b)
       Camenzind, M. J., Hollander, F. J., Hill, C. L., *Inorg. Chem.* **1983**, 22,
       3776.
86.   Spreer, L., Maliyackel, A. C., Holbrook, S., Otvos, J. W., Calvin, M., *J.
       Am. Chem. Soc.* **1986**, *108*, 1949.
87.   Yamaguchi, K., Takahara, Y., Fueno, T., *Appl. Quantum Chem.* **1986**,
       155.
88.   Goddard, W. A., III, Dunning, T. H., Jr., Hunt, W. J., Hay, P. J., *Acc.
       Chem. Res.* **1973**, *6*, 368.
89.   Rappé, A. K., Goddard, W. A., III., *J. Am. Chem. Soc.* **1982**, *104*, 3287.
90.   Laboratory of C. L. Hill, unpublished work.
91.   (a) Reference 2, Chap. 3, p. 47; (b) Sheldon, R. A., Kochi, J. K., *Adv.
       Catal.* **1976**, *25*, 272.
92.   Williamson, M. M., Prosser-McCartha, C. M., Mukundan, S., Hill, C. L.,
       Inorg. Chem. **1988**, *27*, 1061.
93.   Traylor, T. G., Tsuchiya, S., *Inorg. Chem.* **1987**, *26*, 1338.
94.   (a) Hill, C. L., Bouchard, D. A., Kadkhodayan, M., Williamson, M. M.,
       Schmidt, J. A., Hilinski, E. F., *J. Am. Chem. Soc.* **1988**, *110*, 5471;
       see also (b) Prosser-McCartha, C. M., Kadkhodayan, M., Williamson, M.
       M., Bouchard, D. A., Hill, C. L., *J. Chem. Soc., Chem. Commun.*
       1986, 1747; (c) Williamson, M. M., Bouchard, D. A., Hill, C. L., *Inorg.
       Chem.* **1987**, *26*, 1436.
95.   (a) Renneke, R. F., Hill, C. L., *J. Am. Chem. Soc.* **1986**, *108*, 3528; (b)
       Renneke, R. F., Hill, C. L., *New J. Chem.* **1987**, *11*, 763; (c) Schmidt,
       J. A., Hilinski, E. F., Bouchard, D. A., Hill, C. L., *Chem. Phys. Lett.*
       1987, *138*, 346; (d) Renneke, R. F., Hill, C. L., *J. Am. Chem. Soc.*
       1988, *110*, 5461; (e) Hill, C. L., Renneke, R. F., Combs, L.,
       Tetrahedron, **1988**, *44*, 7499; (f) Renneke, R. F., Hill, C. L., *Angew.
       Chem. Int. Ed. Engl.* **1988**, *27*, 1526; (g) Chambers, R. C., Hill, C. L.,

submitted for publication see also (h) Hill, C. L., Bouchard, D. A., *J. Am. Chem. Soc.* **1985**, *107*, 5148.

96. (a) Kozhevnikov, I. V., Matveev, K. I., *Russ Chem. Rev. (Engl Transl.)* **1982**, *51*, 1075; (b) Kozhevnikov, I. V., Matveev, K. I., *Appl. Catal.* **1983**, *5*, 135, and references cited in each.
97. For example, see Sen, A., *J. Am. Chem. Soc.* **1987**, *109*, 8109.
98. (a) Mares, F., Rocek, J., *Coll. Czech. Chem. Commun.* **1961**, *26*, 2370; (b) Rocek, J., Mares, F., *Coll. Czech. Chem. Commun.* **1959**, *24*, 2741; (c) Rocek, J., *Coll. Czech. Chem. Commun.* **1957**, *22*, 1509.
99. Minato, H., Ware, J. C., Traylor, T. G., *J. Am. Chem. Soc.* **1963**, *85*, 3024.
100. Sharpless, K. B., Teranishi, A. Y., Backvall, J.-E., *J. Am. Chem. Soc.* **1977**, *99*, 3120.
101. Benson, S. W., Nangia, P. S., *Acc. Chem. Res.* **1979**, *12*, 223.
102. Wiberg, K. B., in *Oxidation in Organic Chemistry, Part A* (K. B.Wiberg, ed.). Academic Press, New York, 1965, Chap. II and references cited therein.
103. Hamilton, G. A., in *Molecular Mechanisms of Oxygen Activation* (O. Hayaishi, ed.), Chap. 10. Academic Press, New York, 1974.
104. Gassman, P. G., Yamaguchi, R., *J. Am. Chem. Soc.* **1979**, *101*, 1308.
105. (a) Bertram, J., Fleischmann, M., Pletcher, D., *Tetrahedron Lett.* **1971**, 349; (b) Bertram, J., Coleman, J. P., Fleischmann, M., Pletcher, D., *J. Chem. Soc., Perkin II* **1973**, 374; (c) Pitti, S., Herlem, M., Jordan, J., *Tetrahedron Lett.* **1976**, 3221; (d) Fabre, P.-L., Devynck, J., Tremillon, B., *Tetrahedron* **1982**, *38*, 2697.
106. (a) Fleischmann, M., Pletcher, D., *Tetrahedron Lett.* **1968**, 6255; (b) Clark, D. B., Fleischmann, M., Pletcher, D., *J. Chem. Soc., Perkin II* **1973**, 1578; (c) Siegel, T. M., Miller, L. L., Becker, J. Y., *J. Chem. Soc., Chem. Commun.* **1974**, 341; (d) Edwards, G. J., Jones, S. R., Mellor, J. M., *J. Chem. Soc., Chem. Commun.* **1975**, 816; (e) Edwards, G. J., Jones, S. R., Mellor, J. M., *J. Chem. Soc., Perkin II*, **1977**, 505.
107. Roth, R. D., *Acc. Chem. Res.* **1987**, *20*, 343, and references cited therein.
108. Gassman, P. G., Hay, B. A., *J. Am. Chem. Soc.* **1985**, *107*, 4075.
109. (a) Sager, W. F., Bradley, A., *J. Am. Chem. Soc.* **1956**, *78*, 1187; (b) Sager, W. F., *J. Am. Chem. Soc.* **1956**, *78*, 4970; (c) Wiberg, K. B., Foster, G., *J. Am. Chem. Soc.* **1961**, *83*, 423.
110. Mares, F., Rocek, J., Sicher, J., *Coll. Czech. Chem. Commun.* **1961**, *26*, 2355.
111. Wiberg, K. B., Fox, A. S., *J. Am. Chem. Soc.* **1963**, *85*, 3487.
112. Soek, W. K., Dobson, J. C., Meyer, T. J., *Inorg. Chem.* **1988**, *27*, 5.
113. Good review: Kochi, J. K., in *Free Radicals* (J. K. Kochi, ed.), Vol. I, Chap. 11. Wiley, New York, 1973.
114. Rocek, J., *Tetrahedron Lett.* **1962**, 135.
115. (a) Hill, C. L., Smegal, J. A., *Nouv. J. Chim.* **1982**, *6*, 287; (b) Hill, C. L., Smegal, J. A., Henly, T. J., *J. Org. Chem.* **1983**, *48*, 3277.
116. Hill, C. L., Prosser-McCartha, T. P., unpublished work. See also ref. 95.
117. Brown, S. H., Crabtree, R. H., *J. Chem. Soc., Chem. Commun.* **1987**, 970.
118. Groves, J. T., Quinn, R., *J. Am. Chem. Soc.* **1985**, *107*, 5790.

Chapter IX

SELECTIVE FUNCTIONALIZATION OF SATURATED HYDROCARBONS BY THE "GIF" AND "GIF-ORSAY" SYSTEMS

Derek H. R. Barton and Nubar Ozbalik

Department of Chemistry
Texas A&M University
College Station, Texas 77843

I.   Introduction

The selective functionalization of saturated hydrocarbons under mild conditions became a stimulating research area in organic chemistry since the discovery of monooxygenases.[1,2] The existence of an enzymatic system in nature that could catalyze the monooxygenation of nonactivated carbon-hydrogen bonds has prompted the consideration that their organic counterparts could be developed.

Among the many biological systems, the heme-containing monooxygenase, cytochrome P-450 (Cp-450) has been an inspiration for model systems by Crabtree,[3] Shilov,[4] and Sheldon and Kochi.[5] The enzyme, and most of its model systems, involves an Fe^V oxenoid in a porphyrin structure as the active oxidizing species.[6] They show the usual radical selectivity (tertiary > secondary > primary), epoxidize olefins, and oxidize sulfides to sulfoxides.

An ideal system for the oxidation of alkanes would be capable of oxidizing regio- and chemoselectively under mild conditions. In addition, if the principle of Cp-450 oxidation were to be adapted, it would satisfy the requirements of the general Eq. 1 by use of simple and

$$RH + O_2 \xrightarrow[M^{n+}]{2\,H^+,\,2e^-} ROH + H_2O \qquad (1)$$

readily available reagents. Thus, the oxidation of the carbon-hydrogen bond calls for the presence of a reducing agent (electron source), protons (suitably a carboxylic acid), a catalytically active metal ion, oxygen, and a solvent.

II.   Results and Discussion

Preliminary experiments revealed that pyridine in the presence of iron powder, oxygen, and a carboxylic acid (acetic acid, tartaric acid, citric acid, etc.) possessed an unusual oxidizing power and attacked selectively secondary, not tertiary positions.[7] Originally a catalytic amount of hydrogen sulfide was added as a ligand for iron, but in fact this served only to catalyze the dissolution of the iron powder by a surface effect. Simply raising the temperature to 40°C served to initiate the reaction. Later work[8] has shown that the optimum temperature for the reaction is about 30°C. Above 80°C there is no oxidation. Below -20°C the reaction is very slow. The components of this basic system were inert to each other in the absence of the carboxylic acid and hence it was chosen to be the last reagent to be added to start the reaction. A series of experiments established that each component was essential for a successful reaction. We called the most developed form of the iron powder-pyridine oxidative procedure the Gif III system.

     The spherically symmetrical hydrocarbon adamantane **1**, possessing four tertiary and twelve secondary carbon-hydrogen bonds, was chosen as a probe. The ratio of C^2/C^3, where C^2 is the total of the oxidized products at the secondary position, and C^3 is the total of the tertiary alcohol formed, gives a measure of the selectivity of the reaction. Thus, adamantane, if attacked nonselectively, would have a C^2/C^3 of 12/4 = 3. The latest data for the attack of oxygen-based radicals on adamantane give a C^2/C^3 of about 0.15.[9] For a porphyrin-based system similar results were found.[10] In fact the first results using the Gif III system with adamantane gave C^2/C^3 of 3.7, which was evidence for a nonradical mechanism.[7]

     Although there was significant attack at the tertiary position in adamantane **1**, no *tert*-alcohol was detected in the oxidation of isopentane **2** and of methylcyclohexane **3**. It was, of course, shown

       1   **2**   **3**

that the corresponding tertiary alcohols were stable under reaction conditions. The pattern of hydroxylation of **2** and **3** by the Cp-450 enzyme is also different .[11]

An important achievement was the isolation of a crystalline trinuclear organoiron carboxylate cluster (**4**), $Fe^{II}Fe_2^{III} O(OAc)_6pyr_{3.5}$, from the reaction mixture.[12] This cluster had in fact been described in the literature as being accessible through ligand exchange of the aquo complex.[13] The latter, in turn, could be prepared from $Fe^{II}Cl_2$ and $Ca(OAc)_2$ in aqueous acetic acid. With the isolation of this readily available cluster, the hypothesis that such a species could directly or indirectly function as a catalyst was tested. A series of oxidation reactions of adamantane in pyridine-acetic acid, using zinc as a reducing agent, revealed that an efficient catalytic system (Gif IV system) reaching turnover numbers of over 2000 could be attained[12] (Table I).

4

Table I. Oxidation of Adamantane by the Gif IV System in Different Catalyst Concentrations

Catalyst (mmol x 10^{-3})	Products % Adamantane			Total yield %	Catalyst turnover
	1-ol	2-ol	2-one		
0.8	0.7	3.4	5.6	9.7	263
0.5	1.6	4.6	7.7	13.8	584
0.2	0.7	6.3	4.4	11.4	1232
0.1	0.8	5.8	3.4	9.9	2087

After the discovery of this efficient catalytic system, further experiments that demonstrated the regio- and chemoselectivity of the system were performed. Oxidation of *trans*-1,4-dimethylcyclohexane **5**, a symmetrical hydrocarbon having primary, secondary, and tertiary positions, gave mainly ketone **6**, along with small quantities of aldehyde **7**. None of the other possible oxidation products was detected by GLC.[14]

The catalytic activity of nearly all the metals in the periodic table was examined, but only ruthenium salts, and to a limited extent, were found to be also catalytically active. In particular, salts of cobalt, nickel, and manganese were completely inactive.

The chemoselectivity of the Gif system has always been curious. At the outset, the Gif system was oxidizing hydrocarbons selectively in the presence of H_2S, normally considered to quench radicals very efficiently. It was then found that diphenylsulfide was hardly oxidized during a normal oxidation of cyclohexane. Also there was little difference between the reactivity of cyclohexane and cyclohexene, a fact that is surely without precedent. Furthermore, cyclohexene was oxidized in the allylic position and no epoxidation was seen.

Information regarding the fate of oxidation products in the reaction medium would have mechanistic implications. Thus, incubation of cyclohexanol under the usual oxidation conditions gave only 14% of ketone. This indicated that the main pathway leading to ketones did not involve alcohol as an intermediate, which found further support from the failure of simple alcohols (methanol, isopropanol, *tert*-butanol) to alter the course of the reaction, even when present in considerable excess. Oxidation of adamantanone under the usual conditions gave diketones **8** and **9** in total yield of 14.1%. This was an observation against the idea that further oxidation of the initial products could account for the missing hydrocarbon in the mass balance.

A feature of fundamental importance of the Gif system was the dependence of the selectivity ratio (C^2/C^3) on the partial pressure of oxygen.[15] Using air instead of pure oxygen resulted in a higher C^2/C^3 for the oxidation of adamantane. A clear explanation was acquired through analysis of the basic fraction of the reaction mixture.[16] It was found to be a mixture of isomeric bipyridines and pyridine-hydrocarbon

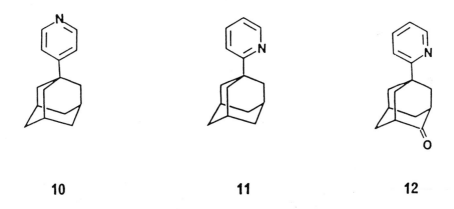

8 X = O , Y = H$_2$

9 X = H$_2$, Y = O

coupled products **10, 11,** and **12** in which pyridine was attached to the tertiary positions of the hydrocarbon. The greater quantity of pyridine-hydrocarbon coupled products found in a high C^2/C^3 reaction (and vice versa) was a clear indication of a competition between oxygen and pyridinium ions (or pyridine itself) for the *tert*-adamantyl radical. Since the ratio of the rate constants for the biomolecular reaction of two different carbon radicals with oxygen is necessarily a constant, the facts demand that if a radical reaction is involved for the tertiary position, there cannot be a similar radical reaction at the secondary position[17] (see further below).

10   **11**   **12**

Pyridine-hydrocarbon coupled products were determined for many other substrates. In every case, only traces of such products (1-2%) were detected, which were mainly *tert*-alkylpyridines. In recent unpublished work, the oxidation of adamantane with limited amounts of oxygen was repeated. When 10% of oxygen in argon is used C^2/C^3 values of more than 100 are observed and the reaction at the tertiary position is almost exclusively coupling to pyridine.

Two other *tert*-carbon-hydrogen bonds show a comparable reactivity to the *tert*-bond in adamantane. They are the tertiary bond in *cis*-decalin and that at the end of the steroidal side chain. In each case where the *tert*-position is oxidized it is a more exposed C-H that is involved. The selectivity of the Gif system (secondary >> tertiary ~ primary) can be explained by a combination of C-H bond strengths (primary > secondary > tertiary) and steric resistance to insertion (tertiary > secondary > primary). The balance normally favors the secondary positions. Insertion into the tertiary position is seen only when the C-H bond is specially exposed.

A number of experiments using perdeuterated cyclohexane[8] revealed that only the carbon center undergoing the oxidation is concerned in the activation process, which is an irreversible process in the sense that the substrate and the oxygenated products (alcohol or ketone) do not undergo hydrogen (deuterium) exchange with the medium. With these experiments, it was also evident that alcohols were not formed from the reduction of ketones.

A kinetic isotope effect of 2.5 was obtained from competitive oxidation of cyclohexane and its perdeutero analog.[16] This value differs from those of Cp-450 and its Fe^V models.[18] However, it is comparable with a number of non-porphyrin-based systems,[19] but not with alkoxy or peroxy radical chemistry.[20]

Among possible candidates, superoxide is highly favored to be the active form of oxygen in the Gif system.[8] This preference was based on the successful replacement of the dioxygen by KO_2 and enhancement of the oxidation yields in the presence of quinones, which increase superoxide concentration by acting as electron transfer reagents.[21] Although H_2O_2 is the expected intermediate to be formed by zinc reduction in a pyridine-acetic acid mixture,[22] it was found to be catalytically destroyed by iron cation in the presence of zinc, without significant oxidation of the hydrocarbon. This experiment has been repeated several times in both the Gif and Gif-Orsay (see later) electrochemical system. The addition of H_2O_2 very slowly to correspond to its possible rate of formation by zinc dust reduction (or electrochemically) does not lead to significant oxidation. A Fenton-like mechanism induced by •OH would be incompatible with the regioselectivity of the Gif system. Furthermore •OH is known to be readily captured by pyridine.[23]

Information concerning the role of pyridine in the Gif system would be of both theoretical and practical importance. In addition to being a good solvent, it could act as a base, a ligand, or a precursor of ligands to iron. Its reduction by zinc could provide efficient means of one electron reduction of oxygen by functioning as an electron transfer agent. One or any combination of these probable roles could operate in the oxidation process. The possibility of being a precursor of a better ligand increased by isolation of a complex of iron ligated to *o*-dipyridyl from the reaction mixture.[8] It was shown to possess the same catalytic activity as the initial iron cluster **4**. A recent experiment of importance has shown that too much *o*-dipyridyl inhibits the oxidation. This suggests reversible binding of *o*-dipyridyl and that reaction cannot occur until at least one or two coordination sites are available. In order to estimate the importance of its basicity, pyridine was replaced by triethylamine, piperidine, and *N,N*-dimethylaniline. In addition dimethyl formamide, acetonitrile, and dimethyl sulfoxide were examined as solvents. In every case, poor yields of oxidized products (about 0.5%) and low selectivity ($C^2/C^3 \sim 1$) were observed. More interestingly, many solvents that bear close structural and chemical resemblance to pyridine were found to be less efficient. The list included 2,3- and 4-methylpyridine, 2,4,6-trimethylpyridine, 2-fluoropyridine, 4-phenylpyridine, quinoline, isoquinoline, and pyridazine. These results emphasized the importance of pyridine in the oxidation process and gave grounds for the extension of this study to the electrochemical version of the Gif system (see later).

In recent publications, the research group of Geletti et al.[24,25] reported that their experimental results confirmed many unusual aspects of the Gif system. Thus, the presence of excess ethanol or acetaldehyde does not inhibit the reaction. Use of pyridine is crucial for good selectivity. They have, however, also been able to oxidize hydrocarbons using copper powder, but not with a copper salt as a catalyst. They propose a new oxygen species capable of selectively oxidizing hydrocarbons at secondary positions. It is the radical cation derived from pyridine-*N*-oxide

$$HO_2^{\bullet} + Py^+H \quad \rightleftharpoons \quad Py^+O^{\bullet} + H_2O \tag{2}$$

1.   Application to Natural Products

Combination of the two unusual features of the Gif system, that is, ready oxidation of secondary positions and preferential oxidation of the unactivated C-H bonds relative to those that exist in oxygenated molecules, could turn out to be a convenient means of modulating regioselectivity in the oxidation of natural products. The first class of compounds chosen for this purpose was steroids.[26] The oxidation of cholestane derivatives by the Gif IV system afforded the industrially important side-chain degradation product, the 20-ketone, as the major isolated product. For example, progesterone **14** can be obtained in one step from cholestenone **13**. The principal mechanism of side-chain degradation has been shown to involve the intermediates of Scheme 1. The terminal *tert*-hydrogen at C_{25} is unhindered and the iron-carbon bond evolves into a carbon radical **15**. Reaction with superoxide (or oxygen) and Fenton type cleavage of the oxygen-oxygen bond affords alkoxy radical **16**. The well known intramolecular 1-5 hydrogen shift then affords carbon radical **17**. Further radical chemistry leads to the carbon radical **18**, which oxidizes further to ketone, and a C-6 fragment **19**. The latter was isolated and characterized.

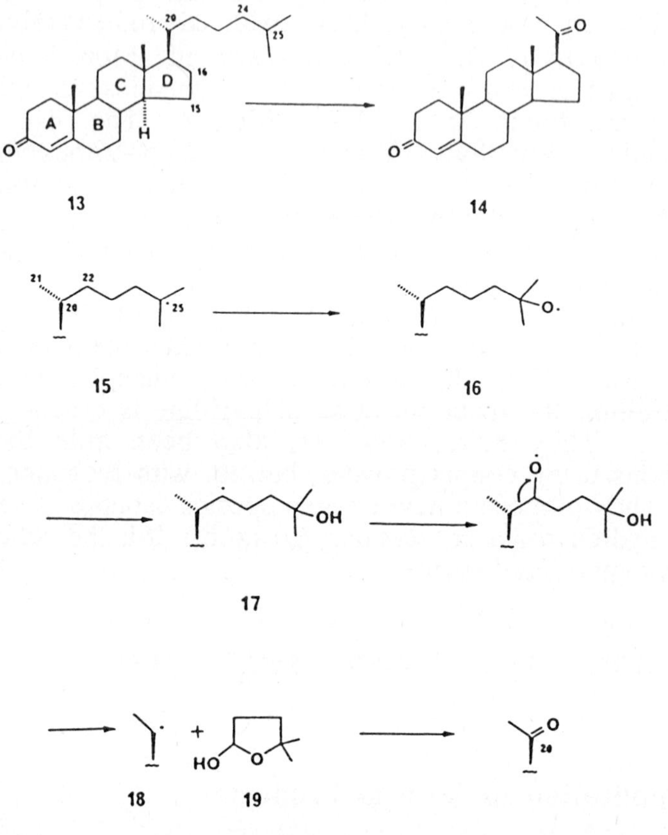

Scheme 1

Gif oxidation of a suitable 25-methylcholesterol derivative would provide clear-cut evidence for the existence of other pathways leading to 20-ketone. If only the proposed mechanism operates, lack of 25-hydrogen necessary for side-chain fragmentation should result in the formation of only the oxo derivatives of the parent compound but none of the 20-ketone. In order to test this idea, 25-methylcholesta-1,4-dien-3-one **28** was synthesized[27] starting from the acetate of lithocholic acid **20** following the steps shown in Scheme 2. The reaction of the acid chloride of **20** with the sodium salt of N-hydroxypyridine-2-thione gave the heterocyclic ester **21**, which upon

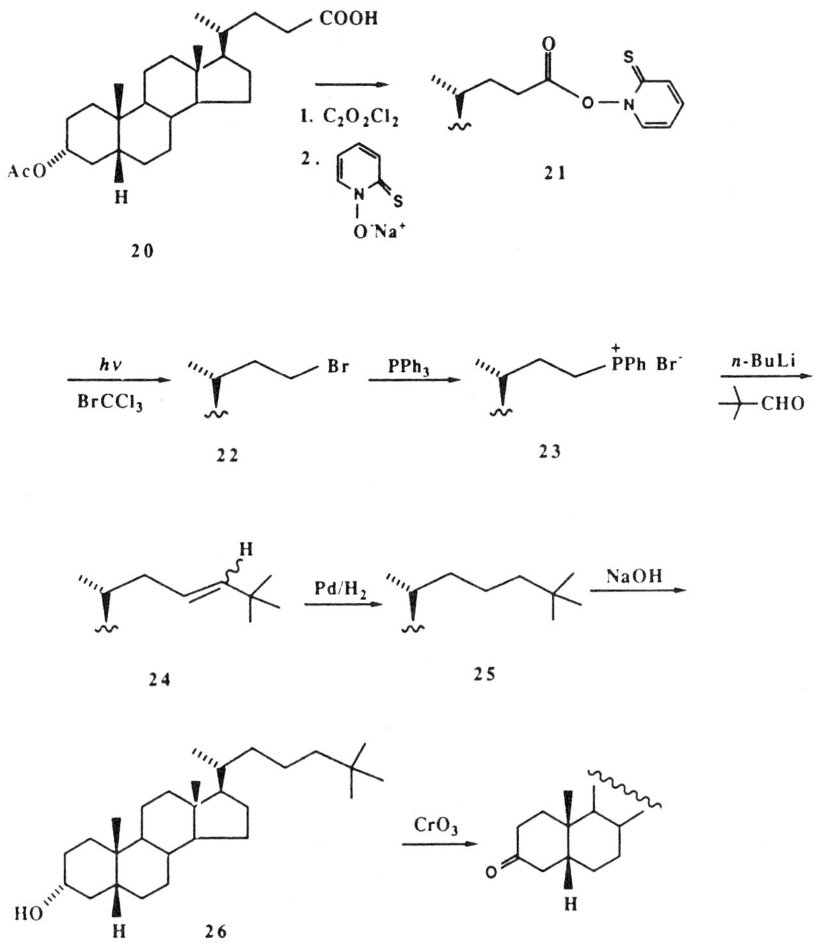

Scheme 2

photolysis in bromotrichloromethane underwent radical decarboxylation to give bromonorcholane derivative **22** in 71% yield. The phosphonium salt **23** derived from **22** was reacted successively with butyl-lithium and dimethylpropanal to afford the olefin **24** as a mixture of isomers. Subsequent hydrogenation of the double bond was followed by ester hydrolysis and Jones' oxidation to give the ketone **27** in 66% overall yield. Finally, dehydrogenation of the ketone to the desired dienone was achieved with benzeneseleninic anhydride.

Oxidation of the dienone by the Gif IV system gave minor amounts of 20-ketone representing 25% of the quantity obtained in oxidation of cholestadien-3-one (**29**). The only other oxidation product isolated from this reaction mixture was the 15-ketone (**30**). Apparently, a second, minor route to the 20-ketone exists.

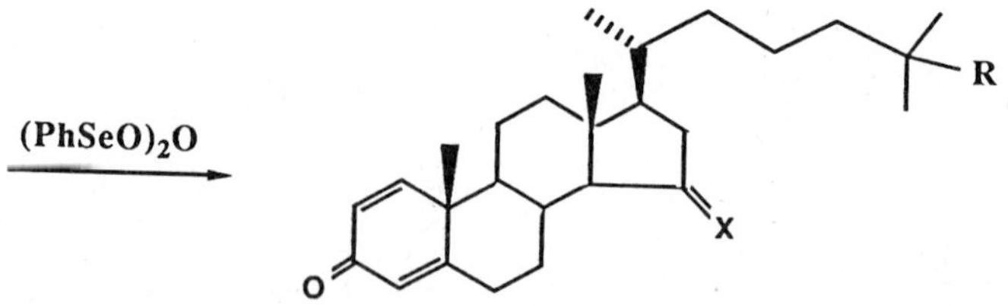

$(PhSeO)_2O$

28	R = CH$_3$, X = H$_2$
29	R = H, X = H$_2$
30	R = CH$_3$, X = O

Independent evidence supporting this conclusion could be obtained from the oxidation of a steroid with a shortened side chain. The dienone **33** was an appropriate substrate since it was readily accessible from the acetate of lithocholic acid **20** by reductive

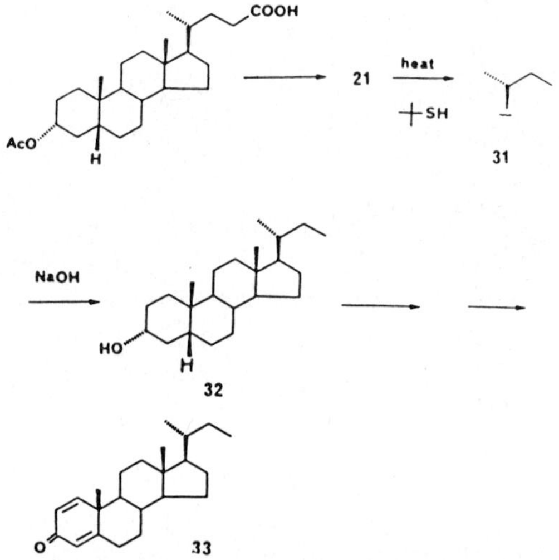

Scheme 3

decarboxylation of the ester **21**. The 3α-acetoxynorcholane **31** thus produced was converted to the target compound following the steps of acetate hydrolysis, Jones' oxidation, and dehydrogenation by benzeneseleninic anhydride as before. Gif oxidation of **33** again yielded the 20-ketone **34** in poor yield. A mechanism for this minor pathway was proposed, which was initiated by hydrogen abstraction from the tertiary 20-position.

Oxidation of steroidal methylene groups to ketones, particularly at positions 15, 16, and 24, is another feature of steroid oxidation.

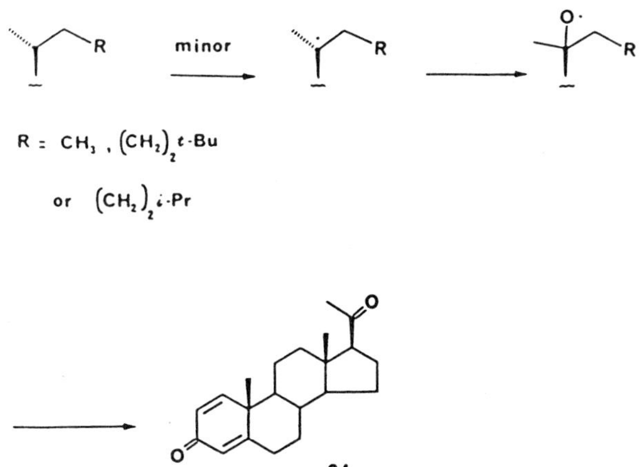

R = CH₃ , (CH₂)₂t-Bu

or (CH₂)₂i-Pr

34

Scheme 4

Comparison of the quantities of oxidation products obtained from a number of suitably functionalized cholestane derivatives[28] showed that hydroxyl or carbonyl groups in ring A deactivated the ring toward oxidation. This effect was transmitted to rings B and C through double bonds situated in conjugation with the carbonyl group. The presence of an enone system in ring B deactivated all the steroidal rings, rendering the side chain susceptible to oxidation. It should be pointed out, however, that long-range conformational transmission effects could contribute to this selectivity.

Oxidation of two well-known examples of sesquiterpenoids was studied for further evaluation of the regioselectivity of the Gif system. The expected oxidation products of caryolan-1-ol **35** and patchouli alcohol **36** are not known and are difficult to obtain by total or partial synthesis. Furthermore, the oxidation products of the latter bear additional importance as a result of their interesting organoleptic properties.

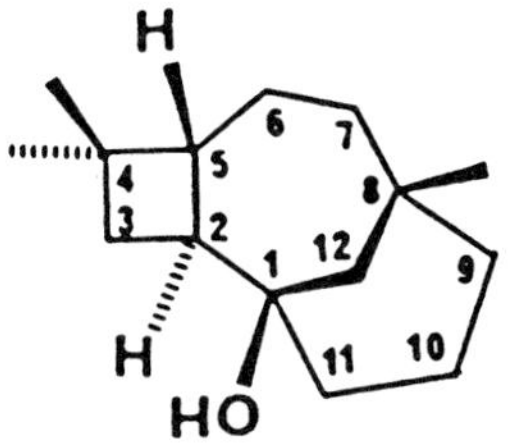

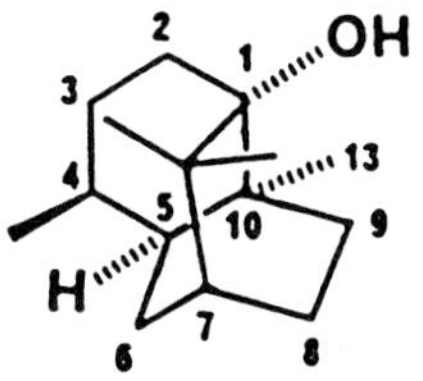

35

36

Oxidation of caryolan-1-ol **35** by the Gif IV system gave 6-(2.5%) and 10-ketone (1.6%) as major products.[29] Its trifluoroacetate derivative was found to be oxidized mainly at C-6 (2.8%) and C-7 (0.9%). The structures of these compounds were established on the basis of their spectral data, particularly by a detailed study of mass and NMR spectra. In the case of patchouli alcohol **36**, products from oxidation of methylenes at positions 3 (1.2%), 8 (3.6%), and 9 (0.9%) were identified.[36] In the latter oxidation GC-mass spectral analysis confirmed that the three ketones were the major oxidation products.

These results again show the high selectivity previously noted for steroidal substrates. They also show once more that oxygen-containing groups deactivate the proximal methylenes towards oxidation.

2.   The Gif-Orsay System

Electroanalytical techniques (cyclic voltammetry, rotating ring-disc voltammetry, etc.) would be very informative regarding the mechanism of the Gif system, which could be described as a complex redox system. However, for dependable conclusions, it was necessary to confirm that the type of chemistry that occurred in zinc or iron powder reductions could be reproduced electrochemically. Furthermore, an efficient electrochemical version of the Gif system would be a positive contribution to its industrial applicability.

Preliminary experiments[30] revealed that a pool of mercury electrode at -0.6 to -0.7 V (versus SCE) could reduce dioxygen (without direct reduction of pyridinium ions), and induce the oxygenation of adamantane with yields close to that of the Gif system. The catholyte at this stage was an oxygen-saturated mixture of pyridine and acetic acid containing adamantane **1**, the iron catalyst **4**, and a tetraethylammonium salt as conducting electrolyte. The extent of protonation of pyridine by trifluoroacetic acid was advantageously used to remove the need for tetraethylammonium salts, which were found to retard oxidation in the chemical system. With this modification, interesting features of the chemical system can be reproduced by the electrochemical system as evidenced by the extent of oxidation of adamantane **1**, cyclododecane **37**, and *trans*-decalin **38** using the

37   **38**   **39**

amount of electricity (3000 C) that roughly corresponds to the quantity of zinc used in the chemical system (Table II).

Table II. Oxidation of Some Hydrocarbons by the Gif-Orsay System

Substrate	Total oxidation (%)	C^2/C^3	Pyridine-hydrocarbon coupled products	Coulombic yield (%)
1	18	8.5	10.5	4.2
3 7	21.1	---	1.0	5.2
3 8	22.2	36	1.4	5.4

In addition to the problems associated with the presence of a ceramic wall between the compartments (diffusion, resistance), the increase in the basicity of the catholyte with electrolysis required continuous addition of trifluoroacetic acid. Moreover, the coulombic yield of the reaction decreases rather sharply at the early stages (1000-1500 C) of the electrolysis. A simple solution for all of these problems was the adaption of an unicellular system, which led us to what we call the Gif-Orsay system.[31] The facts that dioxygen is the most easily reducible species in the system and that its reduction in acidic medium is sufficient for an efficient oxidation allowed us to replace constant potential source (potentiostat) with a simple source of constant current.

A useful observation was the dependence of the coulombic yield on the quantity of substrate. Coulombic yields of 45% could be obtained in the oxidation of cyclohexane (48 mmol) for the first 2000 coulombs. It decreased only to 36% after 8700 C. Further improvements in the coulombic yields could be achieved by the use of electron transfer reagents, such as paraquat or 4,4'-dipyridyl (protonated by trifluoroacetic acid). In the presence of these reagents, electronic yields as high as 49% were obtained after 9400 C.

III. Comparison between Gif and Gif-Orsay Systems

The electron efficiency of zinc in oxidizing the carbon-hydrogen bond, calculated according to the Cp-450 equation, was less than 10% for a 2.0 mmol scale reaction. The positive effect of increased hydrocarbon concentration on the coulombic yield in electrochemical oxidation suggested that the efficiency of zinc as source of electrons in the Gif system could be improved simply by using more hydrocarbon. Thus, a set of hydrocarbons, namely, adamantane, cyclohexane, methylcyclopentane, 3-ethylpentane, and *cis*- and *trans*-decalin, was oxidized by the Gif IV and Gif-Orsay systems at concentrations close to their saturated solutions in pyridine.[32] The following conclusions could be drawn from these experiments:

    1. In the electrochemical system the coulombic yield was higher than that of the chemical system. As opposed to a sharp decrease in the electronic yield toward the end of chemical oxidation, the electrochemical system maintained its efficiency throughout the electrolysis. Better conversions could be achieved simply by extending the duration of electrolysis.

    2. In both systems, ketones were always the major products. Regioselectivities evaluated by the ketone/alcohol and C^2/C^3 ratios were quite similar.

    3. In agreement with the previous results, adamantane gave the highest quantity of pyridine-hydrocarbon coupled products in both oxidation systems.

Cyclohexane, *cis*-**39**, and *trans*-**38** decalins afforded relatively small amounts of this type of compound; they were undetectable in oxidation of methylcyclopentane and 3-ethylpentane.

    As was pointed out, replacement of pyridine in the Gif IV system by other solvents including those closely related to it resulted in a marked decrease both in the yield of oxidation and in the regioselectivity for methylene activation. With the discovery of the highly efficient electrochemical system, the problem of finding a substitute for pyridine became an economic issue. Because of their low cost, and suitable polarity for the dissolution of both hydrocarbons and iron complexes, low-molecular-weight alcohols and ketones seemed to be a good substitute. In addition, these solvents had only slight effect on the Gif system. At first, half of the amount of pyridine was replaced by either methanol, ethanol, isopropanol, or acetone. With the exception of acetone, all of the cosolvents gave rise to an important decrease in the coulombic yield and selectivity. On the other hand, acetone as cosolvent slightly increased the yield and maintained the coulombic yield; however, it gave lower selectivity than pyridine. Complete replacement of pyridine by acetone in the electrochemical system had a profound effect on the distribution of the oxidation products. Ketones were no longer the major products, even though the coulombic yields were not much different.

    Recently, the oxidation of pure *trans*-**38** and *cis*-decalin **39** has been studied in more detail both electrochemically and chemically.[32] The results for the electrochemical system are shown in Table III. *trans*-Decalin shows little oxidation of the *tert*-position and mostly ketones are formed. *cis*-Decalin behaves differently. C^2/C^3 is below the statistical number of 8. There is a major amount of *tert*-alcohol – more than seen in any other hydrocarbons studied. The coulombic yield is satisfactory for both hydrocarbons. In both cases, the ratio *cis* to *trans tert*-alcohol was nearly the same, in agreement with the evolution of the iron-carbon bond (see later) into a radical. The corresponding results for the chemical oxidation of *trans*- and *cis*-decalin are given in Table IV and show the same pattern as the electrochemical results. In particular, the more easier oxidation of *cis*-decalin is confirmed as well as the same ratio of tertiary alcohols whether one starts with the *trans* or the *cis* hydrocarbon.

Table III. Oxidation of *trans*- and *cis*-Decalin by the Gif-Orsay System

38 X = H
40 X = OH

39 X = H
41 X = OH

42 X = O, Y = H₂
43 X = H₂, Y = O

44 X = O, Y = H₂
45 X = H₂, Y = O

trans-Decalin (38)
Oxidation products (mmol)

Q (coulombs)	40	41	42	43	Total	Coulombic Yield (%)	C^2/C^3
1000	0.04	-	0.407	0.357	0.764	29	21.8
2000	0.05	-	0.863	0.828	1.74	33	34.5
3000	0.085	0.053	1.467	1.722	3.265	42	23.1

cis-Decalin (39)
Oxidation product (mmol)

Q (coulombs)	40	41	42	44	45	Total	Coulombic Yield (%)	C^2/C^3
1000	0.28	0.20	0.33	0.24	0.56	1.62	53.1	2.35
2000	0.37	0.22	0.69	0.49	1.08	2.85	49.2	3.85
3000	0.38	0.24	0.96	0.78	1.50	3.86	45.7	5.22

Table IV. Oxidation of *trans*- and *cis*-Decalin by the Gif System

Substrate	Oxidation products (mmol)							f[a]	C^2/C^3
	40	41	42	43	44	45	Total	(%)	
38	0.055	0.042	0.775	0.808			1.68	16.4	16.3
39	0.54	0.35	1.29		0.32	1.39	3.89	34.6	3.37

[a] f = electronic yield; 1.31 g of zinc used corresponds to 3865 C.

The autoxidation of *trans-* and *cis-*decalin by heating at 100°C gave the two known *tert-*hydroperoxides that were reduced to the alcohols by trimethyl phosphite. The *trans* to *cis* ratio from *trans-*decalin was 1.43, from *cis-*decalin 1.89. *cis-*Decalin gave 3.54 mmol of *tert-*alcohols whereas *trans-*decalin gave only 0.309. Thus, in autoxidation *cis-*decalin is oxidized 10 times as fast as *trans-*decalin, but there is little secondary oxidation.

IV.   Conclusions

The Gif system is different from the habitual Cp-450 models in that it gives ketones, not alcohols, attacks preferentially secondary positions, and does not epoxidize olefins or sulfides. A working hypothesis was proposed[16] (Scheme 5) to take account of these facts.

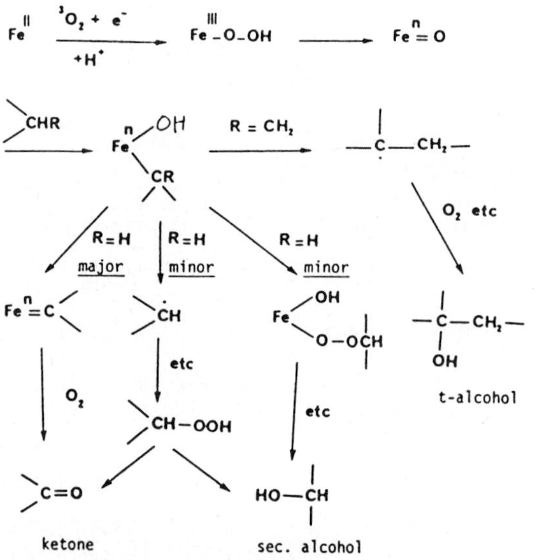

Scheme 5

The formation of ketone was explained by the postulate of an iron-carbene bond. For all positions, the first step was formulated as the formation of an iron-carbon σ-bond. For tertiary positions, this bond can fragment to give radicals. There is a good precedent for this.[33] For secondary positions there is little fragmentation to radicals and the initial bond evolves into a carbene. In this scheme, the valency of the iron is not specified, but is it surely at a different valency than the Fe^V oxenoid postulated in porphyrin models showing typical radical behavior (Crabtree,[3] Shilov,[4] Sheldon and Kochi[5]). Indeed, the characteristic of the Gif system is the flux of electrons that must reduce the iron to at least Fe^{II}.

The variation in the apparent selectivity (C^2/C^3) in the Gif oxidation of adamantane is a key experimental fact in arguments about mechanism. At the tertiary position, the competition between oxygen and pyridine has been confirmed by many recent experiments (Barton, Boivin, and Le Coupanec, unpublished work). For the two reactions (e.g., 3 and 4) the *ratio* of tertiary to secondary oxidation products can

$$R^{\bullet t} + O_2 \xrightarrow{k_t} R^t\text{-O-O}^{\bullet} \qquad (3)$$

$$R^{\bullet s} + O_2 \xrightarrow{k_s} R^s\text{-O-O}^{\bullet} \qquad (4)$$

change with oxygen pressure only if the mechanism of oxidation at the tertiary and secondary positions is *different*. The ratio of k^t and k^s is, by definition, a constant. This is well accounted for in Scheme 5.

Recently, important new evidence has been secured for the postulated iron-carbon bond.[34]

Oxidation of adamantane (2 mmol) under Gif IV conditions in the presence of diphenyldiselenide (2 mmol) provoked a decrease in ketone **46** formation (for example, from 17.2 to 3.5%) with the appearance of major amounts of 2-phenylselenoadamantane **47** (11.5%). A lesser amount of l-phenylselenoadamantane **48** (3.0%) was present, as well as some 1-ol **49** (0.7%) and some 2-ol **50** (2.0%). This was a very interesting result, because many other traps had been used before, but when trapping was seen, it was the tertiary position that was substituted by a radical mechanism. Cyclohexane **51** reacted under the same Gif IV conditions to give phenylselenocyclohexane **52**. In fact, using a large excess of substrate and 1 mmol of diphenyldiselenide we were able to isolate 1.45 mmol of phenylselenocyclohexane **52**, which means that 73% of the initial diphenyldiselenide is incorporated into the hydrocarbon. At the same time 1.0 mmol of cyclohexanol and 2.3 mmol of cyclohexanone were formed.

We then applied Gif III conditions again and the results were again spectacular (Table V). With 1 mmol of adamantane and 2 mmol of diphenyldiselenide, only phenylselenated adamantanes were formed and the unchanged hydrocarbon was recovered quantitatively. The yields were **48** (16.9%) and **47** (18.5%) with recovered adamantane 63%, a total of 98.4%. The selectivity for the reaction (C^2/C^3) was 1.09. This

46	X = O
47	X = H, SePh
50	X = H, OH

| 48 | X = SePh |
| 49 | X = OH |

| 51 | X = H₂ |
| 52 | X = H, SePh |

value is close to that found (1.15) for the total of oxidized and coupled products in the Gif IV system (see above). It is clear that the

Table V. Reaction of Diphenyldiselenide with Adamantane **1** and Cyclohexane **51** under Gif III (Iron Powder) Conditions

Entry	Substrate (mmol)	(PhSe)$_2$ (mmol)	Product yield[a,b] (%)	Total (%)	C^2/C^3	Starting Material Recovered (%)
1	**1** (2)	–	**49** (3.6), **50** (<0.1), **46**, (6.85)	10.45	1.9	
2	**1** (2)	1	**48** (14.8), **47** (17.25)	32.05	1.17	56
3	**1** (2)	2	**48** (16.85), **47** (18.45)	35.35	1.09	63
4	**5 1** (2)	1	**52** (15.05)	15.05		
5	**5 1** (2)	2	**52** (18.15)	18.15		
6[c]	**1** (2)	1	**50** (1.3), **48** (13.0), **47** (15.05)	29.35	1.26	64

[a]Based on substrate.
[b]Yields of **49**, **50**, and **46** were determined by g.l.c. on an aliquot. Yields of **47** and **48** were determined by GLC after removal of excess (PhSe$_2$) by NaBH$_4$ reduction followed by alkaline work-up.
[c]Pyridine-acetone mixture (l/l, v/v, 28 ml) was used as a solvent.

diphenyldiselenide is capturing an intermediate in the Gif system. It is true that diphenyldiselenide is an excellent trap for carbon radicals[35] but we have unimpeachable evidence (see above) that there is no radical involved in reactivity at the 2-position. We consider that the diphenyldiselenide is capturing an iron-carbon bond at the 2-position. The same conclusion for the Gif III system applies to the 1-position, though the Gif IV system at this carbon may involve a mixture of iron-carbon bond capture and radical trapping.

In any case, the functionalization of saturated hydrocarbons by a reagent such as diphenyldiselenide is a truly remarkable reaction. It is hard to conceive of any oxenoid or oxygen radicaloid species that would attack a saturated hydrocarbon faster than it would oxidize diphenyldiselenide because the latter is very easily oxidised. We are forced, therefore, *not* to give further consideration to Shilov's proposed pyridine-*N*-oxide radical cation theory (see above).

The work on the Gif III and Gif IV systems has been accompanied by innumerable blank experiments, some of which have already been mentioned. For the diphenyldiselenide reaction, we showed that without the iron catalyst there was no reaction. More interestingly, there was no reaction in either the Gif III or Gif IV systems without *oxygen*. A number of reactions carried out with the Gif IV system on various natural products[36] give compounds that require that an iron-hydrogen bond be added onto an ethylenic linkage. The most striking example is the oxidation of 1,9-dideoxyforskolin **53**. The major isolated product was the alcohol **54**, which is the formal non-Markovnikov hydration product of the ethylenic linkage. We consider that this alcohol is formed by addition of an iron-hydrogen bond to the ethylenic linkage followed by oxidation of the iron-carbon bond, probably to aldehyde. Aldehydes are reduced to alcohols under Gif IV conditions. This observation does not, of course, prove that selective oxidation reactions of saturated hydrocarbons under Gif conditions involve the chemistry of the iron-hydrogen bond.

53   **54**

It is now clear that the Gif system involves an iron species that is very different from the conventional iron oxenoids studied so far. What the species really is remains a fascinating problem.

References

1.   Hayashi, O., Katagiri, O., and Rothberg, S., *J. Am. Chem. Soc.* **1955**, 77, 5450.
2.   Mason, H. S., Fowlks, W. L., and Peterson, E., *J. Am. Chem. Soc.* **1955**, 77, 2914.
3.   Crabtree, R. H., *Chem. Rev.* **1985**, *85*, 245.
4.   Shilov, A. E., *Activation of Saturated Hydrocarbons by Transition Metal Complexes.* D. Riedel, Dordrecht, 1984.

5.   Sheldon, R. A., Kochi, J. K., *Metal-Catalyzed Oxidations of Organic Compounds*. Academic Press, New York, 1981.
6.   McMurry, T. J., Groves, J. T., in *Cytochrome P450: Structure, Mechanism, and Biochemistry* (P. Ortiz de Mantellano, ed.). Plenum Press, New York, 1985.
7.   Barton, D. H. R., Gastiger, M. J., Motherwell, W. B., *J. Chem. Soc., Chem. Commun.* **1983**, 41.
8.   Barton, D. H. R., Boivin, J., Motherwell, W. B., Ozbalik, N., Schwartzentruber, K. M., Jankowski, K., *Nouv. J. Chim.* **1986**, *10*, 387.
9.   Fossey, J., Lefort, D., Massoudi, M., Nedelec, J. -Y., Sorba, J., *Can. J., Chem.* **1985**, *63*, 678.
10.   Groves, J. T., Nemo, T. E., *J. Am. Chem. Soc.* **1983**, *105*, 6243.
11.   Ullrich, V., *Angew. Chem. Intl. Ed.* **1972**, *11*, 701.
12.   Barton, D. H. R., Gastiger, M. J., Motherwell, W. B., *J. Chem. Soc., Chem. Commun.* **1983**, 731.
13.   Dziobkowski, C. T., Wrobleski, J. T., Brown, D. B., *Inorg. Chem.* **1981**, *20*, 679.
14.   Barton, D. H. R., Boivin, J., Gastiger, M. J., Morzycki, J., Hay-Motherwell, R. S., Motherwell, W. B., Ozbalik, N., Schwartzentruber, K. M., *J. Chem. Soc. Perkin I* **1986**, 947.
15.   Barton, D. H. R., Boivin, J., Ozbalik, N., Schwartzentruber, K. M., *Tetrahedron Lett.* **1984**, *25*, 4219.
16.   Barton, D. H. R., Boivin, J., Ozbalik, N., Schwartzentruber, K. M., *Tetrahedron Lett.* **1985**, *26*, 447.
17.   Minisci, F., Citterio, A., Valeria, F., *J. Org. Chem.* **1980**, *45*, 4752.
18.   White, R. E., Coon. M. J., *Annu. Rev. Biochem.* **1980**, *19*, 315.
19.   Lindsay Smith, J. R., Piggott, R. E., Sleath, P. R., *J. Chem. Soc., Chem. Commun.* **1982**, 55 and references therein.
20.   Russell, G. A., in *Free Radicals* (J. K., Kochi, ed.), p. 312. John Wiley, New York, 1973.
21.   Sawyer, D. T., Gibian, M. J., *Tetrahedron Lett.* **1979**, *35*, 1471.
22.   Cofré, P., Sawyer, D. T., *Anal. Chem.* **1986**, *58*, 1057.
23.   Roberts, J. L., Jr., Morrison, M. M., Sawyer, D. T., *J. Am. Chem. Soc.* **1978**, *100*, 329.
24.   Geletii, Yu. V., Lyubimova, G. V., Shilov, A. E., *Kinet. Katal.* **1985**, *26*, 1019; *C.A.* **1986**, *105*, 60339.
25.   Geletii, Yu. V., Lavreshko, V. V., Shilov, A. E., *Dokl. Akad. Nauk. SSSR* **1986**, *288*, 139.
26.   Barton, D. H. R., Göktürk, A. K., Morzycki, J. W., Motherwell, W. B., *J. Chem. Soc. Perkin I* **1985b**, 583.
27.   Barton, D. H. R., Boivin, J., Crich, D., Hill, C. H., *J. Chem. Soc. Perkin I* **1986**, 1805.
28.   Barton, D. H. R., Boivin, J., Hill, C. H., *J. Chem. Soc. Perkin I* **1986**, 1797.
29.   Barton, D. H. R., Beloeil, J. -C., Billion, A., Boivin, J., Lallemand, J. -Y., Lelandais, P., Mergui, S., Morellet, N., *Nouv. J. Chim.* **1986**, *10*, 439.
30.   Balavoine, G., Barton, D. H. R., Boivin, J., Gref, A., Ozbalik, N., Riviére, H., *Tetrahedron Lett.* **1986**, *27*, 2849.
31.   Balavoine, G., Barton, D. H. R., Boivin, J., Gref, A., Ozbalik, N, Riviére, H., *J. Chem. Soc., Chem. Commun.* **1986**, 1727.
32.   Balavoine, G., Barton, D. H. R., Boivin, J, Gref, A., Le Coupanec, P., Ozbalik, N., Pastena, J. A. X., Riviere, H., *Tetrahedron* **1987**, *44*, 1091.

33.   Bower, B. K, Tennent, H. G., *J. Am. Chem. Soc.* **1972**, *94*, 2512.
34.   Barton, D. H. R., Boivin, J., Le Coupanec, P., *J. Chem. Soc., Chem. Commun.* **1987**, 1379.
35.   Barton, D. H. R., Bridon, D., Zard, S. Z., *Heterocycles* **1987**, *25*, 449.
36.   Barton, D. H. R., Beloeil, J. -C., Billion, A., Boivin, J., Lallemand, J. -Y., Mergui, S., *Helv. Chim. Acta* **1987**, *70*, 273.

Chapter X

ALKANE OXIDATION STUDIES IN DU PONT'S CENTRAL RESEARCH DEPARTMENT

C. A. Tolman, J. D. Druliner, M. J. Nappa, and N. Herron

Central Research and Development Department
E. I. du Pont de Nemours & Company
Experimental Station
Wilmington, Delaware 19880-0328

I. Background

This chapter describes studies that have taken place in Central Research over the past 10 years aimed at understanding partial alkane oxidation and how it might be controlled to give selectivity to desired products. It begins with a discussion of the high-temperature radical chain autoxidation of alkanes with O_2, related to our work on the industrial oxidation of cyclohexane, describes our studies of oxidations by metal porphyrin and phthalocyanine complexes with iodosobenzene as the oxidant, and ends with metal zeolite catalysts that are capable in some cases of selectively hydroxylating linear alkanes at the terminal methyl groups, using a mixture of oxygen and hydrogen at ambient temperature. The latter catalysts are completely inorganic mimics of ω-hydroxylase enzymes of the cytochrome P-450 type.

The oxidation of hydrocarbons is a highly exothermic process, as indicated by the heats of combustion in Table I. This high exothermicity, about 105 kcal/mol for each mole of O_2 consumed, accounts for the use of hydrocarbon combustion to power automobiles,

Table I. Heats of Combustion

Substance	$-\Delta H$ (kcal/mol)^a	$nO_2{}^b$	$\Delta H/n$
Methane	211	2	105
Propane	526	5	105
Cyclohexane	937	9	104
Benzene	782	7.5	104
Carbon(charcoal)	97	1	97
CO	68	0.5	136
H_2	68	0.5	136

^{a}For complete combustion to CO_2 and H_2O (liq.), taken from the *Handbook of Chemistry and Physics*, 41st Ed., pp. 1878-1932. Chemical Rubber Publishing Co., Cleveland, 1959-60.

^{b}The number of moles of O_2 required for complete combustion to CO_2 and H_2O.

airplanes, and ships, as well as to generate much of our electricity. Carbon (coal) is somewhat less exothermic in its combustion, whereas CO and H_2 are somewhat more. (H_2 and O_2 are used in the space shuttle rockets.) The 105 kcal/mol of O_2 (or 52 kcal/g-atom of O) rule of thumb for hydrocarbon combustion works remarkably well, even for unsaturated hydrocarbons such as benzene. Partial oxidations show more variation, as illustrated in Table II, which shows various

Table II. Exothermicity of Partial Oxidation Reactions

Reaction	$-\Delta H$ (kcal/mol)^a
$C_2H_6 + O_2 \longrightarrow C_2H_5O_2H$	19^b
$C_2H_6 + 1/2\,O_2 \longrightarrow C_2H_5OH$	40
$C_2H_5OH + 1/2\,O_2 \longrightarrow CH_3CHO + H_2O$	49
$CH_3CHO + 1/2\,O_2 \longrightarrow CH_3CO_2H$	70

^{a}Calculated from differences in heats of combustion from footnote a in Table I, unless noted otherwise.
^{b}Calculated from group additivity values in Benson, S. W., *Thermochemical Kinetics*, 2nd Ed., pp. 272-275. John Wiley, New York, 1976.

intermediates on the way from ethane to acetic acid. The average deviation from the mean of 53 kcal/g-atom of O for the three nonperoxidic compounds is 11 kcal. Ethyl hydroperoxide formation, however, is exothermic by only 19 kcal/mol of O_2, so that its further conversion to acetaldehyde releases 70 kcal/mol. This stored energy explains why peroxides, particularly, those of low molecular weight, can be quite hazardous.

In spite of the highly favorable thermodynamics for oxidation, flammable materials can exist in our atmosphere in a sea of oxygen because of the normally unfavorable kinetics of oxidation reactions at low temperatures. Table III shows typical bond energies for homolysis

Table III. Typical Bond Energies[a]

$$A-B \longrightarrow A\bullet + B\bullet$$

O-H	C-H	O-O	D (kcal/mol)
HO-H			119
	C_6H_5-H		104
$CH_3CH_2CH_2\overset{\text{O}}{\overset{\|}{C}}$O-H			103
CH_3CH_2O-H; CyO-H[b]			102
	C_2H_5-H		98
	Cy-H		94
	$(CH_3)_3$C-H		91
HO_2-H; CyOO-H[b]			90
	$CH_3\overset{\text{O}}{\overset{\|}{C}}$-H; (cyclohexanone)[b] ; (cyclohexanol)[b]		88
	$C_6H_5CH_2$-H		85
•OO-H[c]			47
		CH_3O-OH	43
		CH_3O-OCH_3	36

[a]Taken from Kerr, J. A., *Chem. Rev.* **1956**, *66*, 465, except as noted otherwise.
[b]Estimated by analogy with compounds of similar structure.
[c]Calculated from $\Delta H°_f$ for H• and HO_2• from Benson, S. W., *Thermochemical Kinetics*, pp. 289-292. John Wiley, New York, 1976.

of various bonds. It is the weakness of the •OO-H bond that provides the key to the normal kinetic stability of hydrocarbons. Direct attack of O_2 on cyclohexane (CyH) in Reaction 1, for example, is endothermic by 47 kcal/mol.

$$O_2 + Cy\text{-}H \longrightarrow \bullet\,O_2\text{-}H + Cy\bullet \qquad (1)$$

From the Arrhenius equation (2), and a typical second order A

$$k = A \exp(-E_a/RT) \qquad (2)$$

factor of 10^{+9} $M^{-1}sec^{-1}$,[1] assuming that the activation energy is no larger than the endothermicity, we can calculate the rate constant at 25°C to be 3×10^{-26} $M^{-1} sec^{-1}$. From the Henry's Law constant,[2] the concentration of O_2 in liquid cyclohexane saturated with air is about 0.004 M. Thus, the pseudo-first-order rate constant for cyclohexane oxidation at 25°C would be about 10^{-28} sec^{-1}, corresponding to a half-life for the oxidation of about 2×10^{20} years! Higher temperature helps; at 300°C the half-life is down to 2 billion years. (Some argue that there is also a spin conservation selection rule against direct reactions of triplet O_2 with singlet hydrocarbons. D. A. Dixon, a theoretical chemist in our department, assures us, however, that this is not a problem.)

Rather than rely on direct oxygen attack, partial oxidation reactions of the type used in the commercial oxidation of cyclohexane rely on the cleavage of the much weaker O-O bonds in peroxides. Once formed, the oxyradicals can form O-H bonds sufficiently strong that they can exothermically abstract hydrogen atoms from alkanes (but not aromatic rings), as indicated in Table III. Peroxyradicals can also attack alkanes, although the reactions are mildly endothermic and have considerably higher activation energies than abstractions by oxyradicals. (Typical values of E_a for attack on cyclohexane are 18 kcal/mol for *tert*-BuO$_2$•[3] and 3 for CyO•[4].) The increasing C-H bond strengths in the series 3° < 2° < 1° account for the generally observed preference for oxidation of 3° C-H bonds over 2°, and of 2° over 1°. The degree of selectivity depends on the strength of the O-H bond that can be formed by removal of the hydrogen, as shown by the first three rows in Table IV. HO•, which can form the strongest O-H bond, is the least selective. Cytochrome P-450 is believed to involve a reactive ferryl (FeO) species with oxyradical character.[5] Table IV suggests that at least for short alkanes (2-methylpentane was used for the P-450 results reported),

Table IV. Relative Reactivities for H Atom Abstraction from Hydrocarbons

Attacking radical	Temperature	Type of C-H bond		
		1°	2°	3°
HO• [a]	17.5°C	1	4.7	9.8
tert-BuO• [a]	40°C	1	10	44
RO_2• [b]	20°C	1	18	134
P-450 [c]	30°C	1	25	150

[a]Trotman-Dickenson, A. F., *Adv. Free Radical Chem.* **1965** *1*, 1.
[b]Bennett, J. E., Brown, D. M., Mile, B., *Trans. Faraday Soc.* **1970**, *66*, 386.
RO_2• = $CH_3CH_2CH_2(CH_2)_2C$•
[c]Ullrich, V., *Angew. Chem., Int. Ed.* **1972**, *11*, 701.

where steric effects are likely to be unimportant, the ferryl acts more like an alkylperoxy radical than a hydroxyl. The remarkable selectivity for terminal methyl groups in the ω-hydroxylases is no doubt the result of the orienting effect of the protein tertiary structure on the *n*-alkane substrate (somewhat like feeding a pencil into a pencil sharpener).

II.   The Radical Chain Oxidation of Cyclohexane

The oxidation of cyclohexane by O_2 of air is carried out commercially on a large scale (about 10^{+9} lb/year by Du Pont) in order to make a mixture of cyclohexanone (K) and cyclohexanol (A), which are then converted to adipic acid using nitric acid as the oxidant.[6] The adipic acid is combined with hexamethylene diamine to produce 6,6-nylon. The weakness of the C-H bonds on the β-carbons of K or on the α-carbon of A (Table III) makes them vulnerable to radical attack, and means that these partial oxidation products are more easily oxidized than cyclohexane itself; this accounts for the commercial practice of limiting cyclohexane oxidation to low conversions to avoid overoxidation of the K and A. Yields are also limited by the ring opening of cyclohexyloxy radicals, to be described in Sec. II.3.A.

1.   Peroxide Decomposition

Hydroperoxides, once formed by hydrogen abstraction by alkyl peroxy radicals, can generate more radicals by homolysis (3); however, a lower

$$CyO_2H \longrightarrow CyO\bullet + \bullet OH \qquad (3)$$

energy pathway is available in the presence of transition metal complexes of variable valence in the so-called Haber-Weiss cycle,[7] shown for cyclohexylhydroperoxide and cobalt in Reactions 4 and 5.

$$CyO_2H + Co(II) \longrightarrow CyO\bullet + Co(III)OH \qquad (4)$$

$$CyO_2H + Co(III)OH \longrightarrow CyO_2\bullet + Co(II) + H_2O \qquad (5)$$

Reaction 4 can be thought of as stabilizing the hydroxyl radical as a hydroxide ion complex of Co(III), whereas in Reaction 5 the Co(III) oxidizes another CyO_2H to give $CyO_2\bullet$ and H^+, the latter reacting with the OH^- to give water. Cobalt can reduce the activation energy for hydroperoxide decomposition from 43 (Table III) to about 23 kcal/mol.
    Another low-energy pathway available for hydroperoxide decomposition is via the perhemiketal **1**, formed as an adduct of CyO_2H and K in Reaction 6.

$$(6)$$

1

    Though not detected as a separate peroxide by TLC, both ^{13}C NMR and IR show the formation of **1** in solutions of CyO_2H and K. Reaction 6 has a small equilibrium constant of ~ 1 M^{-1} in cyclohexane at 25°C. The significance of **1** is that it undergoes homolytic cleavage much more readily than CyO_2H, and leads to extensive ring opening because of the rapid β-cleavage of the 1-hydroxycyclohexyloxy radical formed in Reaction 7. Coupling of the CyO• and linear alkyl radical within the cage gives cyclohexyloxycaproic acid - a kind of fingerprint

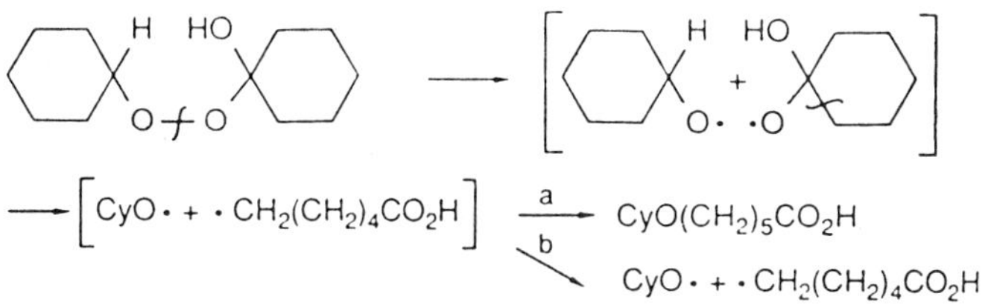

$$\text{(structures shown)} \quad (7)$$

$$\longrightarrow \left[CyO \cdot + \cdot CH_2(CH_2)_4CO_2H \right] \xrightarrow[b]{a} \begin{array}{l} CyO(CH_2)_5CO_2H \\ \\ CyO \cdot + \cdot CH_2(CH_2)_4CO_2H \end{array}$$

for 1.[8] The easy homolysis of 1 is one factor contributing to the autocatalytic decomposition reported for CyO_2H[9]; decomposition produces K, which accelerates further decomposition. Other hydroperoxides no doubt undergo similar adduct formation with K.

Several peroxides are formed in the oxidation of cyclohexane other than CyO_2H, although it is the major one under normal conditions. TLC plates developed with crude oxidation mixtures and visualized by spraying with dimethyl-*p*-phenylenediamine dihydrochloride showed a large number of peroxides (at least 10) with polarities ranging from H_2O_2 (no migration on the silica plate) to dicyclohexyl peroxide (migration with the solvent front). The peroxides isolated and identified included dicyclohexyl peroxide, cyclohexylpentyl peroxide, cyclohexylbutyl peroxide, cyclohexylmethylfuranyl peroxide, CyO_2H, pentyl hydroperoxide, and 6-hydroperoxycaproic acid. The presence of dialkyl peroxides suggests capture of alkyl radicals by the predominant peroxy radical $CyO_2\bullet$, in Reaction 8. The hydroperoxides found are those expected by reaction of carbon centered radicals with

$$CyO_2\bullet + R\bullet \longrightarrow CyO_2R \qquad (8)$$

O_2, followed by hydrogen abstraction, as in Reactions 9 and 10.

$$R\bullet + O_2 \longrightarrow RO_2\bullet \qquad (9)$$

$$RO_2\bullet + CyH \longrightarrow RO_2H + Cy\bullet \qquad (10)$$

Hydrogen peroxide is formed in the oxidation of A to K.[10]

We began our oxidation studies looking for effective catalysts for the decomposition of cyclohexylhydroperoxide.[11] The apparatus, that we called a pulse reactor, is shown schematically in Figure 1. Heating of the reactor, typically to temperatures of 60-150°C, was provided by immersion in an oil bath. Reactions could be followed by

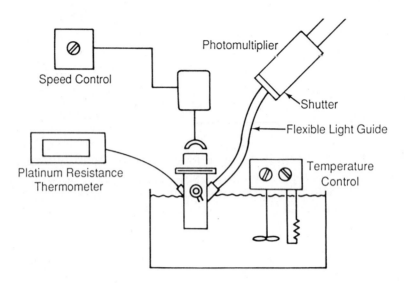

Figure 1. Schematic diagram of the pulse reactor system for monitoring chemiluminescence and temperature changes. From ref. 11.

simultaneously observing both chemiluminescence and temperature changes, after additions by syringe of catalyst to solutions of CyO_2H, or of CyO_2H to catalyst solutions. The chemiluminescence is a consequence of formation of a small fraction of electronically excited ketone molecules in the termination Reaction 11,[12] which is exothermic by 97 kcal/mol. (This can be calculated by the group additivity values in

$$CyO_2{}^{\bullet} \ + \ CyO_2{}^{\bullet} \ \longrightarrow \ K \ + \ A \ + \ O_2 \qquad (11)$$

ref. 13.) The overall Co-catalyzed CyO_2H decomposition is exothermic by about 30-60 kcal/mol, depending on how much of the O_2 released in Reaction 11 is then consumed. Typical traces obtained using $Co(oct)_2$ (oct = 2-ethylhexyloctanoate) are shown in Figure 2. The relatively large temperature rise 10 min after addition of CyO_2H to the catalyst solution is the result of the exothermic quenching Reaction 12 ($\Delta H = -110$ kcal/mol, using refs. 13 and 14), and was used to determine the final extent of decomposition. By comparing the final temperature rise in Figure 2*b* with that in a control experiment without catalyst, we

$$CyO_2H \ + \ P(OMe)_3 \ = \ CyOH \ + \ OP(OMe)_3 \qquad (12)$$

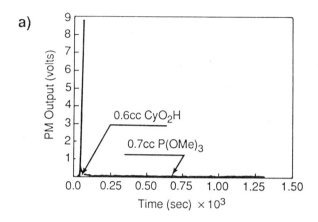

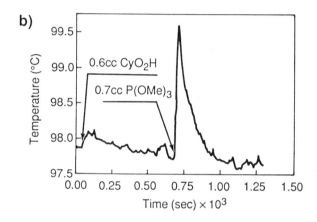

Figure 2. Traces of photomultiplier output (*a*) and temperature (*b*) after additions of neat CyO_2H and $P(OMe)_3$ (10 min later) to $\sim 10^{-5} M$ $Co(oct)_2$. From ref. 11.

could tell that only about half of the hydroperoxide was decomposed in the 10-min period; in fact, the chemiluminescence (Fig. 2*a*) shows that most of the decomposition occurred in the first 10 or 15 sec! It appeared that the Co catalyst was being rapidly deactivated by the products of the decomposition reaction. This was confirmed in an experiment in which the CyO_2H addition was made in three equal portions, only the first of which gave an exotherm, and in subsequent experiments in which K, A, and water were added to the catalyst prior to the CyO_2H.

Reasoning that tightly held ligands that could not be displaced by decomposition products might prolong catalyst activity, we screened many transition metal complexes with chelating ligands, particularly ones with nitrogen donors and with which the metal ion could form stable 5- or 6-membered rings.[15] Early in the search we came across a class of ligands we called BPI [for 1,3-**bis(pyridylimino)**isoindoline **2**] whose Co complexes surpassed anything we subsequently tested.

2

Figure 3 shows the experiment with Co(BPI)$_2$. Rather than a flash in the pan, the catalyst decomposed the hydroperoxide smoothly

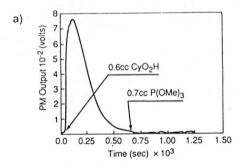

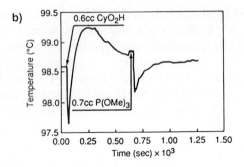

Figure 3. Traces of photomultiplier output (*a*) and temperature (*b*) after additions of neat CyO$_2$H and P(OMe)$_3$ (10 min later) to ~ 10^{-5} *M* Co(BPI)$_2$. From ref. 11.

until it was essentially all gone at the end of the 10-min period, as indicated by the temperature drop when $P(OMe)_3$ was added. (The cooling is because the phosphite was at room temperature.) The temperature dropped when the cool hydroperoxide was added because a full minute was required (see Fig. 3a) before the decomposition reaction reached maximum rate. We attribute this delay to the need to open a coordination position for the hydroperoxide. Six coordination of cobalt in the starting $Co(BPI)_2$ was confirmed by a single crystal X-ray crystal structure determination,[11] which has a structure very similar to that of $Mn(3MeBPI)_2$ which was published earlier.[16] The plausibility of mixed hydroperoxide/BPI complexes as intermediates was subsequently shown by Mimoun's isolation and crystal structure determination of *tert*-$BuO_2CoBPI(OAc)$.[17] A long life for the BPI catalyst during CyO_2H decomposition was indicated by the clean first-order decay of luminescence and by experiments with multiple injections of CyO_2H into catalyst solutions. BPI catalyst activity was found to be insensitive to added hydroperoxide decomposition products.

Spectrophotometric experiments showed that addition of a BPI ligand to a solution of a Co(II) salt gives a clean titration up to a 1:1 mol ratio,[11] indicating that Reaction 13 is rapid and quantitative. (HA is a carboxylic acid and HBPI a free BPI ligand.) Replacement of a second

$$CoA_2 + HBPI = CoA(BPI) + HA \qquad (13)$$

carboxylate by BPI in Reaction 14 is also clean in the sense of showing isosbestic points, but does not have a very high equilibrium constant; it can be reversed by the addition of excess HA.

$$CoA(BPI) + HBPI = Co(BPI)_2 + HA \qquad (14)$$

2. Propagation and Termination

Once released into solution by the decomposition of peroxides, oxygen-centered radicals can attack the cyclohexane solvent as in Reaction 15 to give cyclohexyl radicals, which in turn can react rapidly with O_2 to produce more cyclohexylperoxy radicals. Reactions 16 and

$$CyO\bullet + CyH \longrightarrow CyOH + Cy\bullet \qquad (15)$$

17 provide a radical chain mechanism by which several cyclohexane molecules can be oxidized from a single initiating event. (The ratio of

$$CyO_2\bullet + CyH \xrightarrow{k_p} CyO_2H + Cy\bullet \qquad (16)$$

$$Cy\bullet + O_2 \longrightarrow CyO_2\bullet \qquad (17)$$

oxidation rate to initiation rate is known as the chain length.) Eventually the radicals are removed from solution by bimolecular termination reactions such as 11.

$$CyO_2\bullet + CyO_2\bullet \xrightarrow{k_t} K + A + O_2 \qquad (11)$$

Termination reactions of secondary peroxy radicals have near-zero activation energies, as expected for radical-radical reactions, but unusually low Arrhenius A factors - typically about $2 \times 10^{+6}$ $M^{-1}sec^{-1}$ [18] instead of the normal 10^{+9}. Both the small A factor and the products of the reaction can be explained by the highly constrained tetroxide transition state **3** for the hydrogen transfer, first proposed by Russell.[19] The arrows show the movement of electrons to give the observed products.

3

    The simplified mechanism represented by the six equations 4, 5, 15-17, and 11 can be treated analytically at steady state, by assuming that the [Co(II)/Co(III)] ratio is constant, that the rate of radical generation is equal to the rate of radical removal, and that the rate of CyO_2H generation is equal to its rate of decomposition. Under these conditions the rate of oxidation is given by Eq. 18. For commercially useful oxidation rates (typically 10^{-4} M^{-1} sec^{-1} or higher)

$$-d[CyH]/dt = 3k_p^2[CyH]^2/4k_t \qquad (18)$$

temperatures of 150-175°C are required because of the high activation energy of k_p.

An effect of the differing activation energies of propagation and termination is shown in the amounts of ^{14}C-radiolabeled *K and *A formed at different temperatures when CyO_2H was decomposed in ^{14}C-labeled cyclohexane without added O_2 (Table V). Termination dominates at low temperatures. The converse experiments in Table V - decomposing *CyO_2H in unlabeled CyH - show that the yield of labeled products decreases with increasing temperature, to about

Table V. Yields of ^{14}C-Labeled K and A (*KA) from CyO_2H in *CyH or *Cy_2OH in CyH

Reactants[a]	Metal[b]	Temperature (°C)	Yield of *KA[c] (%)
CyO_2H in *CyH	Co	25	6.4
		60	10.7
		150	33.1
	Mn	25	3.5
		150	27.6
*CyO_2H in CyH	Co	25	96.5
		60	95.7
		150	85.7
	Mn	25	92.9
		150	86.6

[a]1% by weight of hydroperoxide, ~ 0.1 M.
[b]About 20 ppm by weight (~ 2 x 10^{-4} M) as the octoate or naphthenate salts.
[c]100 x counts in *KA product/counts in *CyO_2H reactant for *CyO_2H in CyH or as (100 x counts in *KA/counts in *CyH)/(moles CyO_2H/moles *CyH) for CyO_2H in *CyH.

86% at 150°C. (The 86% yield suggests that about 28% of the CyO•
radicals ring open at this temperature. Since half of the *CyO$_2$H
converted to *CyO$_2$• and half to *CyO• in Reactions 4 and 5, loss of
28% of the *CyO• causes a 14% yield loss.) The best explanation for
the temperature dependence of the yield loss is that the β-cleavage ring
opening of CyO• occurs with a higher activation energy than hydrogen
abstraction (Reaction 15). Another factor contributing is the
nonterminating reaction of cyclohexylperoxy radicals (Reaction 19),
which has a higher activation energy than the termination (Reaction 11),
based on work on analogous s-BuO$_2$• radicals.[20]

$$CyO_2\bullet \ + \ CyO_2\bullet \ \longrightarrow \ CyO\bullet \ + \ O_2 \ + \ CyO\bullet \qquad (19)$$

3.   β-Cleavage of Alkoxy Radicals

After working on CyO$_2$H decomposition, we turned our attention to the
analysis of by-products formed in cyclohexane oxidation reactions, and
the chemistry leading to those by-products. Analysis of nonperoxidic
products showed only very small amounts of ring compounds with
1,3- or 1,4-substitution patterns; the 1,3-cyclohexanediol and
1,4-cyclohexanediol that were isolated were shown by proton NMR to
be equal mixtures of *cis* and *trans* isomers. The vast majority of
by-products are ring opened materials – C$_6$, C$_5$, and C$_4$
α,ω-hydroxyacids, dibasic acids, monobasic acids, and linear alcohols
(*n*-hexanol is not found). There are also small amounts of aldehydes
and esters of the various acids and alcohols present. The complex
by-product mixture (we counted over 100 peaks in one capillary GC
trace) can be vastly simplified by reducing it with LiAlH$_4$, which
converts ketones, aldehydes, acids, and esters to the corresponding
alcohols.[21] Figure 4 shows the result of one such reduction of the
nonvolatile residue remaining after removing most of the K and A by
steam distillation. The major 1,6-hexanediol peak is the result of
reducing adipic acid, 6-hydroxycaproic acid, and several of their
precursors and esters. The 6-CyO-hexanol at longer retention time
comes from the cyclohexyloxycaproic acid formed in Reaction 7. Note
the interesting C$_4$ < C$_5$ > C$_6$ pattern for the monoalcohols in contrast
with the C$_4$ < C$_5$ < C$_6$ pattern for the linear diols. [We attribute the
monoalcohol pattern to ring opening of CyO•, followed by
intramolecular hydrogen transfer and CO loss (Scheme 1)].
        Ring opened by-products suggest an important role for
β-cleavage of cyclohexyloxy radicals. Early in the oxidation, where K
and A are still minor, the major cyclohexyloxy radical will be the

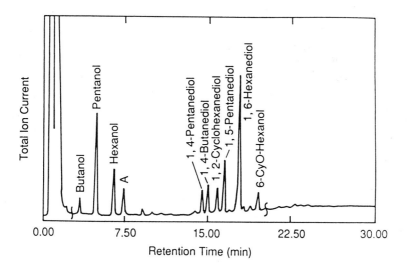

Figure 4. GC/MS trace of a sample of NVR after reduction with LiAlH$_4$.

unsubstituted CyO•; later the 1-hydroxycyclohexyloxy and 2-ketocyclohexyloxy radicals (Section II.3.B), formed via radical attack on A and K at their weakest C-H bonds, become more important sources of ring opening. Thus, we were led into a study of the chemistry of cyclohexyloxy radicals.

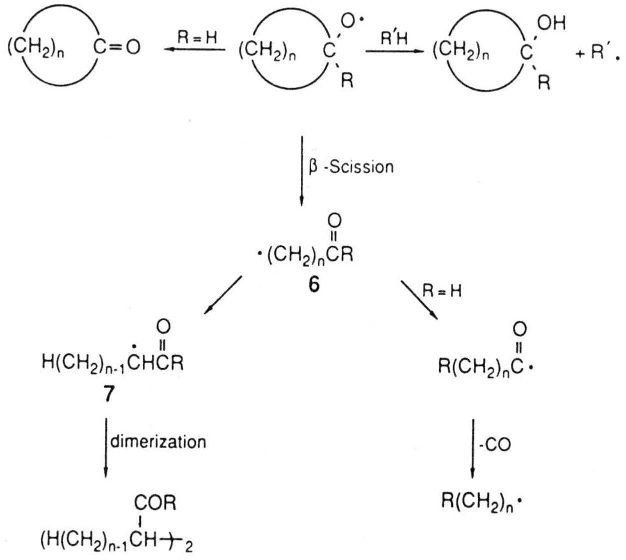

Scheme 1

A.   Cyclohexyloxy Radicals

Ring opening of CyO• radicals at 50°C was established by spin trapping experiments using nitroso-*tert*-butane (NTB) in which the radicals were generated thermally from dicyclohexyl hyponitrite (CyON$_2$OCy).[4] Figure 5 shows the resulting ESR spectrum. The triplet labeled A is the

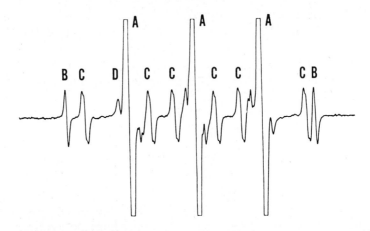

Figure 5. ESR   spectrum   obtained   after   decomposing   CyON$_2$OCy in *o*-dichlorobenzene   containing   NTB.   Lines   A,   B,   and   C   are   assigned   to (*tert*-Bu)$_2$NO•, *tert*-BuN(O•)OCy, and *tert*-BuN(O•)(CH$_2$)$_2$CHO, respectively. From ref. 4.

result of trapped *tert*-Bu• from the NTB; the coupling constant of a(N) = 15.5 G is characteristic of a trapped alkyl radical. The more widely spaced triplet B [a(N) = 28.7 G; the central line is buried] is the result of trapped CyO•. C is a triplet of triplets with 15.4 and 10.4 G coupling constants as a result of trapped •CH$_2$(CH$_2$)$_4$CHO, **4**, with the smaller coupling coming from the two equivalent protons of the terminal methylene. The cleavage Reaction 20 in cyclohexane is in competition with hydrogen abstraction by CyO• in Reaction 15. We have measured

$$\tag{20}$$

4

the rate constant for Reaction 15 by laser flash photolysis and found $k = 8.0 \times 10^{+5}$ M^{-1} sec^{-1} at 25°C, with a small activation energy of about 3 kcal/mol.[4] Although the activation energy of ring opening Reaction 20 is not known, it can be estimated at about 12 kcal/mol from the reported 13.9 kcal for β-cleavage of *tert*-BuO•,[22] taking a smaller value for CyO• because *tert*-BuO• gives the more energetic CH$_3$• radical.

Here we have the central dilemma in radical chain autoxidation reactions: The higher activation energy for propagation than for termination requires a high temperature to have reasonable chain lengths, whereas the higher activation energy for β-cleavage than for hydrogen abstraction by alkoxy radicals causes increasing C-C cleavage as temperature increases.

Other reactions of cyclohexyloxy radicals were considered in addition to hydrogen abstraction and β-cleavage, including disproportionation to K and A (Reaction 21) and dimerization to give dicyclohexyl peroxide (DCHP) (Reaction 22). To ascertain their relative

$$\text{CyO•} + \text{CyO•} \longrightarrow \text{K} + \text{A} \qquad (21)$$

$$\text{CyO•} + \text{CyO•} \longrightarrow \text{DCHP} \qquad (22)$$

importance, a series of experiments was carried out using sources of CyO• radicals with differing numbers of intervening atoms: DCHP (no intervening atoms), dicyclohexyl hyponitrite (DCHN) (two atoms), and dicyclohexylperoxy dicarbonate (DCPD) (four atoms) (Table VI). As expected, the amounts of cage products (DCHP and K) were less from

Table VI. Decomposition Reaction Products from Thermolysis of CyOOCy(DCHP), CyO-N=N-OCy (DCHN), and CyOC(O)OOC(O)OCy DCPD) in Cyclohexane

Compd (Conc. M)	Number of separating atoms[a]	Temperature (°C)	Time[b] (hr)	Reaction products, M				
				K	A	DCHP	Cy$_2$	K/A
DCPD (0.094)	4	80	1	0.0056	0.16	Trace	0.022	0.035
DCHN (0.098)	2	80	1	0.014	0.17	0.004	0.026	0.082
DCHN (0.048)	2	150	1	0.009	0.050		0.006	0.18
DCHP (0.048)	0	150	4	0.009	0.029			0.31

[a]The number of atoms separating incipient CyO radicals.
[b]Experiments carried out in 10-ml glass tubes sealed under vacuum.

from DCPD than from DCHN at 80°C. Also, the amount of noncage product Cy_2 was greater for DCHN than for DCHP, at 150°C. The higher K/A ratio for DCHP than for the other CyO• sources reflects a greater fraction of disproportionation when the radicals are generated adjacent to each other in the solvent cage.

Since CyO• radicals produced under autoxidation conditions are typically present in very low concentrations (our calculations indicate less than $10^{-10}\,M$), they are in close proximity to other CyO• radicals only when they are generated from DCHP, so that their most important reaction pathways are competitive hydrogen abstraction (Reaction 15) and β-cleavage (Reaction 20). At temperatures of 150°C or higher, cleavage becomes quite important. To learn more about the relative importance of these pathways, a series of dicycloalkyl peroxides **5** ($n = 4, 5, 6$; R = H or D) was prepared and decomposed in solution and vapor phases.[23]

$$(CH_2)_n \; C \overset{O-O}{\underset{R \quad R}{\diagdown \; \diagup}} C \; (CH_2)_n$$

5

Possible major reaction pathways considered are shown in Scheme 1. Thermal decompositions of solutions of 1-12% **5** were carried out in F-113®, dodecane, or d_{14}-methylcyclohexane in sealed glass or stainless-steel tubes at 160°C. Product identification was made by GC-MS. Initial experiments were done using undeuterated peroxides to determine the distribution of products obtained under various conditions. The predominant reaction products were cyclic ketones and alcohols, accounting for >90% of the starting peroxides. High O_2 pressures (500 psi) also resulted in efficient O_2 trapping of ring opened radicals to give mainly hydroxyacids and dibasic acids. Thus, all thermolyses of deuterated peroxides were done under N_2 or atmospheric air pressure to minimize O_2 trapping of alkyl radicals in competition with intramolecular H-atom transfer processes. Because the *n*-alkanes were more easily resolved by capillary GC than were the aldehydes or acids, they were chosen for quantification of deuterium isotopic enrichments. The combined amounts of $H(CH_2)_nH$ and $H(CH_2)_nD$, and of $H(CH_2)_{n-1}H$ and $H(CH_2)_{n-2}D$ from **5** are listed in Table VII and show that alkanes resulting from loss of one carbon

Table VII. Amounts of Different Alkane Products Obtained from Thermolysis of Dicycloalkyl Peroxides at 160°C in Solution

Compound	Solvent	Atm	Hydrocarbon product ratios[a] (%)			
			n-Propane	*n*-Butane	*n*-Pentane	*n*-Hexane
5 (*n* = 4)	$CD_3C_6D_{11}$	Air	0	100		
5 (*n* = 4)	$CD_3C_6D_{11}$	Air	0	100		
d$_2$-5 (*n* = 4)	$C_{12}H_{26}$	N$_2$	0	100		
5 (*n* = 5)	$CD_3C_6D_{11}$	Air		2	98	
5 (*n* = 5)	$CD_3C_6D_{11}$	N$_2$		13	87	
5 (*n* = 5)	$C_{12}H_{26}$	Air		11	89	
d$_2$-5 (*n* = 5)	$C_{12}H_{26}$	N$_2$		Trace	100	
5 (*n* = 6)	$CD_3C_6D_{11}$	Air			8	92
d$_2$-5 (*n* = 6)	$C_{12}H_{26}$	N$_2$			6	94

[a]Product ratios determined by comparison of capillary column GC peak area percentage ratios.

as CO) predominate over alkanes formed from loss of two carbons by about a factor of 10. With **5** (*n* = 4) only butane is formed. The absence of any detectable propane suggests that the lower alkanes are not formed by oxidation of $R(CH_2)_n{}^\bullet$ radicals but are derived primarily via intramolecular H-atom abstraction by $\bullet(CH_2)_nCRO$ radicals **6** to give **7** (Scheme 1). Rearrangement of **6** (*n* = 4) to **7** would proceed through a 5-membered transition state, and be energetically less favorable than analogous rearrangements with *n* = 5 or *n* = 6 through 6- and 7-membered transition states.[24]

Percentages of deuterium enrichment in alkane products were determined, both for decomposition of **5** (*n* = 4, 5, 6) in $CD_3C_6D_{11}$ and for d$_2$-**5** (*n* = 4, 5, 6) in dodecane (Table VIII). The results for the d$_0$ compounds show no major differences in the amounts of deuterium enrichment under N$_2$ or air and provide a measure of the reproducibility of the measurements ($\pm$ 10%). The extent of deuterium atom abstraction from $CD_3C_6D_{11}$ solvent was not strongly dependent on chain length of the **6** radicals. The highest amount of intramolecular aldehydic D-atom transfer observed was moderately large (74%), corresponding to a favorable 6-membered transition state for d$_1$-**6**

Table VIII. Percentages of Deuterium Enrichment in $H(CH_2)_nH$ Products from Thermolysis of d_0 and d_2 Dicycloalkyl Peroxides at 160°C in Solution

Compound		Solvent	$H(CH_2)_nD/H(CH_2)_nH$ ratios	
			Air	N_2
5	($n = 4$)	$CD_3C_6D_{11}$	15/85	12/88
5	($n = 5$)	$CD_3C_6D_{11}$	29/71	19/81
5	($n = 6$)	$CD_3C_6D_{11}$	7/93	8/92
d_2-**5**	($n = 4$)	$C_{12}H_{26}$		74/26
d_2-**5**	($n = 5$)	$C_{12}H_{26}$		29/71
d_2-**5**	($n = 6$)	$C_{12}H_{26}$		38/62

($n = 4$, R = D); the amounts for d_1-**6** ($n = 5, 6$, R = D) were about half of that. Studies on the effect of chain length on intramolecular H-atom abstraction have been reported in solution phase studies of *n*-alkyloxy radicals generated by alkyl hypochlorite decomposition at 0°C.[25] A 6-membered transition state was favored over a 7-membered one by a factor of 16, and no rearrangements of the alkyloxy radicals via 8-membered transition states were observed. The percentages of various pathways for H-atom abstraction by radicals **6** ($n = 4, 5, 6$, R = H) that occurred in solution phase reactions at 160°C are shown in Table IX. About half of the H-atom abstraction with $n = 5$ and $n = 6$

Table IX. Extent of Different H-Atom Reaction Pathways for ω-Formyl Radicals in Solution at 160°C under N_2

ω-Formyl radical **6** •$(CH_2)_n CHO$	Reaction pathway (%)		
	Intramolecular aldehydic H-atom abstraction	H-atom abstraction from solvent	Intermolecular H-atom abstraction from other C-H sources[a]
$n = 4$	74	12	14
$n = 5$	29	19	52
$n = 6$	38	8	54

[a]Calcd. as 100% - (% intramolecular aldehydic H-atom abstraction + % H-atom abstraction from solvent).

occurred via intermolecular processes compared with only about 14% in the case of $n = 4$, which has the most favorable transition state for intramolecular transfer.

Vapor phase thermolyses of peroxides d_0 and d_2-5 ($n = 4$ and 5) were done by injection of neat compounds in a glass reservoir heated to 80-110°C from which the compounds could vaporize in the ion source of a mass spectrometer, heated to 200°C to decompose the dialkyl peroxides. The source also contained a small amount of tetracyanoquinodimethane (TCNQ) for trapping and analyzing carbon-centered radicals.[26] The same techniques were not successful in the case of dicycloheptyl peroxide because of insufficient volatility. In the case of 5 ($n = 4$), the major radicals trapped by TCNQ were n-C$_4$H$_9$• and H•, with virtually no C_1, C_2, C_3, or C_5 alkyl radicals. Radical 6 ($n = 5$) likewise gave mostly n-C$_5$H$_{11}$• and H• radicals with only minor amounts of C_1, C_2, C_3, C_4, or C_6 radicals; it also gave an oxygen-containing carbon-centered radical isomeric with 6 ($n = 5$) with an intensity equal to about 10% that of n-C$_5$H$_{11}$•, which we believe to be 7. No significant amount of the corresponding radical was observed starting with 5 ($n = 4$). The energetics of gas phase intramolecular H-atom transfer processes requiring 5- and 6-membered transition states have recently been discussed.[26c] Significantly faster rearrangements involving 6-membered transition states have been noted for both thermally and chemically activated radicals.

Starting with d_2-5 ($n = 4$), the n-butyl radicals trapped by TCNQ were found to contain at least 86% monodeuteration. (Some loss of deuterium enrichment in ω-formyl radicals from H-D exchange with H$_2$O on heated glass and metal surfaces is expected.) Collision induced decomposition (CID) analysis of the trapped n-butyl radicals showed loss of C$_3$H$_6$D• fragments consistent with a location of D in the methyl group. Likewise, d_2-5 ($n = 5$) gave rise to n-pentyl radicals containing at least 95% monodeuteration. CID analysis of the trapped radicals showed the presence of both 2-pentyl and 1-pentyl isomers with D enrichment in the propyl and butyl portions. The high level of deuterium enrichment found in the butyl and pentyl radicals from d_1-6 ($n = 4$ and 5) indicates that gas phase intramolecular H-atom transfer of 6 proceeding through 6- and 7-member transition states is quite favorable.

Early experiments led us to believe that CyO• radicals might react directly with O$_2$ to form K as shown in Reaction 23. One percent solutions of DCHP in Freon F-113® were heated at 160°C under N$_2$ and under 500 psi O$_2$. The amount of K and A under N$_2$ indicated that >90% of the reactions involved disproportionation or hydrogen

$$CyO• + O_2 \longrightarrow K + HO_2• \qquad (23)$$

abstraction from starting materials or products. The dependence of K/A ratios on the presence of O_2 was large: K/A = 1/4 (under N_2) vs K/A = 15/1 with 500 psi O_2. Control experiments run using DCHP in F-113® containing 0.25% cyclopentyl alcohol showed no cyclopentanone formation under N_2, but about 80% conversion to cyclopentanone under O_2. Thus, the corresponding high K/A ratios from DCHP with O_2 can be explained by autoxidation of A to K, rather than by Reaction 23.

B.   1-Hydroxy and 2-Ketocyclohexyloxy Radicals

Hydrogen atom abstraction from the relatively weak C-H bond on the α-carbon of cyclohexanol in Scheme 2 will give a carbon centered

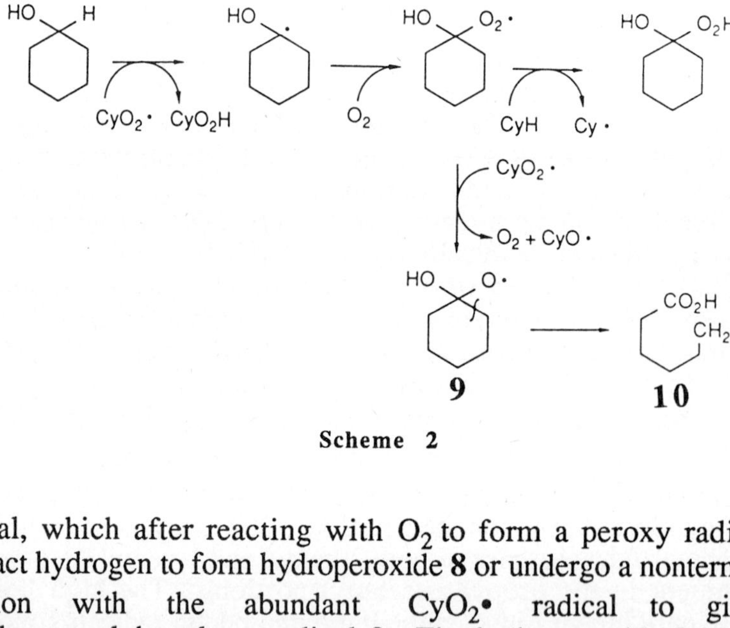

Scheme 2

radical, which after reacting with O_2 to form a peroxy radical can abstract hydrogen to form hydroperoxide **8** or undergo a nonterminating reaction with the abundant $CyO_2\bullet$ radical to give the 1-hydroxycyclohexyloxy radical **9**. The hydroperoxide **8** is nothing more than the perhemiketal adduct of H_2O_2 and K. Ring opening of **9** to give linear radical **10** is very rapid even at low temperatures, and has been used to prepare dodecanedioic acid by radical-radical coupling.[10]

Analogous reactions with cyclohexanone in Scheme 3 form the 2-ketohydroperoxide **11** and ring opened acyl radical **13**. **11** must be very unstable because we have not been able to detect it in cyclohexane oxidation mixtures. (We have, however, been able to prepare and study other 2-ketohydroperoxides.[27]) The 2-ketocyclohexyloxy radical **12** does not β-cleave exclusively, since small amounts of 2-ketocyclohexanol are found.

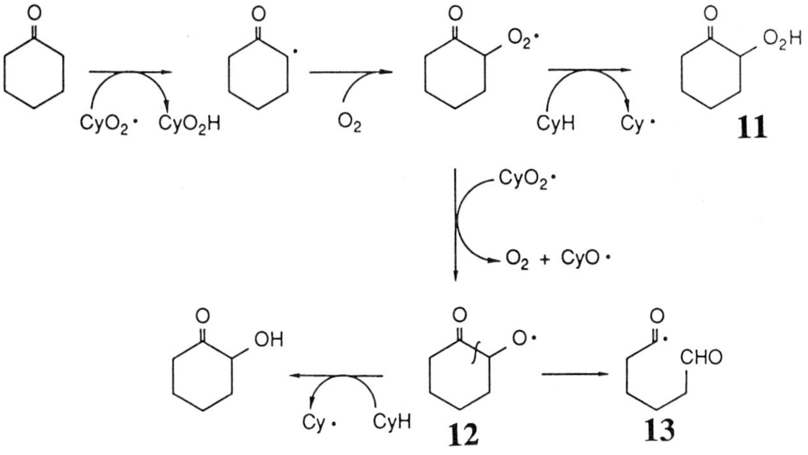

Scheme 3

During our earlier studies of metal carboxylate catalyzed decomposition of $*CyO_2H$ in CyH, we found that though yields of $*KA$ were not strongly dependent on the type of metal used, the $*K/*A$ ratio was, as shown in Table X. The higher ratios could have been

Table X. Decomposition of $*CyO_2H^a$ in Cyane

$$*CyO_2H^b \xrightarrow[\text{30 ppm metal}]{\text{CyH}} *K + *A + K + A$$

Metal	Yield of $*KA^c$ (%)	$*K/A*$
None	76.3 ± 0.8	0.34 ± 0.05
Cr	84.8 ± 0.7	4.0 ± 0.5
Mo	85.7 ± 0.2	1.0 ± 0.1
V	87.0 ± 0.7	1.1 ± 0.2
Co	87.5 ± 0.8	0.58 ± 0.05

[a] ^{14}C label.
[b] About 0.11 M.
[c] 100 x (moles of $*K$ and $*A$)/moles of $*CyO_2H$ decomposed.

the result of Reaction 24 in addition to Reactions 4 and 5 (both heterolytic and homolytic hydroperoxide decomposition pathways have

$$CyO_2H \longrightarrow K + H_2O \qquad (24)$$

been suggested before[28]), or the result of oxidation of *A to *K. To test the latter possibility cyclohexyl hydroperoxide was decomposed in the presence of ^{14}C-labeled *A by various metal carboxylates, and ethylbenzene hydroperoxide was decomposed in the presence of 1-phenylpropanol, indicated schematically by Reactions 25 and 26. The

$$CyO_2H + *A \longrightarrow K + A + *K \qquad (25)$$

$$\underset{\textstyle 14}{\underset{|}{\overset{O_2H}{\underset{|}{PhCHCH_3}}} + \underset{|}{\overset{OH}{\underset{|}{PhCHC_2H_5}}} \longrightarrow \underset{|}{\overset{O}{\underset{||}{PhCCH_3}}} + \underset{15}{\overset{OH}{\underset{|}{PhCHCH_3}}} + \underset{|}{\overset{O}{\underset{||}{PhCC_2H_5}}}} \qquad (26)$$

results in Table XI indicate that Cr, Mo, and V in the presence of hydroperoxides cause considerably more oxidation of the alcohols present than do Mn and Co. This behavior of the first three metals could be related to the tendency of these metals to form metal oxo species.[29]

Table XI. Comparison of Metal Catalysts in the Extent of Oxidation of Alcohols during Decomposition of Cyclohexyl Hydroperoxide[a] or Ethylbenzene Hydroperoxide[b]

Metal[c]	*K / *A	15/14
Cr	0.45	0.27
Mo	0.46	
V	0.45	0.27
Mn		0.07
Co	0.08	0.10
None	0.22	0.11

[a]0.11 M CyO_2H and 0.02 M *A (^{14}C labeled) at 150°C for 2 hr in cyclohexane.
[b]0.06 M $PhCH(O_2H)CH_3$ and 0.01 M $PhCH(OH)C_2H_5$ at 120°C for 1.3 hr in cycloheptane.
[c]Ca. 30 ppm of metal by weight as the 2-ethylhexanoate or naphthanate.

C. Other Oxy Radicals

Just as cyclic peroxy radicals can undergo a variety of reactions as seen in Schemes 2 and 3, linear peroxy radicals formed via reactions of **4** and **10** with O_2 can undergo the reactions shown in Scheme 4. In this case, however, the linear alkoxy radical formed in the nonterminating cross

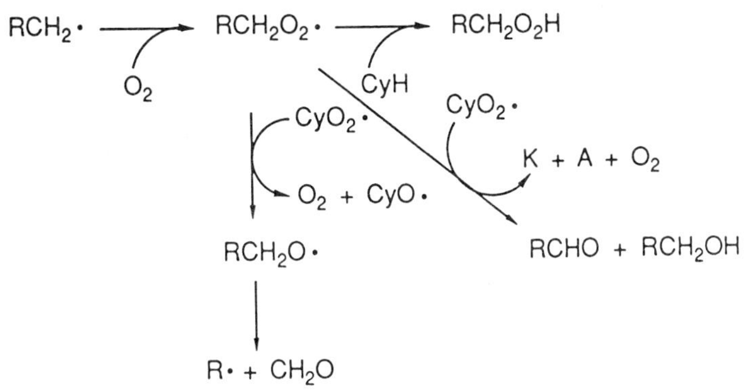

Scheme 4

reaction with $CyO_2\bullet$ β-cleaves to give formaldehyde and a linear carbon-centered radical shorter by one carbon. The cross-termination reaction (down and to the right) gives an aldehyde and A in one case and a linear alcohol and K in the other, depending on the direction of hydrogen atom transfer in the unsymmetrical tetroxide intermediate comparable to **3**. The expected aldehydes and alcohols are observed in the oxidation products, except that the amounts of aldehydes are usually small compared to the corresponding acids to which they are easily oxidized. Though we have observed cyclohexyl formate in the products, the amounts of other esters are normally small because weaker acids form esters more slowly.

Acyl radicals formed by isomerization of **6** in Scheme 1, or **13** in Scheme 3, can lose CO to become one carbon shorter, or react with O_2 and, through a series of steps, eventually lose CO_2. It is the β-cleavage of oxy radicals and the subsequent loss of CO and CO_2 from acyl and carboxyl radicals that gives the complex mixtures observed in cyclohexane oxidation and limits the yields of desired products.

4. Continuous Microreactor Studies

A powerful method for studying autoxidations involves the use of 5 cc continuous stirred microreactors of the type developed in these laboratories. These reactors, which are about an 1 in. in outside diameter and 3 in. high, and are rated for pressures up to 2000 psi, can be stirred with a rotary stirrer and heated in a fluidized sand bath in the laboratory in a 4-in. steel barricade (tested to withstand an explosion of 4 g of TNT); they are much more convenient to use than the typical walk-in barricades used for reactions with potentially explosive mixtures. Figure 6 shows a schematic arrangement we have used for

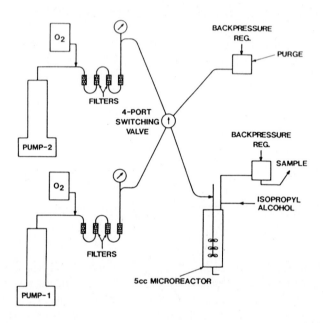

Figure 6. Schematic diagram of the 5cc microreactor system showing filters to speed O_2 dissolution and a 4-port valve for switching reactor feeds.

oxidation studies. The dual pumps and four-port switching valve make it possible to switch feed solutions rapidly; one or both can be presaturated with air at 1000 psi, giving an O_2 concentration in solution of about 0.15 M. The filters just after the air inlets break up bubbles and speed up dissolution. Backpressure regulators maintain the 1000 psi system pressure. Isopropyl alcohol is added just prior to the regulator on the reactor outlet line to dissolve any organic solids (such as adipic acid) and water formed during the oxidation, thus maintaining a single phase. In the experiments described here, the catalyst was generally Co(oct)$_2$ in the feed, though tert-BuO$_2$CoBPI(OAc)[17] was used in some cases.

In one experiment in which a solution of cyclohexanone in cyclopentane was switched in after reaching steady state in the oxidation of cyclopentane at 150°C, comparison of the K concentration before and after the addition showed that 14% of the K was gone at the new steady state. (Cyclopentane was used so that the added ketone would be different from that produced by oxidizing the solvent.) From this experiment, in which the cyclopentane conversion was about 1.5%, we determined that K was oxidized about 10 times as fast as cyclopentane. From a similar experiment with added A, we found that it reacted 12.5 times as fast as the hydrocarbon, with 80% going to K and 20% to ring opened products, consistent with the mechanism shown in Scheme 2.

In another experiment we switched the feed from cyclohexane to d_2-cyclohexane (abbreviated *C, easily prepared by the deuteration of cyclohexene). The formation of labeled products was followed by GC-MS of samples taken at various times after switching feeds. The GC traces appeared unchanged, but the mass spectra showed the growth of product peaks at *m/e* ratios larger by +2. Figure 7 shows the time

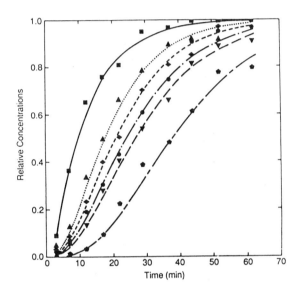

Figure 7. The time course of relative concentrations of *C and various labeled products after switching from d_0-cyclohexane to d_2-cyclohexane. The symbols are the observed points for *C (◼), *P (▲), *A (◆), *K ( ), *HY (▼), and *AA (✦). The curves were calculated using the rate constants of Scheme 5.

evolution of the *C and five of the products exiting the reactor, all normalized to their final steady-state values. The first-order approach of *C to its final value gives the hold-up time of the experiment, about 10 min. The appearance in turn of labeled peroxide (*P), alcohol (*A), ketone (*K), 6-hydroxycaproic acid (*HY), and adipic acid (*AA) could be fit rather well (Fig. 7) by the simple nonradical model shown in Scheme 5 (with first order rate constants in units of sec^{-1}). The relative rate constants from *C imply that at very low conversion the *C goes to *P, *A, *K, and *HY, in the ratios of 59/28/12/1. Direct

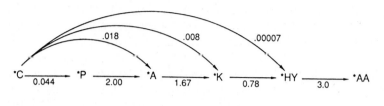

Scheme 5

pathways to *A and *K as well as to *P are consistent with radical
models in which CyO_2• can go to CyO_2H (in Reaction 16) and also to
K and A in Reaction 11. The ring opened product
6-hydroxycaproaldehyde can come from **4** via the reactions shown in
Scheme 4, then be easily oxidized to the hydroxycaproic acid observed.

One of the most interesting experiments in the continuous
microreactor was one in which the feed (cyclohexane presaturated with
air at 1000 psi) was kept constant but the reactor temperature was cycled
up and down between 145 and 165°C. Figure 8 shows the temperature

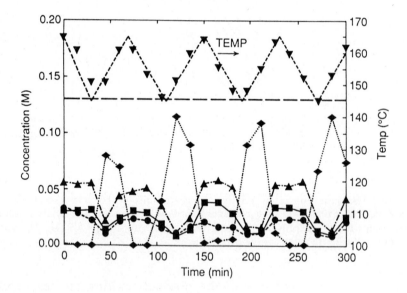

Figure 8. The observed temperature (▼) and concentrations of O_2 (◆), CyO_2H (▲),
A (■), and K (●) on cycling the temperature to a 5cc CSTR. Predissolved O_2 at 0.13 *M*
was fed.

ramp at the top and the response of the exit O_2, peroxide, A, and K in
the curves at the bottom.   The horizontal dashed line at 0.13 *M* is the
concentration of O_2 in the feed, that is the exit O_2 concentration in the

absence of chemical reaction. Note that the O_2 leakage goes through maxima soon after the temperature minima; product concentrations at those points are correspondingly low. Soon after the temperature maxima, the O_2 leakage is at its lowest, with an average value of about 0.5 mM. Figure 9 shows calculated curves for the same experiment,

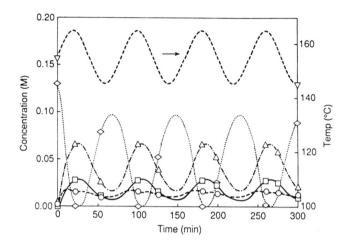

Figure 9. The calculated temperature and concentrations of O_2 (◇), CyO$_2$H (··), A (▱), and K (○) on cycling the temperature to a 5 cc CSTR with a predissolved 0.13 M O_2 feed, using a 74 reaction radical model.

calculated with a radical model with some 70 reactions[30] that account for the formation of CO, CO_2, cyclohexyloxycaproic acid (Reaction 7) and the C_6, C_5, and C_4 aldehydes, alcohols, dialdehydes, and hydroxyaldehydes formed in the cross termination reactions of the various linear peroxy radicals. For ease in calculation a continuous sine function was used for the temperature. Clearly, in a temperature cycling experiment of this type, both activation energies and rate constants have to be varied in order to fit the observed concentrations. Cycling temperatures or feed compositions can be a powerful way to study complex reaction systems, as described for biological systems by Finkelstein and Carson.[31]

5. Computer Modeling

Computer modeling was carried out in conjunction with the cyclohexane oxidation studies. For a complex system of this sort, the value of modeling, both as a guide for experimental work and as a repository of accumulated knowledge, cannot be overemphasized. Models evolved in complexity as the work progressed, with nonradical models as simple as Reaction 27 (where KAP represents the sum of K, A, and P, and X is

$$C \longrightarrow KAP \longrightarrow X \tag{27}$$

all other products) through the one shown in Scheme 5, and with radical models as simple as the six reaction model described in Sec. II.2 to models with over 150 reactions that include the oxidation of aldehydes to acids and the formation and decomposition of DCHP and H_2O_2. Rate constants and activation energies were measured for individual reactions or taken from the literature where possible; in other cases they were fit using laboratory cyclohexane oxidation results from continuous reactors, or even plant data in some cases. Finally, the full set of reactions was incorporated into a reactive distillation model that simulates the countercurrent flow of air and cyclohexane in the commercial reactors, and includes heat and mass balances, mass transfer of O_2 from the vapor to the liquid phase, and vapor/liquid equilibria of the volatile components. The model, run on a Cray computer, is being used to optimize plant performance.

III. Alkane Oxidations by Soluble Metalloporphyrin Complexes

Iron porphyrins had been known to catalyze the autoxidation of hydrocarbons,[32] but not until a report by Groves and co-workers in 1979,[5] which showed that simple iron-porphyrins could function as models for the cytochrome P-450 monooxygenase enzymes, did metalloporphyrins begin receiving much attention as selective oxidation catalysts. Since then a large number of metalloporphyrin-oxidant combinations have been reported to selectively oxidize a wide range of hydrocarbons.[33] Iron and manganese porphyrins seem to be the most active, particularly when used with iodosoarenes as the oxidant, and are the metals of choice for oxidizing saturated hydrocarbons. Much emphasis has been placed on understanding the mechanism of oxygen atom transfer with the hope that a better understanding would lead to the development of more selective and more reactive metalloporphyrin oxidation systems. We and others have undertaken studies to understand the relationships between porphyrin structure and reactivity-selectivity of the oxidation reactions, and have concentrated on the iron[33h] and manganese systems.[34]

The biological monooxygenases are intriguing because of their ability to activate molecular oxygen and oxidize hydrocarbon substrates selectively under mild conditions. The cytochromes P-450 are a class of these enzymes that catalyzes the monooxygenation of a variety of organic substrates using both dioxygen and NADPH, or single oxygen atom donors such as iodosoarenes or alkylhydroperoxides.[35] This class of enzymes has been the focus of much research, and much is known about the catalytic cytochrome P-450 cycle, shown in Scheme 6. The enzyme, with the iron in the +3 oxidation state, will react with oxygen

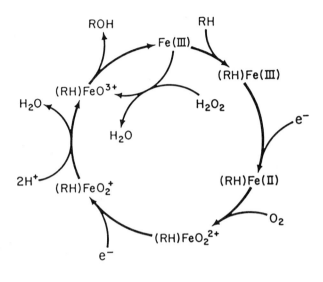

Scheme 6

transfer reagents such as iodosobenzene or alkylhydroperoxides anaerobically to form the active ferryl intermediate, by a pathway known as the "peroxide shunt." Simple metalloporphyrins react similarly with oxygen transfer reagents to generate an active species capable of oxidizing hydrocarbons, however, the selectivity in cytochrome P-450 oxidations is unparalleled by the metalloporphyrin model systems.

1.   Iron Porphyrins

A.   Cyclohexane Oxidation Ligand Effects

Under mild conditions iron tetraphenylporphyrins catalyze oxygen transfer from iodosobenzene to saturated hydrocarbons, resulting in oxygen atom insertion into C-H bonds. In the oxidation of cyclohexane, the yield of cyclohexanol is critically dependent on the concentrations of all reactants, on other reaction parameters, and on phenyl ring substituents on the catalysts. The reactions between cyclohexane, iodosobenzene, and iron tetraarylporphyrin **16** give the yields of cyclohexanol (based on iodosobenzene) shown in Table XII for reactions carried out under standard conditions.

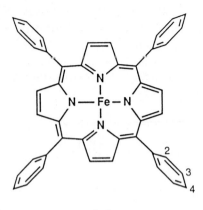

16

Table XII. Cyclohexane Oxidation: Cyclohexanol Yields for Different Iron Porphyrin Derivatives

Iron porphyrin	Cyclohexanol yield[a]	Iron porphyrin	Cyclohexanol yield[*a]
$Fe(T_{4-Me}PP)Cl$	6.9 ± 0.4	$Fe(T_{2-F}PP)Cl$	48.8 ± 0.2
$Fe(T_{3-Me}PP)Cl$	7.9 ± 0.1	$Fe(TF_5PP)^{b}Cl$	66.6 ± 1.0
$Fe(T_{2-Me}PP)Cl$	31.7 ± 1.0	$Fe(FPP-3)^{c}Cl$	39.4 ± 0.4
$Fe(T_{3,4,5-MeO}PP)Cl$	23.1 ± 0.5	$Fe(TPP)Cl$	10.1 ± 0.1
$Fe(T_{2,4,6-MeO}PP)Cl$	13.0 ± 1.2		

[a]Based on iodosobenzene charged; 30% cyclohexane in CH_2Cl_2 was stirred vigorously for 2 hr at 25°C.
[b]Tetra(pentafluorophenyl)porphyrin.
[c]Tetra(2-perfluoropropylamidophenyl)porphyrin.

Steric and electronic effects on the cyclohexanol yield are distinguished by a plot shown in Figure 10, in which the yield term $[\log (Y_X/Y_H)]$ is plotted against the electrochemical potential for ring oxidation of the Fe(III) chloride complexes. The plot is linear for the Hammett series and a fluoro-pocket[33h] porphyrin complex. Electron-withdrawing groups have two effects: to increase the reactivity of the ferryl intermediate toward hydrocarbons more than toward other possible substrates (iodosobenzene or the iron porphyrin catalyst), and to reduce the susceptibility of the catalyst to oxidation, resulting in less wasted oxidant and longer catalyst lifetimes. In Figure 10, the vertical distances of points above the line give a measure of steric effects. It appears as though they are approximately the same for any ortho-substituent used. The increased yields suggest that these substituents prevent the bimolecular destruction of the catalyst (in a face to face approach), and even a small substituent such as fluorine is effective.

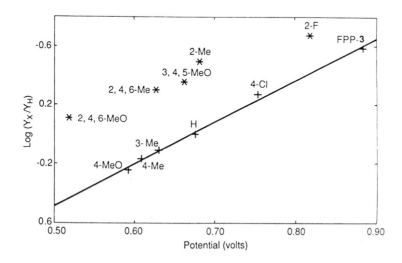

Figure 10. Cyclohexane oxidation. Steric and electronic effects on the yield: $\log(Y_X/Y_H)$ vs the porphyrin ring oxidation potentials, measured in CH_2Cl_2 (0.2 M tetrabutylammonium tetrafluoroborate) vs SCE and referenced to internal ferrocene.

Varying the axial ligand in FeTPPX causes the yields of cyclohexanol to increase in the series ClO_4^-, Cl^-, CH_3O^-, Br^-, F^- (Table XIII). The yield correlates with the amount of catalyst decomposition; less decomposition means higher yield. The anion

Table XIII. Effect of Axial Ligand on Cyclohexanol Yield for Fe(TPP)X Catalysts

X^-	ClO_4^-	MeO^-	Br^-	Cl^-	F^-
Cyclohexanol yield (%)[a]	<1	8	8	10	12
$E_{1/2}(Fe^{III}/Fe^{II})$, V	0.22		-0.21	-0.29	-0.50

[a]Based on iodosobenzene, ± 1%.

effects on the iron porphyrin catalysts, shown here, are markedly different from the effects on the manganese porphyrin catalysts (Sec.

III.2). The main reason for this is that although a weakly coordinating anion may actually activate the ferryl species as in the manganese system, the yield of oxidized hydrocarbon is very low because of the anion's effect on catalyst lifetime.

Although iron porphyrins are simple models for the cytochromes P-450, the mechanism of hydrocarbon oxidation indicated by Reactions 28-31 is complicated by undesirable side Reactions 30 and 31. The

$$PhIO + Fe \longrightarrow PhI + OFe \qquad (28)$$

$$OFe + CyH \longrightarrow Fe + CyOH \qquad (29)$$

$$OFe + PhIO \longrightarrow Fe + PhIO_2 \qquad (30)$$

$$OFe + Fe \longrightarrow Fe + D \qquad (31)$$

initial step is oxygen transfer from iodosobenzene to the iron(III) porphyrin to generate a high oxidation state ferryl intermediate. The next three involve competitive reactions of this intermediate. D represents destroyed catalyst. Bimolecular catalyst destruction is supported by kinetic studies.[33h]

B.   *n*-Alkane Oxidations: Regioselectivity and Substrate Selectivity

We oxidized pentane using a variety of substituted iron tetraphenylporphyrins under standard conditions to give a mixture of 1-, 2-, and 3-pentanols and traces of 2- and 3- pentanones. The 2-/3-pentanol ratio can be used as a measure of the shape selectivity of a particular catalyst. (Little or no 1-pentanol was observed in most cases.) This ratio varies from 1.61 to 3.82, depending on the nature and location of the substituents; a ratio of 2.0 (4/2) is expected for totally random attack on pentane. Steric and electronic effects can be distinguished by the plot shown in Figure 11. Electron-withdrawing groups on the porphyrin ring make the ferryl intermediate more reactive and less selective. Inductive effects on the alkane favor reactivity at the 3-position over the 2-position for a more selective H-atom abstractor. As the ferryl becomes more reactive and less selective, the ratio approaches a value of 2.0 (1.84) with FPP-3. In addition to the electronic effect there is the steric effect, whose magnitude is indicated by the vertical distance between the starred points and the line. As expected, based on the sizes of the substituents, the order observed for the steric effect is 2,4,6-Me > 2,4,6-MeO > 2-Me > 2-F > 3-Me.

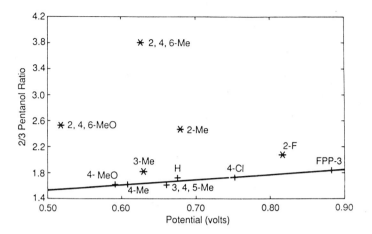

Figure 11. Pentane Oxidation. Steric and electronic effects on regioselectivity: 2-/3-pentanol ratio vs the porphyrin ring oxidation potentials, measured in CH_2Cl_2 (0.2 M tetrabutylammonium tetrafluoroborate) vs SCE and referenced to internal ferrocene.

Substrate selectivity is observed in competition experiments between pentane and octane. Figure 12 shows the total pentanol/total octanol ratio for different iron porphyrin catalysts. Porphyrins having substituents at both 2- and 6-positions show the highest substrate selectivity because both faces are protected.

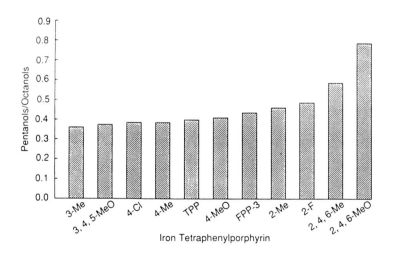

Figure 12. Competition experiment using cooxidation of pentane (1.08 M) and octane (1.08 M) showing ratios of total pentanols/octanols for different iron porphyrin catalysts.

2.   Manganese Porphyrins

Manganese porphyrin systems give higher yields and longer catalyst lifetimes than the corresponding iron systems, however, the reactions display considerable radical character, leading to by-products and losses in selectivity.[33x,y] Anion exchange in simple manganese porphyrin systems can lead to both a reduction in radical character and an increase in overall reactivity in oxidations using iodosobenzene. These changes have been rationalized in terms of anion effects on the electronic structure of the oxo-intermediate.[34]

A.   Axial Ligand Effects

We expected that Lewis acid $Ph_3SnO_2C_8F_{15}$ (**17**) would coordinate to pyridyl groups in MnTpyPOAc (structure **16** with N instead of CH at position 4) and thus affect cyclohexane oxidations when using it with iodosobenzene. The effects shown in Figure 13 and Table XIV are, however, the result of anion metathesis. The UV/VIS spectral changes

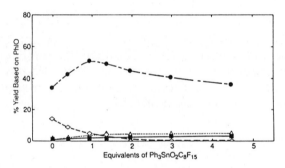

Figure 13. Yields of cyclohexanol (●), cyclohexyl chloride (◇), cyclohexene oxide (▲), and cyclohexanone (■) as a function of **17** added to cyclohexane oxidations using MnTpyPOAc. $[PhIO]_0 = 0.05\ M$, $[MnTpyPOAc]_0 = 0.0125\ M$ in CH_2Cl_2. Yields are based on PhIO.

for the addition of **17** to MnTPPOAc are consistent with a stoichiometric metathesis of the -OAc by $-O_2C_8F_{15}$. Authentic $MnTPPO_2C_8F_{15}$, prepared from MnTPPCl and $AgO_2C_8F_{15}$, gives the same results for cyclohexane oxidation as with the $MnTPPOAc/Ph_3SnO_2C_8F_{15}$ system (see Table XIV, entry 17 versus entries 12 and 13), confirming that anion metathesis is the cause of the observed changes in product distribution.

To probe the nature of the change in product distribution with anion metathesis, we measured the deuterium isotope effect for cyclohexane oxidation using MnTpyPOAc with different tin Lewis acids (Table XV). There is a reduction in k_H/k_D when one

Table XIV. Cyclohexane Oxidation Yields[a] with Lewis Acid Additions

Entry number	Manganese porphyrin	Equivalents of additive[b]	Yield (%)			
			Cyclo-hexanol	Cyclo-hexanone	Cyclohexyl chloride	Cyclohexene oxide
1	MnTpyPPOAc		34.3	1.5	14.3	0.9
2	MnTpyPPOAc	0.39	42.8	1.3	9.0	2.1
3	MnTpyPPOAc	0.94	51.1	1.8	4.9	3.3
4	MnTpyPPOAc	1.36	49.3	1.8	3.1	4.0
5	MnTpyPPOAc	1.96	44.9	2.5	1.4	4.4
6	MnTpyPPOAc	2.97	40.9	3.1	0.5	4.6
7	MnTpyPPOAc	4.46	36.3	3.4	0.3	4.7
8	MnTpyPPOAc	1.09[c]	38.1	10.8	1.7	3.9
9	MnTpyPPOAc	2.85[c]	49.2	7.8	0.2	2.7
10	MnTPPOAc		22.4	2.8	13.3	0.6
11	MnTPPOAc	0.49	28.5	2.3	11.9	0.7
12	MnTPPOAc	0.86	32.2	2.6	10.5	1.1
13	MnTPPOAc	1.35	29.9	2.6	6.7	1.4
14	MnTPPOAc	1.85	28.0	3.0	4.5	2.3
15	MnTPPOAc	3.00	26.2	2.6	1.8	2.1
16	MnTPPOAc	4.84	18.6	1.9	0.7	1.6
17	$MnTPPO_2C_8F_{15}$		31.8	1.9	8.4	0.9
18	$MnTPPO_2C_8F_{15}$	1.64	28.6	1.5	2.9	2.0
19	$MnTPPO_2C_8F_{15}$	2.81	22.2	2.6	1.4	1.6

[a]Yields are based on iodosobenzene and are the average of at least two runs; all yields are ± 0.5%.
[b]$Ph_3SnO_2C_8F_{15}$ (**17**) was used unless otherwise noted.
[c]$Ph_3SnO_3SC_6F_{13}$.

equivalent of **17** is added (anion metathesis is complete), and the second equivalent has no effect on k_H/k_D, indicating that, even though

Table XV. Deuterium Isotope Effects for Cyclohexane Oxidation

Catalyst System	k_H/k_D
MnTpyPOAc	11.8
MnTpyPOAc + 1 equiv. of $Ph_3SnO_2C_8F_{15}$ (**17**)	5.7
MnTpyPOAc + 2 equiv. of $Ph_3SnO_2C_8F_{15}$ (**17**)	5.7
MnTpyPOAc + 1 equiv. of $Ph_3SnO_3SC_6F_{13}$	3.7
FeTTPCl[a]	12.9

[a] TTP, tetra(2-methylphenyl)porphyrin; ref. 33p.

the product distribution changes when more than one equivalent of **17** is added (Table XIV), it is not the result of altering the mechanism. When one equivalent of $Ph_3SnO_3SC_6F_{13}$ is used, replacing the acetate by an even weaker anion, k_H/k_D is 3.7, an indication that the mechanism changes as the Mn-X bond in MnTPPX weakens.

We looked at the oxidation of norcarane that was used by Groves et al. in manganese porphyrin-catalyzed oxidations to probe the radical character of the oxidation reaction;[33y] our results using this substrate are shown in Table XVI. There is an overall increase in the

Table XVI. Norcarane Oxidation

Catalyst System	Relative yields[a]			
	18	**19**	**20**	**21**
MnTpyPPOAc	1.00	.27	.14	.16
MnTpyPPOAc + 1 equiv. $Ph_3SnO_2C_8F_{15}$	1.77	.42	.01	.18

[a] Yields are based on iodosobenzene; for absolute yields multiply the relative yields by a factor of 11.3.

yield of alcohols produced, as well as an increase in the norcaranols/cyclohexenyl methanol[33y] ratio from 9.1 to 220. The suppression of the radical rearrangement pathway is consistent with our cyclohexane results.

B. Theoretical Results

An extreme case for the decreasing donor strength beyond that of the perfluoroanions is the removal of the axial ligand altogether. Semiempirical molecular orbital calculations were performed using four hypothetical geometries: (Fig. 14a) the high-valent oxo-manganese porphyrin acetate, Mn-O (oxo) = 1.59 Å, Mn-O (acetate) = 2.201 Å;

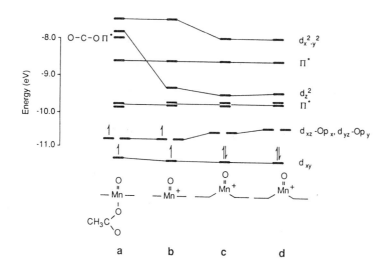

Figure 14. Correlation diagram mapping *d*-orbital energies for four geometeries of a proposed monomeric oxo-manganese intermediate. For *a*, Mn-O (oxo) = 1.59 Å, Mn-O (⁻OAc) = 2.201 Å with Mn in the porphyrin plane; for *b*, Mn-O (oxo) = 1.59 Å; for *c*, Mn-O (oxo) = 1.59 Å with Mn 0.3 Å out of the plane; for *d*, Mn-O (oxo) = 1.55 Å with Mn 0.3 Å out of the plane.

Mn is in the plane of the four nitrogen atoms; (Fig. 14b) oxo-manganese porphyrin, same as *a* with the acetate ligand removed; (Fig. 14c) same as *b* with the manganese atom 0.3 Å out of the plane of the four nitrogen atoms toward the oxo ligand retaining a Mn-O (oxo) bond distance of 1.59 Å; and (Fig. 14d) same as *c* with the Mn-O (oxo) bond distance shortened to 1.55 Å. High valent oxo-manganese porphyrin reactive intermediates are proposed to be formally Mn(V), with a d_2 electron configuration. For a spin paired configuration, this would make orbital d_{xy} the HOMO and Mn d_{yz}-Op_y the LUMO. Because the HOMO-LUMO gap is relatively small (0.59 eV) for geometry *a*, we may expect a considerable contribution from spin unpaired configurations, thereby giving the oxo-manganese intermediate oxy-radical character (consistent with experimental observations). As the acetate ligand is removed, the molecule relaxes by moving the oxo-manganese moiety 0.3 Å out of the plane of the porphyrin and by shortening the Mn-O (oxo) bond distance by 0.04 Å (geometry *d*). The substantial increase in the HOMO-LUMO gap in going from *a* to *d* (0.59 to 0.89 eV) would be expected to reduce the contribution from spin unpaired configurations and thereby significantly reduce the oxy-radical character of the oxo-manganese intermediate.

The effect of changing the axial ligand on the product distribution in cyclohexane oxidation is best discussed with respect to the mechanism in Scheme 7, which is similar to that proposed by Smegal and Hill[33r] and which is essentially identical to that proposed for

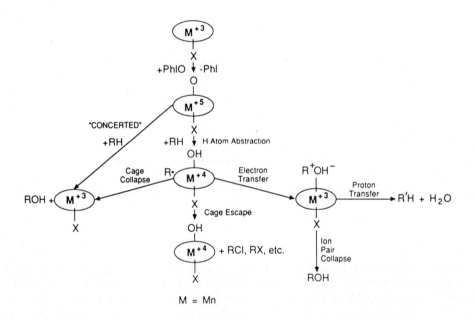

Scheme 7

iron porphyrins.[33h] The mechanism invokes both a concerted pathway to explain the reduction in radical character and a proton transfer pathway to explain the formation of cyclohexene oxide.

The initial step is the oxygen atom transfer from iodosobenzene to Mn(III)(porphyrin)X to generate a high-valent oxo-manganese porphyrin complex, Mn(V)=O(porphyrin)X. It has been proposed for both manganese[33r] and iron porphyrins[33h] that this high-valent oxo-metal intermediate reacts with a hydrocarbon substrate to generate Mn(IV)-OH(porphyrin)X and a carbon-centered organic free radical. The high degree of free-radical character in manganese porphyrin-catalyzed oxidations compared to iron has been attributed to the relative stability of Mn(IV)-OH(porphyrin)X compared to Fe(IV)-OH(porphyrin)X. As a result, cage escape and electron transfer with Mn effectively compete with cage collapse.

We rationalize the reduction in radical character with anion to changes in the electronic structure of the oxo intermediate, which in turn results in a more concerted oxygen atom insertion into C-H bonds. This is supported by the norcarane results and the deuterium isotope effects in Table XV. A linear transition state, depicted for the stepwise H-atom abstraction pathway, should result in a large deuterium isotope effect, since there is considerable C-H bond breaking in the transition state. On

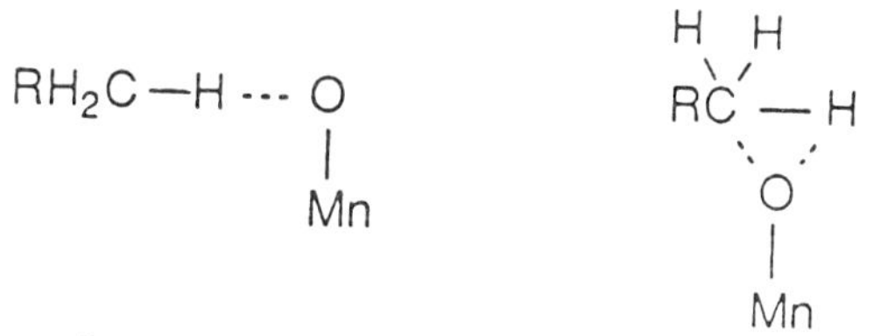

$$RH_2C-H \cdots O$$
$$|$$
$$Mn$$

Stepwise

$$\begin{array}{cc} H & H \\ \backslash & / \\ RC & -H \\ & \ddots \\ & O \\ & | \\ & Mn \end{array}$$

Concerted

the other hand, a concerted oxygen atom insertion into C-H bonds should have a small or possibly an inverse isotope effect. A continuum of transition-state geometries differing little in energy may be visualized between the extremes pictured above, and the reduction of k_H/k_D to 3.7, by removing the axial ligand, suggests that we are moving along this continuum, in which the transition state now looks more triangular than linear, thus explaining our cyclohexane, norcarane, and deuterium isotope results. A k_H/k_D value of 11.8 for MnTPPOAc is consistent with the stepwise process and is in agreement with the value measured by Groves, and Nemo for an iron porphyrin system (Table XV).[33p]

As shown in Table XIV, there is also an overall increase in the yield of products derived from cyclohexane as the coordinating ability of the axial ligand is reduced. The active intermediate in the catalytic cycle for cytochrome P-450 is proposed to be RS-Fe(IV)=O(porphyrin+•). The actual electron configuration could be RS•-Fe(IV)=O(porphyrin)X, with the thiolate axial ligand having transferred an electron to the oxidized porphyrin ring. This differs from closely related model compounds studied by Balch et al. [B-Fe(IV)=O(porphyrin), B= pyridine or *N*-methylimidazole], which have good sigma-donating axial ligands.[36] These model complexes are not reactive enough to oxidize olefins, but removing the axial ligand, as in the case proposed for cytochrome P-450, which may have a thiyl radical in the axial position *trans* to the oxo ligand, will activate the oxo-iron toward less reactive substrates.

3.   Summary

In the case of iron porphyrins, those with bulky electron-withdrawing substituents close to the iron center are more active and more selective oxidation catalysts. An inherent problem with the iron porphyrin systems is their propensity to undergo self-destruction. This can be overcome, to some extent, by using manganese porphyrins as catalysts. In doing so it is important to use a weakly coordinating axial ligand because reducing the donor ability of the axial ligand on manganese reduces the oxy-radical character of the oxo-manganese intermediate. This results in a shift from stepwise to more concerted processes for oxygen atom transfer, and also increases the oxidizing power of the catalyst.

IV.   Alkane Oxidations inside Metal/Zeolite Catalysts: "Ships in Bottles"

Examination of cytochrome P-450 shows that it consists of a very potent, but nonselective, oxidizing center (the ferryl porphyrin prosthetic group, Table IV) embedded within a very size- and shape-selective environment (the protein tertiary structure), which is the source of selectivity in the subsequent oxidation chemistry.[37] With this as our model, we have developed two related alkane-oxidizing systems using zeolites to provide the size- and shape-selective environment for potent iron-based oxidation centers.

Although zeolites have found applications in many chemical transformations of interest in petrochemical refining (cracking, isomerization, alkylation, etc.),[38] their use for selective alkane oxidations was unknown prior to our work. The open framework of zeolites contains channels and cavities of molecular dimensions [from 3 to 13 Å (Figure 15)][39] whereas the aluminosilicate composition imparts great robustness and makes them inert to oxidizing conditions. The aluminum leads to a net anionic charge on the framework that is compensated by loosely attached cations, giving the well-known ion-exchange properties of zeolites. This combination of properties makes these porous materials attractive inorganic replacements for the protein tertiary structures of the natural system (previous work suggested this idea[40]) and offers an alternative approach to controlling oxidation selectivities that can be compared with the organic modification of porphyrins described in Sec. III and elsewhere in this volume.

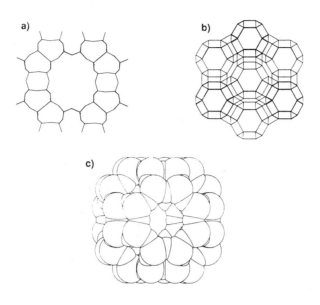

Figure 15. Structures of representative zeolites. (*a*) Mordenite. (*b*) Y. (*c*) A (space-filling model).

1. Iron Phthalocyanine Complexes in Large Pore Zeolites

One example of a transition metal complex capable of hydrocarbon oxidation inside the pore structure of a zeolite is iron phthalocyanine within zeolite X or Y.[41] This type of supported complex catalyst is typically prepared by a stepwise procedure: introducing the metal ion by ion exchange, followed by absorption of ligand components that react to generate the desired complex. We have coined the term "ship-in-a-bottle" complexes in cases such as this in which the metal-ligand complex, once constructed inside, is too large to get back out.[42] The porphyrin-related iron phthalocyanine molecule **22** is a molecular "ship" approximately 16 Å in diameter - much too large for direct absorption into even the largest pore zeolites (X and Y), which have channels only ~8 Å in diameter. Prior reports have indicated, however, that such a species can be readily made.[40]

Iron(II) ion-exchanged zeolite X or Y is mixed and heated with dicyanobenzene above 200°C in an inert atmosphere. Over a period of about an hour, the pale-green zeolite becomes increasingly blue-green as the dicyanobenzene is absorbed and reacts with the Fe(II) in a template condensation. Removal of the excess dicyanobenzene by sublimation and acetone extraction leaves a deep blue-green material. The

22

macrocyclic product on the exterior surface of the zeolite must be removed before shape-selective catalysis by complexes inside the pores can be observed. (Because of the greater accessibility of surface catalyst, those molecules, if present, dominate product distributions.) Extended (48 hr) soxhlet extraction with 1-chloronaphthalene removes unencapsulated complex to give a deep blue solution. When no further color is extracted from the zeolite, the catalyst is collected and vacuum dried, leaving a deep blue-green powder.

ESCA analysis indicates no residual surface iron; powder X-ray patterns show no lines associated with crystalline iron phthalocyanine but do confirm the retention of zeolite crystallinity. Chemical analysis shows the Fe:N ratio of 1:8 expected if all of the entrained iron is present as phthalocyanine complex, which is also consistent with the diffuse reflectance spectra. This now presents a problem: we have a complex of diameter 16 Å whereas the zeolite supercage available for it is at most 13 Å in diameter. This paradox is resolved by considering Figure 16. The complex can fit into the zeolite by giving it a saddle

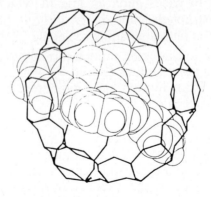

Figure 16.   Saddle-deformed FePc in the zeolite Y supercage.

distortion and allowing its arms to protrude through the supercage windows. Figure 16 also shows that a maximal loading level of something less than one complex per supercage is all that is possible and that access to the central iron atom is restricted by partial blocking of the windows into the supercage in which the complex is located.

Using the popular iodosobenzene oxidizing agent, iron phthalocyanine itself is a poor but operable catalyst for alkane oxidation in methylene chloride, but suffers rapid (<15 min) deactivation, presumably by bimolecular oxidative degradation as is the case with iron porphyrin complexes.[33h] After a turnover of close to one, the originally blue oxidizing solutions became dark brown and inactive. The zeolite-entrapped material also produces oxidized products in an identical reaction procedure but with some marked differences (Table XVII). Most significant is the increased turnover based on iron. This is

Table XVII. Products from the Partial Oxidation of Methylcyclohexane Using Iron Phthalocyanine/Zeolite Catalysts and Iodosobenzene

Catalyst[a]	Turnover	Alcohol/ketone
FePc	1.1	3.1
FePc/X 20%	0.5	1.8
FePc/X 2%	4.1	2.1
FePc/Y10 1%	2.5	2.9
FePc/Y20 1%	5.6	3.0

[a]The percentage refers to the extent of exchange of cations by Fe(II); 20% corresponds to one Fe per supercage.

probably a result of the very effective site isolation within the pores that prevents any bimolecular pathways to catalyst destruction. The recovered zeolite materials remain blue-green, and we find that a recovered catalyst can be reactivated by simply baking to remove water and entrained organics, giving back the original activity. The reaction appears to stop because the pore system of the zeolite becomes clogged with reaction products - preventing further access of reagents and substrate to the iron sites. By acid dissolution of the zeolite, we have identified (by GC) iodobenzene as the major product arising from acid decomposition of iodoso- and/or iodoxy-benzene trapped in the pores. One other notable feature of the zeolite oxidations is their very slow rates; the reactions take ~24 hr to go to completion.

Substrate selectivity (Fig. 17a) is the simplest kind of selectivity to expect from the zeolite-based catalysts since zeolites are well known for their molecular sieving action. We chose to examine pairs of

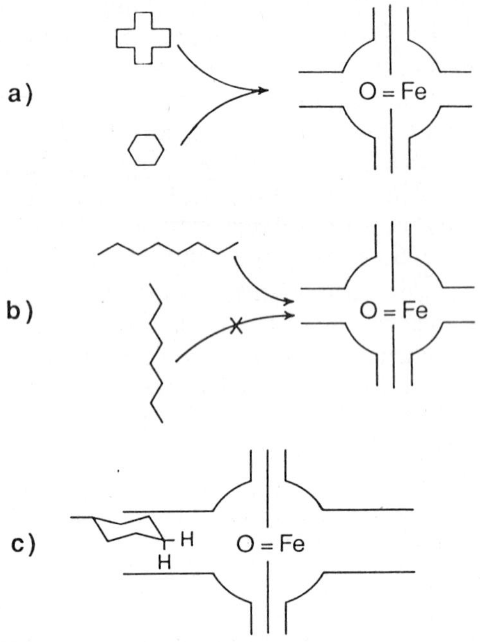

Figure 17. Schematic representations of (*a*) substrate, (*b*) regioselectivity and (*c*) stereoselectivity.

similarly reactive substrates of different sizes in competitive oxidations with these catalysts. Figure 18*a* shows the results of such an experiment with cyclohexane and cyclododecane and Figure 18*b* shows pentane and octane. In both cases the selectivity with the zeolite catalysts favors the smaller substrate, but the effect is small. We thought the reason for the small size of the effect was the flexibility of the substrates and the large pore size of the zeolites used. Since our synthetic approach required large pore zeolites, we sought to reduce the pore size by partial pore blockage with the exchangeable cations. The results of this approach are illustrated in Figure 19 in which the substrate selectivity between cyclohexane and cyclododecane is shown as a function of the cations present. There is a clear selectivity optimum with the medium sized cations Rb^+ and NH_4^+, and the rate of oxidation declines continuously as the cation size increases. Apparently increasing cation size helps differentiate between the substrates by narrowing the channels until further blockage leads only to exclusion of both. These trends are exactly analogous to those observed in competitive hydrogenation of pairs of olefins over rhodium-exchanged zeolites, in which water of hydration served to narrow the pores and change selectivity.[43]

The natural monooxygenases have structures that tend to orient incoming substrates with respect to the active center by a combination of

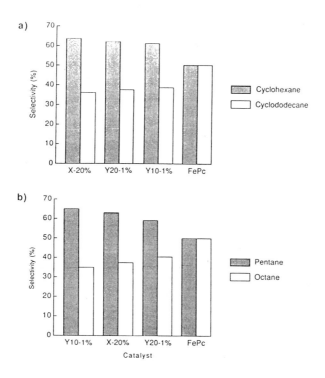

Figure 18. Bargraph representation of substrate selectivities in (*a*) cyclohexane/cyclododecane and (*b*) octane/pentane competitive oxidations.

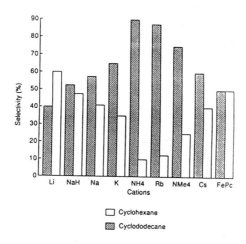

Figure 19. Bargraph representation of substrate selectivities in competitive cyclohexane/cyclododecane oxidations over various cation-exchanged FePc zeolite catalysts.

hydrophobic, steric, and hydrogen bonding effects. In this way they introduce regioselectivity (Fig. 17*b*) and stereoselectivity (Fig. 17*c*) into their oxidations. In the case of the iron phthalocyanine/zeolite catalysts, since the substrate must diffuse to the iron center through narrow windows of the zeolite it might be expected that this would tend to orient it in favor of oxidation nearer the ends of its long molecular axis. Figure 20 clearly shows a trend favoring oxidation near the ends of the *n*-octane chain. However, as with substrate selectivity the effects are small.

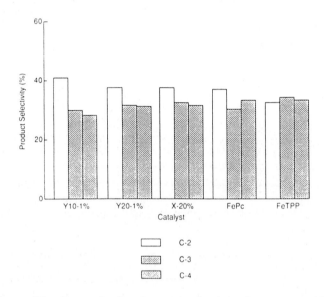

Figure 20.   Bargraph showing regioselectivity in octane oxidation.

Stereoselectivity requires differentiation between diastereotopic C-H bonds so as to generate one particular stereoisomer from a possible pair. Again this depends on the ability of the structure to direct one of the pair of diastereotopic hydrogen atoms selectively toward the active site. Such selectivity is observed in the oxidation of methyl (or *tert*-butyl) cyclohexane and norbornane with the zeolite catalysts as seen in Table XVIII.

Although the results show that iron phthalocyanine/zeolite-based oxidation systems are a reality, they also point to a number of problems with these materials:

1. Low rates are a result of the need to diffuse substrates and oxidant in and products out through narrow pores that are partially blocked by the phthalocyanine molecules themselves. This problem of counterdiffusion should be minimized by using the smallest possible oxidants and active complexes so as to maximize the space available for substrate and product diffusion.

Table XVIII. Stereochemistry of Alkane Oxidations with Iron Phthalocyanine/Zeolite Catalysts and Iodosobenzene

Catalyst		*trans/cis* Ratio 4-methylcyclohexanol	*endo/exo* Ratio of norborneols
FePc		1.2	0.11
FePc/Y20	1%	1.6	0.16
FePc/X	2%	1.9	0.18
FePc/Y10	1%	2.0	0.17
FePc/X	20%	2.0	0.17

2. Low turnovers are the result of pore blockage by products of oxidation and oxidant disproportionation - not phthalocyanine degradation. This problem should also be minimized by using a small oxidant and minimizing pore blockage at the catalyst site.

3. The small selectivities in all but the large cation-exchanged systems are a result of having to use large pore zeolite hosts where the molecular sieving and orientation properties are not optimal for high selectivities. A smaller pore zeolite would be highly desirable as the host.

Combining these observations led to the development of the second generation oxidation catalyst system described below.

2. Iron/Palladium Zeolites with O_2/H_2 as Completely Inorganic P-450 Mimics

By combining the ability of colloidal Pt or Pd metal to directly combine molecular hydrogen and oxygen into hydrogen peroxide[44] with the ability of iron ions to use hydrogen peroxide to hydroxylate organic materials[45] - all at room temperature - we have prepared a bimetallic Pd/Fe oxidizing system and embedded it inside the framework of a zeolite.[46] In contrast to the phthalocyanine system previously described, the new catalysts can be prepared with ease with zeolites of any pore size and so should be capable of high selectivities. In addition, the use of molecular oxygen as the primary oxidant minimizes diffusion or pore blocking problems, whereas the bare metal ions or atoms provide the minimum pore restrictions themselves.

Zeolite 5A[39] (Si/Al ~1.2) was first ion exchanged with iron(II) and then with palladium(II) tetramine chloride, to give a material containing ~1 wt% Fe and ~0.7 wt% Pd. Calcination in oxygen (400°C) followed by reduction in hydrogen (150°C) generates a well-dispersed

Pd(0) phase.[47] A typical oxidation run used 500 mg of such a catalyst in 10 ml hydrocarbon under 40 psig oxygen and 15 psig hydrogen at 25°C for 4 hr. Addition of water followed by conc. sulfuric acid (1 ml) (with stirring for 1 hr to dissolve the zeolite) released any oxidation products on the exterior surface or trapped inside the pore system. As a means of estimating the selectivity imposed by the zeolite a control catalyst of Fe/Pd on an amorphous silicoaluminate (Si/Al ~1.2) was also investigated.

When a 1:1 mixture of *n*-octane and cyclohexane is subjected to this oxidation, the results are as shown in Figure 21. The amorphous

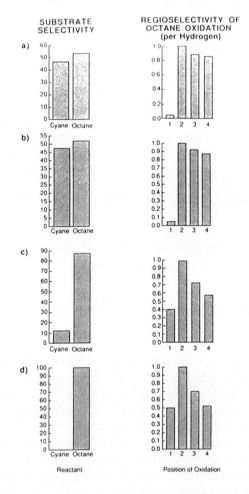

Figure 21. Selectivities in the oxidation of a 1:1 (volume) mixture of *n*-octane and cyclohexane. Substrate selectivities are on a molar basis. Regioselectivities along the chain have been normalized for the relative number of hydrogens at each chain position, with the activity at the 2 position in the chain assigned a nominal activity of 1.0. Bars represent the total of all products (alcohols plus ketones) derived from a given reactant or at a given position within each reactant. (*a*) Amorphous silicoaluminate support; (*b*) zeolite 5A support after water addition; (*c*) zeolite 5A after acid dissolution; (*d*) zeolite 5A with 2,2'-bipyridine as poison after acid dissolution.

silicoaluminate control catalyst shows moderate activity (~75% yield based on iron , ~205% based on palladium, and ~0.5% based on hydrogen at 0.15% conversion), with no substrate selectivity (octane/cyclohexane oxidation = 0.9), and a regioselectivity of octane oxidation products (primary H/secondary H oxidation = 0.05) that is atypical of hydroxyl radical oxidation in an aqueous system [*prim/sec* ~0.2 (Table IV)]. The detected products were not different in either quantity or distribution regardless of whether only water was added or the catalyst was dissolved in acid. With the 5A catalyst, the organic phase after initial water addition showed a small quantity of oxidized products with a distribution (Fig. 21*b*) almost identical to the control. This represents the oxidation that occurred at the exterior surface of the zeolite particles. Following dissolution in acid, the product quantity and distribution were dramatically changed (Fig. 22), with a 5-fold increase in products being entirely the result of octane oxidation products (Fig. 21*c*). The substrate selectivity now shows an octane/cyclohexane

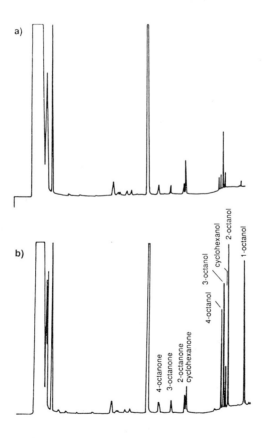

Figure 22. GC traces of oxidation reaction mixture over zeolite 5A (*a*) after water addition and (*b*) after acid dissolution of zeolite. Peak identification as marked based on retention times of genuine compounds.

ratio of 6 whereas the regioselectivity of octane products shows a *prim/sec* oxidation ratio of 0.54. All of the additional oxidized products come from the zeolite interior in which they were trapped because of their physical size. (Zeolite 5A has 5 Å 8-ring windows that are extremely selective for adsorption or release of unsubstituted linear alkanes, to the exclusion of cyclic or branched species.[39])

The addition of a small amount of 2,2'-bipyridine to the oxidation medium acts to poison the iron activity, presumably as a result of coordination to produce a redox inactive complex. The poison's size dictates that only exterior surface iron sites are eliminated, and there is no evidence of oxidation products prior to the acid dissolution of the zeolite when bipyridine is used. This provides a means of establishing the intrinsic selectivity of the interior zeolite sites (Fig. 21*d*), with a substrate selectivity of octane/cyclohexane of >190 and a regioselectivity in *prim/sec* oxidation ratio of 0.67. Such regioselectivity is superior to that reported for our sterically hindered porphyrin systems[33h] or Suslick's best iron porphyrin (0.3) or manganese systems (0.53)[33b] using iodosobenzene as the oxidant.

Regioselectivity with this zeolite A system (Fig. 23) is almost independent of the *n*-alkane chain length; in all three cases, the *prim/sec* oxidation ratio is very close to 0.6 for the first three atoms in the chain. This is in contrast to the results of Suslick and co-workers[33b]

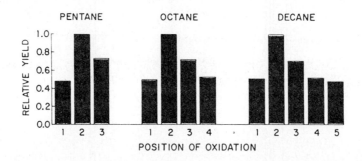

Figure 23. Regioselectivities along the chain of *n*-pentane, *n*-octane, and *n*-decane normalized for the relative number of hydrogens at each chain position (with activity at the 2 position normalized to 1.0).

who noted a marked chain-length dependence in regioselectivity with hindered porphyrin systems. This implies that the rigid zeolite framework exerts a much tighter control over the substrate conformations during oxidation of short-chain alkanes than the flexible periphery of the porphyrin complexes can impose.

The oxidation with our catalysts appears to involve the generation of hydrogen peroxide at Pd(0)[44] rather than reduction of an iron-coordinated dioxygen species by a Pd-hydride,[48] since a physical mixture of a Pd zeolite and a Fe zeolite is also capable of hydrocarbon oxidation in solvents in which hydrogen peroxide is soluble and hence able to migrate.[49] In addition, experiments using simple iron zeolites and hydrogen peroxide give product selectivities identical to those observed with the O_2/H_2 system. Whether the actual oxidizing species or anhydrous hydroxyl radicals or high-valent iron-oxo units (which we prefer) is unknown. As discussed in detail above, we believe the substrate selectivity observed is a consequence of the sorption selectivity of the 5A zeolite whereas the ω-oxidation selectivity arises from the very close fit of alkane to pore size that essentially constrains it to have an extended "linear" conformation. Thus the methyl end-groups are the first to encounter (and so be oxidized by) the iron active sites in the zeolite supercages. This concept has previously been demonstrated for severely hindered porphyrin systems[33g] and in micellar media.[50]

The overall activity of the 5A zeolite system is lower than the control with 30% yield based on iron in this batchwise experiment. However, a recovered, calcined, and dried catalyst can be reused with the same initial activity as virgin material. As was the case with the iron phthalocyanine/zeolite catalysts, we believe that a combination of organic oxidation products and the water by-product of the O_2/H_2 reaction fill the zeolite interior and eventually stop access of further substrate to the active sites. It should be noted that, based on the pressure drop during the course of the reaction, we estimate that >95% of the O_2/H_2 mixture that is consumed gives water rather than oxidized organics. Other zeolites, particularly ZSM-5,[39] also display good activity in this type of reaction but with the advantage that the oxidation products and water by-product are readily released from these more hydrophobic materials by simple washing while selectivities can remain high.

These systems thus take molecular oxygen at room temperature and pressure and use a reducing cofactor (molecular hydrogen/Pd) to generate an iron-based potent oxidant. This oxidant is embedded in a very size- and shape-selective site in which it can selectively hydroxylate unactivated alkanes. This description could equally apply to the natural system, and we feel that our present systems, although consisting entirely of inorganic components, provide excellent mimics of the long-sought monooxygenase chemistry. In this regard we feel that they represent significant progress toward the elusive goals of selective partial oxidation of alkanes and verify the feasibility of this novel approach to oxidation selectivity using zeolites as protein replacements.

V.   Activation of O_2 by Reduction - A   Thermodynamic Scale for O Atom Transfer Reagents

One feature of the P-450 enzymes that seemed curious at first was the necessity for a reducing agent (usually NADPH) to activate the O_2 for oxidations under mild conditions. A number of nonbiological systems capable of oxidizing hydrocarbons with O_2 at room temperature, including Barton's Gif and Gif-Orsay systems, also require reducing agents or their electrochemical equivalent as electrons.[51] O atoms are powerful oxidizing agents capable of reacting rapidly with hydrocarbons even below room temperature.[52] The effect of the reducing agent can be thought of as reducing one of the O atoms of O_2 to water and using the energy produced   to   pump the second thermodynamically uphill. The

Table XIX. Heats of Oxidation[a]

Reaction	ΔH (kcal/mol)
$1/2\ O_2 \longrightarrow O$	59
$O_2\ +\ 1/2\ O_2 \longrightarrow O_3$	34
$CH_3CO_2H\ +\ 1/2\ O_2 \longrightarrow CH_3CO_3H$	34
$H_2O\ +\ 1/2\ O_2 \longrightarrow H_2O_2$	24
$N_2\ +\ 1/2\ O_2 \longrightarrow N_2O$	17
$HCl\ +\ 1/2\ O_2 \longrightarrow HClO$	8
$C_2H_4\ +\ 1/2\ O_2 \longrightarrow$ ethylene oxide	-25[b]
$Me_2S\ +\ 1/2\ O_2 \longrightarrow Me_2SO$	-32[c]
$C_2H_6\ +\ 1/2\ O_2 \longrightarrow C_2H_5OH$	-40[d]
$H_2\ +\ 1/2\ O_2 \longrightarrow H_2O$	-68
$CO\ +\ 1/2\ O_2 \longrightarrow CO_2$	-68
$PPh_3\ +\ 1/2\ O_2 \longrightarrow OPPh_3$	-70[c]

[a]For compounds in their standard states at 25°C, taken from the *Handbook of Chemistry and Physics*, 41st Ed., pp. 1800-1808.   Chemical Rubber Company, Cleveland, 1959-60; pp 1800-1808, unless otherwise noted.

[b]Calculated from $\Delta H°_{f300}$ values from Benson, S. W., *Thermochemical Kinetics*, 2nd Ed., pp. 295-296. John Wiley, New York, 1976.

[c]From Stully, D. R., *The Chemical Thermodynamics of Organic Compounds*. John Wiley, New York, 1969.

[d]From Table II.

activated O atom will not generally be free, but will be partly stabilized by coordination to some two-electron donor, such as the Fe(III) in a porphyrin. The better the donor, the more stabilized the O atom, and the less capable it will become for oxidizing other substrates. Table XIX shows that the electron donor ability increases in the sequence $O_2 < N_2 < HCl < Me_2S < PPh_3$ as the exothermicity of the oxidation reaction increases. Ozone, though a weaker oxidant than an O atom by 25 kcal/mol, is stronger than N_2O by 17, which in turn is stronger than HClO by 9. Reduction of one atom of O_2 to water by H_2 has more than enough exothermicity to produce any of these oxidants. It is thermodynamically downhill to move an oxygen atom from any XO species to one lower on the list; $OPPh_3$ represents the thermodynamic sink of the XO species shown.

The question naturally arises as to where to put PhIO and ferryl FeO species on the thermodynamic scale. Unfortunately, experimental data are not available. The fact that iodine is less electronegative than chlorine suggests that PhI should be a better two-electron donor than HCl, so that Reaction 32 should have a ΔH less than 8 kcal/mol. The

$$PhI + 1/2O_2 \longrightarrow PhIO \qquad (32)$$

ability of Fe(III) porphyrin complexes to use PhIO to epoxidize olefins indicates that ΔH for Reaction 32 must be greater than -25 and that the Fe(III) porphyrin must be intermediate in its electron donor ability between PhI and ethylene. After we developed insight into the role of reducing agents in activating dioxygen, and the thermodynamic scale for O atom transfer reactions,[53] we learned that Holm had independently devised the same kind of scale as a result of his work on O atom transfer reactions involving Mo enzymes, and has extensively reviewed the available data on metal-centered oxygen atom transfer reactions for a variety of metal centers and oxidants.[54]

Acknowledgments

We are indebted to R. N. McGill for measuring the activation energy for cyclohexylhydroperoxide decomposition by cobalt 2-ethylhexanoate, to A. Mical and J. B. Sieja for the separation and isolation of a number of peroxides from cyclohexane oxidation mixtures, to S. D. Ittel for BPI ligand and catalyst synthesis, and to I. D. Williams for determining the single crystal X-ray structure of $Co(BPI)_2$. We are also indebted to O. Van Buskirk for the microreactors, to W. C. Seidel for performing many of the continuous microreactor studies, to W. T. Robbins for fitting the data shown in Figure 7, and to D. L. Filkin and D. W. Drew for assistance with computer modeling of the cyclohexane oxidation.

We are also grateful to R. H. Holm for sending a copy of his review of metal-centered oxygen atom transfer reactions in advance of its publication.

References

1. Benson, S. W., Golden, D. M., in Eyring, H., ed., *Physical Chemistry - An Advanced Treatise*, (H. Eyring, ed.), Chap. 2. Academic Press, New York, 1975.

2. Wild, J. D., Sridhar, T., Potter, O. E., *Chem. Eng. J.* **1978**, *15*, 209.

3. Korcek, S., Chenier, J. H. B., Howard, J. A., Ingold, K. U., *Can. J. Chem.* **1972**, *50*, 2285.

4. Druliner, J. D., Krusic, P. J., Lehr, G. F., Tolman, C. A., *J. Org. Chem.* **1985**, *50*, 5838.

5. Groves, J. T., Nemo, T. E., Meyers, R. S., *J. Am. Chem. Soc.* **1979**, *101*, 1032.

6. Weissermel, K., Arpe, H.-J., in *Industrial Organic Chemistry*, Verlag Chemie, New York, 1978. p. 212.

7. Sheldon, R. A., Kochi, J. K., *Adv. Catal.* **1976**, *25*, 272.

8. Duynstee, E. F., J., Hennekens, J. L. J. P., *Recueil* **1970**, *89*, 769.

9. Farkas, A., Passaglia, E., *J. Am. Chem. Soc.* **1950**, *72*, 3333.

10. Brown, N., Hartig, M. J., Roedel, M. J., Anderson, D. W., Schweitzer, C. E., *J. Am. Chem. Soc.* **1955**, *77*, 1756.

11. Tolman, C. A., Druliner, J. D., Krusic, P. J., Nappa, M. J., Seidel, W. C., Williams, I. D., Ittel, S. D., *J. Mol. Catal.*, **1988**, *48*, 129.

12. Kellogg, R. E., *J. Am. Chem. Soc.* **1969**, *91*, 5433.

13. Benson, S. W., *Thermochemical Kinetics*. John Wiley, New York, 1976.

14. Stully, D. R., in *The Chemical Thermodynamics of Organic Compounds*; John Wiley & Sons, New York, 1969.

15. (a) Hancock, R. D., Marsicano, F., *J. Chem. Soc., Dalton Trans.* **1976**, 1096; (b) Beck, M. T., *Chemistry of Complex Equilibria*, p. 253. Van Nostrand, New York, 1970.

16. Domaille, P. J., Harlow, R. L., Ittel, S. D., Peet, W. G., *Inorg. Chem.* **1983**, *22*, 3944.

17. Saussine, L., Brazi, E., Robine, A., Mimoun, H., Fischer, J., Weiss, R., *J. Am. Chem. Soc.* **1985**, *107*, 3534.

18. McCarthy, R. L., MacLachlan, A., *Trans. Faraday Soc.* **1961**, *57*, 1107.

19. Russell, G. H., *J. Am. Chem. Soc.* **1957**, *79*, 3871.

20. Cowley, L. T., Waddington, D. J., Woolley, A., *J. Chem. Soc., Faraday Trans.* **1982**, *78*, 2535.

21. Jensen, R. K., Korcek, S., Mahoney, L. R., Zinbo, M., *J. Am. Chem. Soc.* **1979**, *101*, 7574.

22. Howard, J. A., *Adv. Free Radical Chem.* **1972**, *4*, 49.

23. Druliner, J. D., Kitson, F. G., Rudat, M. A., Tolman, C. A., *J. Org. Chem.* **1983**, *48*, 4951.

24. Kochi, J. K., in Kochi, J. K., ed. *Free Radicals*; (J. K. Kochi, ed.), Vol. 1, p. 378. John Wiley, New York, 1973.

25. Walling, C., Padwa, A., *J. Am. Chem. Soc.* **1963**, *85*, 1597.

26. (a) McEwen, C. N., Rudat, M. A., *J. Am. Chem. Soc.* **1979**, *101*, 6470; (b) *J. Am. Chem. Soc.* **1981**, *103*, 4343; (c) Rudat, M. A., McEwen, C.

N., *J. Am. Chem. Soc.* **1981**, *103*, 4349; (d) McEwen, C. N., Rudat, M. A., *J. Am. Chem. Soc.* **1981**, *103*, 4355.

27. Druliner, J. D., Hobbs, F. W., Seidel, W. C., *J. Org. Chem.* **1988**, *53*, 700.

28. Kochi, J. K., Sheldon, R. A., In *Metal-Catalyzed Oxzidation of Organic Compounds*, p. 48. Academic Press, New York, 1981.

29. Sharpless, K. B., Teranishi, A. Y., Backvall, J. E., *J. Am. Chem. Soc.* **1977**, *99*, 3120.

30. Tolman, C. A., Seidel, W. C., *A Radical Model for Cyclohexane Oxidation*, presented at the Florida Catalysis Conference, Sheraton Palm Coast, April 27-May 2, 1987.

31. Finkelstein, L., Carson, E. R., *Mathematical Modelling of Dynamic Biological Systems*, 2nd Ed. John Wiley, New York, 1985.

32. (a) Paulson, D. R., Rudiger, U., Sloan, R. B., *J. Chem. Soc., Chem. Commun.* **1974**, 186; (b) Ohkatsu, Y., Tsuruta, T., *Bull. Chem. Soc. Jpn.* **1978**, *51*, 188.

33. (a) Mansuy, D., Battioni, P. *Bull. Chem. Soc. Belg.* **1986**, *95*, 959; (b) Cook, B. R., Reinert, T. J., Suslick, K. S., *J. Am. Chem. Soc.* **1986**, *108*, 7281; (c) Che, C. M., Chung, W. C., *J. Chem. Soc., Chem Commun.* **1986**, 386; (d) Meunier, B. *Bull. Chem. Soc. Fr.* **1986**, 578; (e) Battioni, P., Renaud, J. P., Bartoli, J. F., Mansuy, D., *J. Chem. Soc., Chem Commun.* **1986**, 341; (f) De Poorter, B., Meunier, B., *J. Chem. Soc., Perkin Trans.* **1985**, 1735; (g) Suslick, K., Cook, B., Fox, M., *J. Chem. Soc., Chem. Commun.* **1985**, 580; (h) Nappa, M. J., Tolman, C. A. *Inorg. Chem.* **1985**, *24*, 4711; (i) Renaud, J. P., Battioni, P., Bartoli, J. F., Mansuy, D., *J. Chem. Soc., Chem. Commun.* **1985**, 888; (j) Yuan, L. C., Bruice, T. C., *J. Chem. Soc., Chem. Commun.* **1985**, 868; (k) Dicken, C. M., Lu, F. L., Nee, M. W., Bruice, T. C., *J. Am. Chem. Soc.* **1985**, *107*, 5776; (l) Mansuy, D., Battioni P., Renaud, J. P., *J. Chem. Soc., Chem. Commun.* **1984**, 1255; (m) Khenkin, A. M., Shteinman, A. A., *J. Chem. Soc., Chem. Commun.* **1984**, 1219; (n) Fontecave, M., Mansuy, D., *Tetrahedron* **1984**, *40*, 4297; (o) Collman, J. P., Brauman, J. I., Meunier, B., Raybuck, S. A., Kodadek, T., *Proc. Natl. Acad. Sci. U.S.A.* **1984**, *81*, 3245; (p) Groves, J. T., Nemo, T. E., *J. Am. Chem. Soc.* **1983**, *105*, 6243; (q) Mansuy, D., Fontecave, M., Bartoli, J. F., *J. Chem. Soc., Chem. Commun.* **1983**, 253; (r) Smegal, J. A., Hill, C. L., *J. Am. Chem. Soc.* **1983**, *105*, 3515; (s) *J. Am. Chem. Soc.* **1983**, *105*, 2920; (t) Ando, W., Tajima, R., Takata, T. *Tetrahedron Lett.* **1982**, *23*, 1685; (u) Mansuy, D., Bartoli, J. F., Momenteau, M., *Tetrahedron Lett.* **1982**, *23*, 2781; (v) Nee, M. W., Bruice, T. C., *J. Am. Chem. Soc.* **1982**, *104*, 6123; (w) Perree-Fauvet, M., Gaudemer, A., *J. Chem. Soc., Chem. Commun.* **1981**, 874; (x) Hill, C. L., Schardt, B. C., *J. Am. Chem. Soc.* **1980**, *102*, 6374; (y) Groves, J. T., Kruper, W. J., Haushalter, R. C., *J. Am. Chem. Soc.* **1980**, *102*, 6375.

34. Nappa, M. J., McKinney, R. J., *Inorg. Chem.* **1988**, *27*, 3740.

35. (a) Coon, M. J., White, R. E., in *Metal Ion Activation of Dioxygen*, (T. G. Spiro, ed.), Chap. 2. John Wiley, New York, 1980; (b) Sligar, S. G., Gunsalus, I. C., *Adv. Enzymol. Rel. Areas Mol. Biol.* **1979**, *47*, 1; (c) Groves, J. T., *Adv. Inorg. Biochem.* **1979**, *1*, Chapter 4; (d) Chang, C. K., Dolphin, D., *Bioorg. Chem.* **1978**, *4*, 37.

36. Balch, A. L., Chan, Y-.W., Cheng, R-J., La Mar, G. N., Latos-Grazynski, L., Renner, M. W., *J. Am. Chem. Soc.* **1984**, *106*, 7779.

37.   (a) White, R. E., Coon, M. J., *Annu. Rev. Biochem.* **1980**, *49*, 315; (b) Gunsalus, I. C., Sligar, S. C., *Adv. Enzymol.* **1978**, *47*, 1.
38.   (a) Maxwell, I. E., *J. Incl. Phen.* **1986**, *4*, 1; (b) Weisz, P. B., *Chem. Tech.* **1973**, *3*, 498.
39.   Breck, D. W., *Zeolite Molecular Sieves.* Krieger Publishing Co., Malabar, Florida, 1984.
40.   Romanovsky, B. V., *Proc. Int. Symp. on Zeolite Catal. Siofok,* 13-16 May, 1985.
41.   Herron, N., Stucky, G. D., Tolman, C. A., *J. Chem. Soc., Chem. Commun.* **1986**, 1521.
42.   (a) Herron, N., *Inorg. Chem.* **1986**, *25*, 4714; (b) Herron, N., Stucky, G. D., Tolman, C. A., *Inorg. Chim. Acta* **1985**, *100*, 135.
43.   Corbin, D. R., Seidel, W. C., Abrams, L., Herron, N., Stucky, G. D., Tolman, C. A., *Inorg. Chem.* **1985**, *24*, 1800.
44.   (a) Hooper, G. W., British Patent 1,056,121 (1967) to ICI; (b) Dyer, P. N., Mosely, F., U.S. Patent 4,128,627 (1978) to Air Products; (c) Kim, L., Schoenthal, G. W., German Patent 2,615,625 (1976) to Shell; (d) Izumi, Y, Miyazaki, H., Kawahara, S., U.S. Patent 4,009,252 (1977) to Tokuyama Soda; (e) Eguchi, M., Morita, N., *Kogyo Kagaku Zasshi* **1968**, *71*, 783 .
45.   (a) Groves, J. T., van der Puy, M., *J. Am. Chem. Soc.* **1974**, *96*, 5274; (b) Udenfriend, S., Clark, C. T., Axelrod, J., Brodie, B. B., *J. Biol. Chem.* **1954**, *208*, 731; (c) Hamilton, G. A., *J. Amer. Chem. Soc.* **1964**, *86*, 3391.
46.   Poutsma, M. L., Elek, L. F., Ibarbia, P. A., Risch, A. P., Rabo, J. A. *J. Catal.* **1978**, *52*, 157.
47.   Herron, N., Tolman, C. A. *J. Am. Chem. Soc.* **1987**, *109*, 2837.
48.   (a) Tabushi, I., Koga, N., in *Biomimetic Chemistry;* D. Dolphin, C. McKenna, Y. Murakami, I. Tabushi, eds., American Chemical Society, Washington, D.C. 1980; Adv. Chem. Ser. No. 191, p. 291; (b) James, B. R., Adv. Chem. Ser. No. 191, p. 253; (c) Tabushi, I., Yazaki, A., *J. Am. Chem. Soc.* **1981**, *103*, 7371.
49.   Rebek, J., *Tetrahedron* **1979**, *35*, 723.
50.   Sorokin, A. B., Khenkin, A. M., Marakushev, S. A., Shilov, A. E., Shteinman, A. A., *Doklady Akad. Nauk SSSR* **1984**, *279*, 939.
51.   (a) Barton, D. H. R., Gastiger, M. J., Motherwell, W. B., *J. Chem. Soc., Chem. Commun.* **1983**, 41 and 731; (b) Balavoine, G., Barton, D. H. R., Boivin, J., Gref, A., Ozbalik, N., Riviere, H., *J. Chem. Soc., Chem. Commun.* **1986**, 1727.
52.   Varkony, T. H., Pass, S., Mazur, Y., *J. Chem. Soc., Chem. Commun.* **1975**, 457.
53.   Tolman, C. A., Herron, N., in *Oxygen Complexes and Oxygen Activation by Metal Complexes* (D. T. Sawyer, Martell, A. E. eds.). Plenum, New York, 1988.
54.   (a) Holm, R. H., Berg, J. M., *Acc. Chem. Res.* **1986**, *19, 363.* (b) Holm, R. H., *Chem. Rev.* **1987**, *87*, 1401.

INDEX